Graphene Oxide in Enhancing Energy Storage Devices

The world is filled with electronic devices that use batteries and supercapacitors, such as laptops, cellphones, and cameras, creating the need for the efficient and effective production of good energy storage devices. The depletion of fossil fuels demands alternative sources of energy, which prompted the creation of solar cell (photovoltaic) and fuel cell technologies. The introduction of graphene oxides to these technologies helps improve the performance of various energy storage and conversion devices.

This book provides a broad review of graphene oxide synthesis and its applications in various energy storage devices. The chapters explore various fundamental principles and the foundations of different energy conversion and storage devices with respect to their advancement due to the emergence of graphene oxide, such as supercapacitors, batteries, and fuel cells. This book will enable research toward improving the performance of various energy storage devices using graphene oxides and will be a valuable reference for researchers and scientists working across physics, engineering, and chemistry on different types of graphene oxide-based energy storage and conversion devices.

Graphene Oxide in Enhancing Energy Storage Devices

Edited by
Fabian I. Ezema, Ishaq Ahmad,
and Tingkai Zhao

CRC Press
Taylor & Francis Group
Boca Raton London New York

CRC Press is an imprint of the
Taylor & Francis Group, an **informa** business

First edition published 2022

by CRC Press
6000 Broken Sound Parkway NW, Suite 300, Boca Raton, FL 33487-2742

and by CRC Press
4 Park Square, Milton Park, Abingdon, Oxon, OX14 4RN

CRC Press is an imprint of Taylor & Francis Group, LLC

© 2023 selection and editorial matter, Edited by Fabian I. Ezema, Tingkai Zhao and Ishaq Ahmad; individual chapters, the contributors

Library of Congress Cataloging-in-Publication Data
Names: Ezema, Fabian I., editor. | Ahmad, Ishaq, editor. | Zhao, Tingkai, editor.
Title: Graphene oxide in enhancing energy storage devices / edited by Fabian I. Ezema and Tingkai Zhao and Ishaq Ahmad.
Description: First edition. | Boca Raton : CRC Press, 2022. | Includes bibliographical references and index.
Identifiers: LCCN 2022023211 | ISBN 9781032062181 (hbk) | ISBN 9781032104102 (pbk) | ISBN 9781003215196 (ebk)
Subjects: LCSH: Graphene. | Electric batteries–Materials. | Capacitors–Materials. | Energy storage–Materials.
Classification: LCC TK2945.C37 G73 2022 | DDC 621.31/242–dc23/eng/20220818
LC record available at https://lccn.loc.gov/2022023211

ISBN: 9781032062181 (hbk)
ISBN: 9781032104102 (pbk)
ISBN: 9781003215196 (ebk)

DOI: 10.1201/9781003215196

Typeset in Times
by codeMantra

Contents

Editors

Fabian I. Ezema is a Professor at the University of Nigeria, Nsukka. He obtained a PhD and an MSc in Physics and Astronomy from University of Nigeria, Nsukka and a BSc from the then Anambra State University of Science and Technology, Enugu. His researches are on several areas of Materials Science: synthesis and characterizations of particles and thin-film materials through chemical routes with emphasis on energy applications. He is interested in materials for thin-film solar cells fabrication and nanoparticle synthesis. For the last 15 years, he has been working on the nano-/submicron-sized materials for energy conversion and storage (cathodes, anodes, supercapacitors, thin-film solar cells, DSSC, etc.), including novel methods of synthesis, characterization and evaluation of the electrochemical and optical properties.

Prof. Dr. Ishaq Ahmad currently holds the positions of Director General, Centre of Excellence in Physics, National Centre for Physics, Islamabad. He obtained his PhD of Engineering in Nuclear Technology and Applications in Materials Science from Shanghai Institute of Applied Physics, Graduate University of CAS (Chinese Academy of Science), Beijing, China. He has co-authored over 180 research papers in internationally reputed journals and contributed 6 books and 8 book chapters in reputed publishers. He supervised several Master's, PhD, and Postdoctoral local and abroad students. His research interests are focused on ion beam applications in materials/nanomaterials, synthesis of nanomaterials/thin films and ion beam analysis of materials. He was TWAS-UNESCO-iThemba LABS research Associate from 2014 to 2016. He has established international nano center called "NPU-NCP Joint International Research Center on Advanced Nanomaterials & Defects Engineering" at Northwestern Polytechnical University, Xi'an, Shaanxi, China with the collaboration of Prof. Dr. Zhao Tingkai. He has many awards and hours on his credit.

Dr. Tingkai Zhao is currently working as a Full Professor in the School of Materials Science and Engineering of Northwestern Polytechnical University (NPU, Xi'an China) and obtained his PhD from Materials Science and Engineering of Xi'an Jiaotong University (XJTU, Xi'an China) in 2005. He is a Fellow of International Association of Advanced Materials (IAAM, Sweden) and Vebleo Fellow. As a director of NPU-NCP Joint International Research Center on Advanced Nanomaterials & Defects Engineering and Vice-director of Shaanxi Engineering

Laboratory for Graphene New Carbon Materials & Applications, his research topics include the synthesis of graphene, carbon nanotubes (single-wall, multi-wall CNTs and amorphous CNTs), flexible graphite, 2D nanomaterials (like MXene), the growth mechanism, and applications of carbon nanomaterials in composites, energy conversion (solar cell, supercapacitor, and Li-ion batteries), smart devices, and biosensors. He has published 5 English books, 19 Chinese patents, and more than 150 SCI Papers at journals, and also served as reviewer of many international journals. He has been awarded 25 Honors & Rewards including the first prize of the Science and Technology of Shaanxi Province in 2013. Dr. Zhao's achievements have also been reported four times in *Science Daily*, the famous newspaper media of China. He held international conferences every year as Chair of International Symposium on Graphene Advanced Nanomaterials & Defects Engineering (GANDE) and Chair of International Conference on Graphene Novel Nanomaterials (GNN).

Contributors

Ada C. Agbogu
Department of Physics and Astronomy
University of Nigeria
Nsukka, Enugu State, Nigeria

Ishaq Ahmad
National Center for Physics
Quaid-i-Azam University
Islamabad, Pakistan
NPU-NCP Joint International Research
    Center on Advanced Nanomaterials
    and Defects Engineering
Northwestern Polytechnical University
Xi'an, China
Nanosciences African Network
    (NANOAFNET)
iThemba LABS-National Research
    Foundation
Somerset West, Western Cape Province,
    South Africa

Manickam Anusuya
Registrar, Indra Ganesan Group of
    Institutions
Tiruchirappalli, Tamilnadu, India

Malachy N. Asogwa
Department of Physics and Astronomy
University of Nigeria
Nsukka, Enugu State, Nigeria

Assumpta C. Nwanya
Department of Physics and Astronomy
University of Nigeria
Nsukka, Enugu State, Nigeria

Chinedu P. Chime
Agricultural & Bioresources
    Engineering
University of Nigeria
Nsukka, Enugu State, Nigeria

M. Dakshana
Sathyabama Institute of Science and
    Technology
Chennai, Tamilnadu, India

Chandrasekar Divya
CITE
Manonmaniam Sundaranar University
Tirunelveli, India

Eugene O. Echeweozo
Department of Physics with
    Electronics
Evangel University
Akaeze, Ebonyi State, Nigeria

Jeyachandran Eindhumathy
Department of ECE
Saranathan College of
    Engineering
Tiruchirappalli, Tamilnadu, India

A. B. C. Ekwealor
Department of Physics and Astronomy
University of Nigeria
Nsukka, Nigeria

Blessing N. Ezealigo
Dipartimento di Ingegeneria
    Meccanica, Chimica e
    dei Materiali
Universita Degli Studi di Cagliari via
    Marengo
Cagliari, Italy

Fabian I. Ezema
Department of Physics
University of Nigeria
Nsukka, Enugu State, Nigeria
NPU-NCP Joint International Research
Center on Advanced Nanomaterials
and Defects Engineering
Northwestern Polytechnical University
Xi'an, China
Nanosciences African Network
(NANOAFNET)
iThemba LABS-National Research
Foundation
Somerset West, Western
Cape Province,
South Africa
UNESCO-UNISA Africa Chair in
Nanosciences/Nanotechnology,
College of Graduate Studies
University of South Africa (UNISA)
Pretoria, South Africa
Africa Centre of Excellence for
Sustainable Power and Energy
Development
University of Nigeria
Nsukka, Enugu State, Nigeria

Chinedu Iroegbu
Department of Physics
Federal University of
Technology
Owerri, Nigeria

David C. Iwueke
Department of Physics
Federal University of
Technology
Owerri, Nigeria

Gabriel N. Kasozi
Department of Chemistry, College of
Natural Sciences
Makerere University
Kampala, Uganda

Moses Kigozi
Department of Chemistry, Faculty of
Science and Education
Busitema University
Tororo, Uganda

John Baptist Kirabira
Department of Mechanical Engineering
Makerere University
Kampala, Uganda

Venkatesh Koushick
Department of ECE
Vel Tech Rangarajan Dr. Sagunthala
R&D Institute of Science and
Technology
Chennai, India

Malik Maaza
Nanosciences African Network
(NANOAFNET)
iThemba LABS-National Research
Foundation
Somerset West, Western Cape Province,
South Africa
UNESCO-UNISA Africa Chair in
Nanosciences/Nanotechnology,
College of Graduate Studies
University of South Africa (UNISA)
Pretoria, South Africa

M. Malarvizhi
K.S. Rangasamy College of Technology
Tiruchengode, Tamilnadu, India

Sylvester M. Mbam
Department of Physics and Astronomy
University of Nigeria
Nsukka, Enugu State, Nigeria

Tariq Mehmood
College of Environmental
Hohai University
Nanjing, China

S. Meyvel
Department of Physics
Chikkaiah Naicker College
Erode, Tamilnadu, India

Agnes Chinecherem Nkele
Department of Physics and
   Astronomy
University of Nigeria
Nsukka, Nigeria
&
Department of Physics
Colorado State University
USA

Hope E. Nsude
Department of Physics and
   Astronomy
University of Nigeria
Nsukka, Enugu State, Nigeria

Kingsley U. Nsude
Department of Physics and
   Astronomy
University of Nigeria
Nsukka, Nigeria

Onyekachi Nwakanma
Department of Physics and
   Astronomy
University of Nigeria
Nsukka, Enugu State, Nigeria

Assumpta C. Nwanya
Department of Physics and
   Astronomy
University of Nigeria
Nsukka, Enugu State, Nigeria
SensorLab
University of the Western Cape Sensor
   Laboratories
Cape Town, South Africa

Chinwe Nwanya
Department of Physics and
   Astronomy
University of Nigeria
Nsukka, Enogu State, Nigeria

Raphael M. Obodo
Department of Physics and Astronomy
University of Nigeria
Nsukka, Enugu State, Nigeria
Department of Physics
University of Agriculture and
   Environmental Sciences
Umuagwo, Imo State, Nigeria
National Center for Physics
Quaid-i-Azam University
Islamabad, Pakistan
NPU-NCP Joint International Research
   Center on Advanced Nanomaterials
   and Defects Engineering
Northwestern Polytechnical University
Xi'an, China

Izunna S. Okeke
Department of Physics and Astronomy,
   Faculty of Physical Sciences
University of Nigeria
Nsukka, Enugu State, Nigeria
Medical Biotechnology Department
National Biotechnology Development
   Agency
FCT Abuja, Nigeria

Chinemerem Jerry Ozoude
Department of Metallurgical &
   Materials Engineering
&
Africa Centre of Excellence for
   Sustainable Power and Energy
   Development (ACE-SPED)
University of Nigeria
Nsukka, Enogu State, Nigeria

Swati N. Pusawale
Department of Sciences & Humanities
Rajarambapu Institute of Technology,
    affiliated to Shivaji University
Uran Islampur, Maharashtra, India

P. Sathya
Department of Physics
Salem Sowdeswari College
Salem, Tamilnadu, India

Emmanuel Tebandeke
Department of Chemistry, College of
    Natural Sciences
Makerere University
Kampala, Uganda

Ikechukwu Ifeanyi Timothy
Department of Civil Engineering
University of Nigeria
Nsukka, Enogu State, Nigeria

Ericdavid E. Ugochukwu
Department of Physics and Astronomy
University of Nigeria
Nsukka, Enugu State, Nigeria

Sabastine E. Ugwuanyi
Department of Physics and Astronomy
University of Nigeria
Nsukka, Enugu State, Nigeria

Raju Venkatesh
Department of Physics
PSNA College of Engineering &
    Technology
Dindigul, TamilNadu, India

Balasubramaniam Yogeswari
Department of Physics
Sri Eshwar College of Engineering
    (Autonomous)
Coimbatore, TamilNadu, India

1 Elementary Concept of Graphene Oxide Additive in Nanoscience and Nanotechnology

Sylvester M. Mbam, Ericdavid E. Ugochukwu, and Sabastine E. Ugwuanyi
University of Nigeria

Raphael M. Obodo
University of Nigeria
University of Agriculture and Environmental Sciences
Quaid-i-Azam University
Northwestern Polytechnical University

Assumpta. C. Nwanya
University of Nigeria
iThemba LABS-National Research Foundation
UNESCO University of South Africa (UNISA)

A. B. C. Ekwealor
University of Nigeria

Ishaq Ahmad
Quaid-i-Azam University
Northwestern Polytechnical University

Malik Maaza
iThemba LABS-National Research Foundation
UNESCO University of South Africa (UNISA)

Fabian I. Ezema
University of Nigeria
iThemba LABS-National Research Foundation
UNESCO University of South Africa (UNISA)

DOI: 10.1201/9781003215196-1

CONTENTS

1.1   INTRODUCTION

Recently, there has been an increasing interest in the research to strategically tune the functionalities and properties of nano-sized materials into a hybrid nanostructure to obtain a distinct property that can be specifically employed in several applications such as spin electronics, energy, memory, and microwave devices, catalysis, sensor, and bio-medical (where biocompatible nanomaterials are applied directly or used to substitute natural materials to function/be in contact with the living systems) (Obodo et al., 2020a, 2020b, 2021a,).

Thus, nanotechnology has overwhelmingly offered increasing innovations such as environmental-friendly technology, renewable energy technology, and nanomedicine. Nanomaterials are generally of the dimension ranging from 1 to 100 nm, and they lie in the region between molecules (which are of the order of 1 Å–1 nm) and micromaterials (with a dimension of 100 nm–100 μm). When likened to bulk materials, nanostructures possess distinct and improved behaviors due to their higher surface area, and significant confinement effect where the energy bands that appear separately in molecules and continuous in bulk structures hereby turn discrete, and the density of states thus becomes proportional to the low size of the material (Mbam et al., 2021a; Meng & Xiao, 2014; Ozin & Arsenault, 2009).

Materials such as quantum dots (Zrazhevskiy et al., 2010), DNA structures (Goodman et al., 2005), nanoparticles (Gupta & Gupta, 2005; Jain et al., 2007), graphene (Obodo et al., 2020c; Geim, 2009), fullerenes (Popov et al., 2013), carbon nanotubes (Terrones, 2003), and zeolites (Meng & Xiao, 2014) have all been reported to exist in the nanoscale dimension, together with such structures synthesized through atomic layer deposition (Fischer & Wegener, 2013), electron beam lithography (Altissimo, 2010), and direct laser inscription (Zeng et al., 2014). In the olden days, nanoscale materials have been adequately used, for example, the plasmonic nanoparticles which have been earlier as a colorant in glasses (Freestone, Meeks, Sax, & Higgitt, 2007), carbon black which has previously served as a raw material for producing tires (Medalia, 1978), and amorphous silica which has also served for producing toothpaste and paint (Merget et al., 2002). These olden nanostructures, therefore, require low specifications and functionality.

Graphene is a flat lattice with a honeycomb structure comprising one layer of carbon atoms being chained together by the support of sp^2 hybrid bonds. The graphene structure, therefore, characterizes a new category of material that is one atom thick, which is also known as two-dimensional (2D) materials (as they can only expand in length and width, the third-dimension height is zero). Other graphitic materials such as graphite, fullerene, and carbon nanotube are structured from the graphene mono-layer (Mbam et al., 2021b).

This review presents the functional properties of graphene, effective techniques for engineering the band gap of graphene to boost its diverse applications, the super-conducting properties of graphene, and various techniques (including flash Joule heating) for synthesizing high-quality and large-scale graphene oxide (GO). Also, the applications of graphene in various fields and specifically in energy storage systems were reviewed.

1.2 FUNCTIONAL PROPERTIES OF GRAPHENE

Graphene being a new structure of carbon discovered in 2004 (Mayorov et al., 2011) has been widely reported to possess excellent intrinsic properties (Mbam et al., 2021b). The exceptional properties of graphene originate from the π state bands that delocalize over its constituent carbon layers. (Table 1.1).

1.3 BAND GAP ENGINEERING IN GRAPHENE

Graphene being a semimetal has zero energy gap (Liu & Shen, 2009), which makes it not applicable to logic circuits. It thus requires a bandgap engineering to enhance it for applications in electronics. Modifying the properties of graphene involved the covalent infusion of heteroatoms into the graphitic structure as earlier demonstrated by Wang et al. (2009). Various techniques for enhancing the electronics proper-ties of graphene have been reported by many authors. To fabricate graphene-based field-effect transistors, most researched techniques have involved tuning graphene armchair-edged nanoribbon (Yamaguchi et al., 2020), doping graphene (Wei et al., 2009), and also introducing a matrix such as boron nitride to act as support (Britnell et al., 2012). Graphene with semiconducting properties (Chang et al., 2013; Xu et al., 2008) was obtained through co-doping nitrogen and boron with carbon nanotubes and graphene, employing the chemical vapor deposition technique (Fukushima et al., 2013). Recently, the atoms of boron, nitrogen, and sulfur have been employed in creating the heterostructures of a graphene nanoribbon with enhanced proper-ties (Kawai et al., 2015; Zhang et al., 2014; Nguyen et al., 2016). In this case, the bottom-up synthesis approach has been effectively employed in obtaining the hetero-structure where the dopant atoms are attached only at the edges without infusing into the graphene nanoribbon structure (Bronner et al., 2013; Rizzo et al., 2019; Mateo et al., 2020). Among the dopants, nitrogen atoms have attracted greater interest due to their electron affinity in graphene, more feasibility, coupled with their tendency to also create enough ferromagnetic property in graphene, especially when doped at high concentrations (Babar & Kabir, 2019; Blonski et al., 2017). A heteroatom dop-ing of graphene using boron/nitrogen was carried out by Jung et al. (2014). Through a solvothermal process, the reaction involved carbon tetrachloride (CCl_4), boron tri-bromide (BBr_3), and nitrogen (N_2) in the presence of potassium (K). The fabricated

TABLE 1.1

Functional Properties of Graphene

Properties	Values	Explanation	References
Electron mobility	200,000 cm^2 (Vs)$^{-1}$	Electron in graphene is the most mobile among any known material (100 times more high than silicon).	Chen et al. (2008)
Electrical resistivity	1×10^{-8} Ωm	The most conductive known material, having very low resistivity at room temperature (about 65% more conductive than copper).	Chen et al. (2008); Morozov, et al. (2008)
Surface area	2,630 m^{2}g^{-1}	Due to the intrinsic two-dimensional structure of graphene, its particles are so much tiny.	Peigney et al. (2001)
Atomic thickness	About (0.4–1.7 nm)	Graphene is only one atom thick (monolayer), thus known as a two-dimensional material.	Shearer et al. (2016)
Thermal conductivity	5,300 W^{-1} mK	At room temperature, graphene conducts heat even far higher than diamond.	Balandin et al. (2008)
Strength	42 N^{-1}m	A pure graphene single layer is the strongest known material, and with an equivalent intrinsic strength of about 130 GPa, it is many times stronger than steel.	Lee et al. (2008)
Stiffness	1.0 TPa (Young's modulus)	Young's modulus value of the single-layer graphene is among the highest when compared to other known materials.	Lee et al. (2008); Papageorgiou et al. (2017)
Flexibility and toughness	About 25% stretchability	Though graphene is slightly brittle, but can withstand pressure and thus flexible.	Lee et al. (2008); Liu et al. (2017)
Impermeability	100% impermeable	The honeycomb lattice structure formed by the sp^2-hybridized carbon atoms has offered it an intrinsic highly elastic and impermeable layer. Not even helium atom can pass through it.	Liu et al. (2013)
Transparency	About 97.7% optically transparent	Graphene sheet absorbs a minimal value (2.3%) when exposed to a reflecting light, and this can be linked to its non-interacting Dirac fermions.	Sheehy and Schmalian (2009)

boron/nitrogen co-doped graphene (BCN-graphene) possessed 2.66% and 2.38% contents of nitrogen and boron, respectively, which consequently originate into a field-effect transistor that has an on/off ratio of 10.7 and a bandgap of 3.3 eV.

Pawlak et al. (2020) used a brominated tetrabenzophenazine through a silver-aided Ullmann polymerization reaction to synthesize nitrogen-doped graphene nanoribbons. The one-dimensional metastable compound catalyzed the formation of the porous nitrogen-doped graphene nanoribbons, which have a semiconducting bandgap of about 2.2 eV on the silver (Ag) substrate and a significant donor-acceptor heterostructure. These were enhanced by the orderly paired nanopores that have a diameter of 4.6 Å and a high concentration of nitrogen atoms. Analysis of the transport properties in porous graphene had earlier been recorded by Moreno et al. (2018). Wang et al. (2017) and Li et al. (2015) which also yielded similar results. Talirz et al. (2017) illustrated the basic techniques in the periodic penetration of an armchair graphene nanoribbon doped with the nitrogen atom. In the contrast, this involved a top-down technique, as the benzene ring needs to be periodically displaced along the graphene nanoribbon longitudinal axis, while four unstable carbon bonds are removed by atoms of hydrogen, and nitrogen atoms instantly replace the rest of the carbon atoms.

The other alternative approaches for enhancing the magnetic and electronic properties of graphene also involved creating pores or defects into the heterostructure. Proton movement through single-sheet graphene was analyzed by Actyl et al. (2015), and it was found that monolayer graphene on fused silica substrate subjected to a series of different pH levels, can permit a reversible transfer of protons in the aqueous phase across the graphene layer to the opposite side. Using a computer simulation, the authors identified that the proton transfer uses a small energy barrier of 0.61–0.75 eV from the aqueous phase through naturally arising atomic defects. Pederson et al. (2008) reported graphene sheets that have orderly arranged holes, which results in an antidote lattice. The antidote lattice is, thus, a triangular cluster of holes in a graphene layer. In this case, the antidote lattices consequently induce a semiconducting property with a tunable energy gap into the semimetallic graphene layer. Also, whenever the holes dissociate, defect states/pairs of entangled defects states are always formed, and this can host quantum information (DiVincenzo et al., 2000). The induced defects, however, enhance the creation of confined electron bands in the graphene antidote lattice.

1.4 SUPERCONDUCTIVITY IN GRAPHENE

Despite the excellent electrical conductivity of graphene, researchers have recently developed it into a superconducting material that seamlessly allows the flow of electrons without any resistance. This was actualized when a sheet of graphene is being stacked over another sheet, then turning one sheet to a certain angle (otherwise known as the 'magic angle') and allowing the set-up to cool at a temperature little above zero kelvin (0 K). This twisting can tremendously alter the properties of the graphene bilayer, changing into an insulator at first, and then into a supercapacitor when enough electric field is applied (Phong et al., 2021; Cao et al., 2018, 2020). The double-layer structure when rotated forms a highly ordered structure taken as a superlattice which has a broader unit cell (He et al., 2021). Electrons, therefore, flow in between the stacked layers, which consequently display a strange property never seen in graphene before. When warped to a particular 'magic angle' (usually 1.1°), the superlattice, in turn, becomes a superconductor. Nevertheless, Chittari et al. (2019) earlier reported a superconductive property in multiple-stacked graphene sheets without any rotation. This implies that patterning the three layers in such a definite alignment forms a superlattice geometry that is analogous to that of magic-angle-tuned double layers, and this also demonstrated a comparable strongly correlated interaction.

Theoretical studies by Morell et al. (2010) and Bistritzer and MacDonald (2011) earlier evidenced the presence of a superconducting property in 2D materials at a certain little twist ('magic angles'), and this is because the fundamental structure of the superlattice can hugely alter its electrons' configurations, and thus fully enhance the electronics property of the materials. However, there is still hidden information on the exact twist angles that can initiate such changes, and how deep the properties can fluctuate.

However, the previous attempt to induce superconductivity into graphene had involved merging it with other materials that are good superconductors or using other elements (Uchoa & Neto, 2007; Nandkishore et al., 2012; Chen et al., 2019). Intriguingly, the non-superconducting thin layers of graphene can now be successfully engineered to offer significant superconducting behaviors. But researchers have pointed out that the key challenge in this process has been on getting the perfect twist angle (about 1.1°), and this is because twisted graphene layers when slightly feigned can possess a similar structure as graphite which has all its subsequent layers aligned in one direction (Li et al., 2010), also yet to fully understood , is whether the underlying superconducting mechanism in twisted graphene involves the same mechanism as that of a high-temperature superconductor (Lee et al., 2006) of which when explored will initiate the development of materials that transmits electricity with no resistance, even at room temperature. Consequently, this will open wider applications for superconductors, especially for efficient power transmission at a much-reduced cost.

A superconducting mechanism had always involved electrons coupling in a manner that permits them to move with zero hindrance. In the case of conventional superconductors, electrons can couple up in no specified manner due to the interaction involving the particles of the material and vibrations in its atomic lattice—(phonon). This implies that the electrons will slightly move apart, but at the end, they will pair up in a pattern that allows them transverse through the lattice with no resistance at a temperature little above absolute zero. While for unconventional superconductors, electrons can couple up using a direct and more compact interaction, thus conducting with no resistance even at a high temperature. Graphene having the same band structure as metals is just a conductor, where the mobile electrons only intermingle with the atomic lattice and never with each other. Thus, the insulating phenomenon also observed shortly in bilayer graphene originates as the bilayer structure can also block the flow of electrons, and this scenario is the evidence of underlying interactions in the bilayer graphene. Understanding explicitly the fundamentals of strongly correlated systems (Avsar et al., 2020; Shen et al., 2020) as observed in twisted bilayer graphene at the magic angle can open a clear concept of many other peculiar and fascinating states of matter yet to be deeply explored. Such as quantum spin liquids (odd/chaotic states where there is no alignment in the magnetic field) (Fujihala et al., 2020), fractional quantum hall states/effect (Stormer, 1999; Sanchez-Yamagishi et al., 2012; Cysne et al., 2018), and topological phase (Zaletel & Khoo, 2019; Yin et al., 2020). Clear studies on these systems are of great potential applications in a high technological innovation like very powerful quantum computing systems (Wolf et al., 2006).

Graphene has been confirmed to exhibit both conventional and unconventional superconducting mechanisms. Cao et al. (2018) recorded an unconventional behavior in bilayer graphene twisted at a magic angle of 1.1° which first initiated a correlated insulating state on half-filling. A superconducting state was recorded in the superlattice at a critical temperature of about 1.7 K when introduced with an electric field (electrostatic doping). However, the temperature-carrier density phase plots of the twisted double-layer graphene were observed to display the same structure as those of cuprates (Tsuei & Kirtley, 2000) (that is, a dome-shaped area that signifies superconductivity). It was also deduced that the superconducting behavior at such a fairly high critical temperature, coupled with a little Fermi surface (about $1,011\,cm^{-2}$ carrier density) of the graphene superlattice, is an indication of a very tough pairing power among the electrons. On the other hand, the studies on the correlated insulating and superconducting phases in twisted bilayer graphene by Codecido et al. (2019) recorded a low twist angle of 0.93° and a near low critical temperature of 0.3–0.5 K. The strength of the electrons' interactions in the magic angle of twisted graphene has been observed to be influenced by a tunable electric field (Gonzalez-Arraga et al., 2017; Yankowitz et al., 2014). Also, introducing pressure on the superlattice can consequently enhance the interlayer coupling/induced Josephson coupling in the perpendicular planes (Yankowitz et al., 2018; Anđelković et al., 2018; Mao et al., 2020).

1.5 METHODS OF SYNTHESIS

Scientists had previously argued the impossibility of 2D crystalline structure since they are thermodynamically not stable. However, in 2004, Andre Geim and Konstantin Novoselov from the University of Manchester were able to extract a single layer of graphite (graphene) out of the bulk graphite in an ordinary pencil. This was possible using an adhesive tape which successfully removed a flake of carbon that is only one atom thick. This technique (mechanical exfoliation) is the most simple technique to first made a stand-alone graphene possible (which fetched Andre Geim and Konstantin Novoselov a Nobel prize in physics in 2010) (Geim & Novoselov, 2010). Several techniques have subsequently been used to produce quality graphene on large scale. These can be broadly categorized into bottom-up techniques (such as chemical/physical vapor deposition, arc discharge, unzipping of carbon nanotubes, and epitaxial growth on silicon substrates) or top-down techniques (chemical reduction of exfoliated graphite, thermal expansion of graphite, electrochemical exfoliation, and liquid-phase exfoliation of graphite) (Mbam et al., 2021b).

The most widely utilized technique for producing GO was first established by Hummers and Offeman (Hummers' method) (Hummers Jr & Offeman, 1985). This method involves the oxidation of graphite flake in a concentration of H_2SO_4, $KMnO_4$, and $NaNO_3$. This technique has significant advantages over earlier methods reported by Brodie (1859), where $KClO_3$ and HNO_3, which are environmentally harsh, are replaced with $KMnO_4$ and $NaNO_3$, respectively. Hummers' method also has a short reaction time; however, it has some defects which include the evolution of toxic NO_2 and N_2O_4 gases, and the formation of insoluble Na^+ and NO^{3-} ions during the synthesis process. Marcano et al. (2010) improved Hummers' technique by increasing the ratio of $KMnO_4$ while removing $NaNO_3$. The reaction was also carried out in a 9:1 solution of H_2SO_4/H_3PO_4, and this alteration in the reactants successfully increased the reaction yield and also reduced the amount of harmful gas emitted from the reaction process. Hummers' method was also modified by Chen et al. by synthesizing GO without the use of $NaNO_3$. This improvement prevented the production of poisonous gases and also reduced the cost of the synthesis. However, the GO samples synthesized using both the conventional and modified Hummers' method have almost the same chemical and physical structures (Chen et al., 2013; Nwanya et al., 2020; Obodo et al., 2020d). The X-ray diffraction (XRD) pattern of the GO synthesized through this technique recorded a diffraction peak around $2\theta = 11°$ as shown in Figure 1.2a. Likewise, it was found that the omission of $NaNO_3$ did not reduce the yield of the entire reaction, but rather simplified the synthesis process.

Luong et al. (2020) employed a new process known as flash Joule heating to synthesize graphene. In this technique, amorphous carbon black is casually compacted in between the electrodes of a quartz tube. The electrode might be a copper or graphite material. Introducing a high voltage into the system skyrockets the temperature of the amorphous carbon to above 3,000 K within a very short time, thereby changing it to turbostratic flash graphene as illustrated in Figure 1.1. The XRD pattern of the flash graphene recorded an intense diffraction peak at angle $2\theta = 26.1°$ as shown in Figure 1.2b. The result recorded by the high-resolution transmission electron microscopy showed disoriented layers of the flash graphene with the usual Moire patterns as shown in Figure 1.3. The quality of the flash graphene synthesized from the carbon black was studied using Raman spectroscopy. It was observed that the D band exhibited an extremely low intense peak which is also evidenced in the $I_{2D/G}$ map showing the multiple sites of the intensity ratio (Figure 1.4). The high value of the intensity ratio ($I_{2D/G} = 17$) indicates a low amount of defects available in the flash graphene. However, the high temperature utilized in the flash Joule technique therefore successfully eliminated other non-carbon materials and impurities from the sample.

1.6   APPLICATIONS OF GRAPHENE OXIDE

Graphene has been applied in various fields of science and technology. These include electronics, biology, material science, tissue engineering, ultra-high sensors, and nanomedicine. In the field of electronics, graphene was successfully used to fabricate a tunable single-electron transistor. The charging energy of the transistor

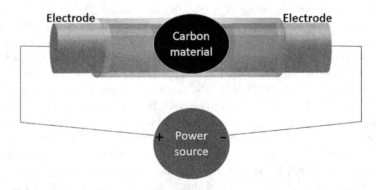

FIGURE 1.1 Synthesis of graphene using the flash Joule heating process. (Adapted from Luong et al. (2020), Copyright 2020, Nature.)

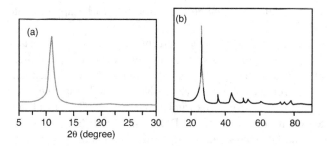

FIGURE 1.2 (a) The XRD pattern of the graphene oxide synthesized using the improved Hummers' method (without $NaNO_3$) (Chen et al., 2013). (Reprinted with permission from Chen et al. (2013), Copyright 2013, Elsevier). (b) The XRD pattern of the flash graphene synthesized using flash Joule heating technique (Luong et al., 2020). (Reprinted with permission from Luong et al. (2020), Copyright 2020, Nature.)

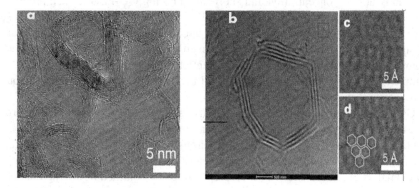

FIGURE 1.3 (a) The TEM image and (b) HR-TEM image of the flash graphene synthesized using flash Joule heating technique (Luong et al., 2020). (Reprinted with permission from Luong et al. (2020), Copyright 2020, Nature.)

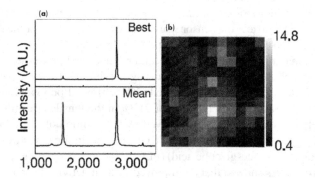

FIGURE 1.4 (a) Raman spectra of the flash graphene and (b) Raman ((I_2D/G) map of the flash graphene synthesized using flash Joule heating technique (Luong et al., 2020). (Reprinted with permission from Luong et al. (2020), Copyright 2020, Nature.)

was estimated by coulomb diamond measurements to be approximately 3.4 MeV (Stampfer et al., 2008).

Graphene has proven potent in the biotechnology field in areas such as biosensing, disease diagnosis, cancer-targeting, drug delivery, and photothermal therapy. Dembereldorj et al. reported on the in vitro and in vivo intracellular glutathione (GSH)—triggered releases of polyethylene glycol-coated graphene oxide (PEG-GO) loaded on the anticancer drug, doxorubicin (DOX). PEG-GO did not show cytotoxicity up to 100 mg mL^{-1}, while the DOX-embedded PEG-GO reduced the cancer cells above the DOX concentration of 10–5 M. Their study showed promising real-time release of DOX from PEGylated GO after an external triggering by GSH (Dembereldorj et al., 2012). PEG-functionalized GO was also prepared using the modified Hummers' method for biocompatibility and stability and then used to improve the operating rate of paclitaxel (PTX), another anticancer drug. The GO-PEG bound well to 11.2 wt% of the PTX via π-π stacking and hydrophobic interactions. GO-PEG/PTX exhibited a higher cytotoxicity effect compared to free PTX, especially at low concentration and short time proving GO-PEG to be a promising nanomaterial in medical areas (Xu et al., 2014). Derivatives of graphene when fused with biopolymers have been observed to be good candidates for use as functionalizing agents in drug delivery procedures. For example, gelatin-functionalized graphene nanosheet (GNS) gave evidence of its potential for drug delivery, and GNS–DOX complex also exhibited high toxicity toward U251, 1,800, and A-5RT$_3$ cells through endocytosis (Liu et al., 2011).

GO has been utilized effectively in sensors, as in the detection of dangerous gases and greenhouse gases such as nitrogen dioxide, hydrogen, and carbon dioxide. Studies have shown that a resistance structure with a layer of reduced GO can react to NO$_2$ already at a concentration of parts per billion in an atmosphere of carrier gas (Drewniak et al., 2016). Prezioso et al. prepared to GO obtained from graphite flakes using Hummers' method and used for NO$_2$ sensing. The resistance behavior of the GO was observed to depend on the temperature. They concluded that 150°C was the best temperature for sensor operation, and as-deposited GO was superior to annealed GO with the conductivity of the GO increase in exposure of NO$_2$ (Esfandiar et al., 2012). Making use of reduced graphene oxide (rGO) obtained from GO and doped on lead, Pandey et al. reported on the sensing of hydrogen gas. The thickness of the rGO was reported to affect the sensor's sensitivity as an increase in the thickness of rGO led to a decrease in the sensor response (Pandey et al., 2013).

Due to its amazing mechanical, chemical, and electrical properties and stabilizing property, graphene application in tissue engineering has been reported by many researchers. Wang et al. (2011) synthesized GO-reinforced polyvinyl alcohol composite films which recorded an increase of 212% in the tensile strength, and a 34% increase in elongation at break was achieved by the addition of 0.5 wt% of GO. Fu et al. also reported on the engagement of GO for the improvement of the tensile strength of poly(l-lactic-co-glycolic acid) (PLGA) and hydroxyapatite (HA) (Fu et al., 2017). In vitro studies showed that an improved cellular activities and the osteogenic markers expression observed on PLGA/GO/HA nanofibrous could be ascribed to the synergistic additive effect between GO and HA which highly enabled a successful proliferation and osteogenic differentiation of MC$_3$T$_3$-E$_1$ cells. Graphene-based biohybrids have been effectively used for neural tissue engineering, possibly due to their electrical coupling with human neural stem cells, and have been reported by many authors (Park et al., 2011).

1.7 SPECIFIC APPLICATIONS OF GRAPHENE OXIDE IN ENERGY STORAGE SYSTEMS

The excellent electrical conductivity, thermal conductivity, and good mechanical properties of graphene have made it uniquely invaluable for energy storage applications. These energy storage devices could be graphene-based—supercapacitors, lithium batteries, lithium-sulfur batteries, and even lithium-air/batteries (Zhu et al., 2014).

Natorianni et al. reported the use of double-wall carbon nanotubes films with graphene films as electrodes in the fabrication of supercapacitors. The capacitance of this fabricated supercapacitor was noted to be higher than that of the supercapacitor which made use of the cathode nanotube film as both current collector and electrode. Increasing the electrode (graphene film) thickness was observed to improve the areal-specific capacitance of the fabricated supercapacitor (Notarianni et al., 2014). Kannappan et al. successfully employed Hummers' method and tip sonication for the synthesis of graphene-based supercapacitors. Scanning electron microscopy images showed the graphene to be highly porous which improved the charge storage ability. The crippled and wrinkled nature of the prepared graphene, as observed by transition electron microscopy, proved important as it helped prevent the stacking of the graphene sheets together. The supercapacitors were found to be very porous and possessed high-energy density, power density, and specific capacitance (Kannappan et al., 2013). rGO was used to coat magnesium nickel (Mg_2Ni) storage alloy by Du et al., employing the electrostatic adsorption method. Coating with rGO improved the capacity-maintenance property as well as stability of the alloys (Du et al., 2017).

Another class of energy storage device is graphene-based lithium-ion batteries, and these too have received ample interest among researchers. Yoo et al. pioneered the investigation of the use of GNS materials for purpose of increasing lithium storage capacities. They found the specific capacity of the GNS to be 540 mAhg^{-1} (Yoo et al., 2008). Using graphene foam (GF) network as both a conductive pathway for electrons and current collector, Li et al. assembled thin, lightweight, and flexible full lithium-ion batteries. In terms of high discharge rates, they noted that batteries with GF performed better than those without it. The capacity of the assembled flexible battery was very high at 117 mAh g^{-1} (Li et al., 2012).

Lithium-sulfur (Li-S) batteries have received tremendous interest because theoretically, and they can possess high specific energy if there is a complete reaction of lithium and sulfur. Graphene-based materials have been used to improve the mechanical flexibility and electrical conductivity of Li-S batteries. For example, the layered graphene-based porous carbon-sulfur composite material was synthesized and used as the cathode material in Li-S material (Yang et al., 2014). Using electrochemical impedance spectroscopy, the electrochemical performance of the Li-S batteries was shown to be outstanding. Wang et al. performed a similar procedure and obtained a reversible specific capacity of 400 mAh g^{-1} after 600 cycles at 1.0°C.

Herein, the hierarchical carbon pores together with graphene successfully give the Li-S battery long cycling reversibility (Wang et al., 2019).

Among rechargeable batteries, Li-air batteries have been reported to possess the highest energy density. To improve the efficiency and cycle life stability of these classes of lithium batteries, different approaches have been employed which include the use of nanostructured transition metal oxide, mesocellular carbon foam, superfine platinum particles, and graphene-based materials. Sun and his group reported on the use of GNSs as a cathode catalyst for Li-air batteries using an alkyl carbonate electrolyte. The cycling performance of the GNSs was reported to be better when compared to the Vulcan XC-72 electrode used in Li-air batteries (Sun et al., 2012).

1.8   CONCLUSION

Graphene being a new structure of carbon has been widely reported to possess excellent intrinsic properties. The exceptional properties of graphene originate from the π state bands that delocalize over its constituent carbon layers. Graphene being a semimetal has zero energy gap, which makes it not applicable in logic circuits. Various techniques for enhancing its electronics properties include tuning graphene armchair-edged nanoribbon, doping graphene, intercalation of matrix ions, and creating pores or defects into the heterostructure. Graphene was also reported to exhibit both conventional and unconventional superconducting mechanisms. Synthesis of quality graphene involves any of chemical/physical vapor deposition, arc discharge, unzipping of carbon nanotubes, epitaxial growth on silicon substrates, chemical reduction of exfoliated graphite, thermal expansion of graphite, electrochemical exfoliation, and liquid-phase exfoliation of graphite, among others. The excellent electrical conductivity, thermal conductivity, high mechanical strength, and high surface area of graphene have made it very useful in energy storage systems.

REFERENCES

Achtyl, J. L., Unocic, R. R., Xu, L., Cai, Y., Raju, M., Zhang, W., … & Geiger, F. M. (2015). Aqueous proton transfer across single-layer graphene. *Nature Communications, 6*(1), 1–7.

Altissimo, M. (2010). E-beam lithography for micro-nanofabrication. *Biomicrofluidics, 4*(2), 2–7.

Anđelković, M., Covaci, L., & Peeters, F. M. (2018). DC conductivity of twisted bilayer graphene: Angle-dependent transport properties and effects of disorder. *Physical Review Materials, 2*(3), 034004.

Avsar, A., Ochoa, H., Guinea, F., Özyilmaz, B., Van Wees, B. J., & Vera-Marun, I. J. (2020). Colloquium: Spintronics in graphene and other two-dimensional materials. *Reviews of Modern Physics, 92*, 021003.

Babar, R., & Kabir, M. (2019). Ferromagnetism in nitrogen doped graphene. *Physical Review B, 99*, 115442.

Balandin, A. A., Ghosh, S., Bao, W., Calizo, I., Teweldebrhan, D., Miao, F., & Lau, C. N. (2008). Superior thermal conductivity of single-layer graphene. *Nano Letters, 8*(3), 902–907.

Bistritzer, R., & MacDonald, A. H. (2011). Moiré bands in twisted double-layer graphene. *Proceedings of the National Academy of Sciences, 108*(30), 12233–12237.

Blonski, P., Tucek, J., Sofer, Z., Mazanek, V., Petr, M., Pumera, M., ... & Zboril, R. (2017). Doping with graphitic nitrogen triggers ferromagnetism in graphene. *Journal of the American Chemical Society, 139*, 3171–3180.

Britnell, L., Gorbachev, R. V., Jalil, R., Belle, B. D., Schedin, F., Mishchenko, A., ... & Ponomarenko, L. A. (2012). Field-effect tunneling transistor based on vertical graphene heterostructures. *Science, 335*(6071), 947–950.

Brodie, B. C. (1859). On the atomic weight of graphite. *Philosophical Transactions of the Royal Society of London, 149*, 249–259.

Bronner, C., Stremlau, S., Gille, M., Brausse, F., Haase, A., Hecht, S., & Tegeder, P. (2013). Aligning the band gap of graphene nanoribbons by monomer doping. *Angewandte Chemie International Edition, 52*, 4422–4425.

Cao, Y., Fatemi, V., Demir, A., Fang, S., Tomarken, S. L., Luo, J. Y., ... & Jarillo-Herrero, P. (2018). Correlated insulator behaviour at half-filling in magic-angle graphene superlattices. *Nature, 556*(7699), 80–84.

Cao, Y., Fatemi, V., Fang, S., Watanabe, K., Taniguchi, T., Kaxiras, E., & Jarillo-Herrero, P. (2018). Unconventional superconductivity in magic-angle graphene superlattices. *Nature, 556*(7699), 43–50.

Cao, Y., Rodan-Legrain, D., Rubies-Bigorda, O., Park, J. M., Watanabe, K., Taniguchi, T., & Jarillo-Herrero, P. (2020). Tunable correlated states and spin-polarized phases in twisted bilayer–bilayer graphene. *Nature, 583*(7815), 215–220.

Chang, C. K., Kataria, S., Kuo, C. C., Ganguly, A., Wang, B. Y., Hwang, J. Y., ... & Chen, K. H. (2013). Band gap engineering of chemical vapor deposited graphene by in situ BN doping. *ACS Nano, 7*(2), 1333–1341.

Chen, G., Jiang, L., Wu, S., Lv, B., Li, H., & Watanabe, K, ... & Wang, F. (2019). Gate-Tunable mott insulator in trilayer graphene-boron nitride moiré superlattice. In *APS March Meeting Abstracts*, vol. 2019, pp. S14–008.

Chen, J. H., Jang, C., Xiao, S., Ishigami, M., & Fuhrer, M. S. (2008). Intrinsic and extrinsic performance limits of graphene devices on SiO_2. *Nature Nanotechnology, 3*, 206–209.

Chen, J., Yao, B. L., & Shi, G. (2013). An improved Hummers method for eco-friendly synthesis of graphene oxide. *Carbon, 64*, 225–229.

Chittari, B. L., Chen, G., Zhang, Y., Wang, F., & Jung, J. (2019). Gate-tunable topological flat bands in trilayer graphene boron-nitride moiré superlattices. *Physical Review Letters, 122*(1), 016401.

Codecido, E., Wang, Q., Koester, R., Che, S., Tian, H., Lv, R., ... & Lau, C. N. (2019). Correlated insulating and superconducting states in twisted bilayer graphene below the magic angle. *Science Advances, 5*(9), eaaw9770.

Cysne, T. P., Garcia, J. H., Rocha, A. R., & Rappoport, T. G. (2018). Quantum Hall effect in graphene with interface-induced spin-orbit coupling. *Physical Review B, 97*, 085413.

Dembereldorj, U., Kim, M., Kim, S., Ganbold, E., & Lee, S. Y. (2012). A spatiotemporal anti-cancer drug release platform of PEGylated graphene oxide triggered by glutathione in vitro and in vivo. *Journal of Materials Chemistry, 22*, 23845.

DiVincenzo, D. P., Bacon, D., Kempe, J., Burkard, G., & Whaley, K. B. (2000). Universal quantum computation with the exchange interaction. *Nature, 6810*(408), 339–342.

Drewniak, S., Muzyka, R., Stolarczyk, A. I., Pustelny, T., Kotyczka-Mora´nska, M., & Setkiewicz, M. (2016). Studies of reduced graphene oxide and graphite oxide in the aspect of their possible application in gas sensors. *Sensors, 16*(1), 103.

Du, Y., Li, N., Zhang, T., Feng, Q., Du, Q., Wu, X., & Huang, D. (2017). Reduced graphene oxide coating with anticorrosion and electrochemical property-enhancing effects applied in hydrogen storage system. *ACS Applied Materials & Interfaces, 9*(34), 28980–28989.

Esfandiar, A., Ghasemi, S., Irajizad, A., Akhavan, O., & Gholami, M. R. (2012). The decoration of TiO_2/reduced graphene oxide by Pd and Pt nanoparticles for hydrogen gas sensing. *Journal of Hydrogen Energy, 37*(20), 15423–15432.

Fischer, J., & Wegener, M. (2013). Three-dimensional optical laser lithography beyond the diffraction limit. *Laser & Photonics Reviews, 7*(1), 22–44.

Freestone, I., Meeks, N., Sax, M., & Higgitt, C. (2007). The Lycurgus cup - a roman nanotechnology. *Gold Bulletin, 40*(4), 270–277.

Fu, C., Bai, H., Zhu, J., Niu, Z., Wang, Y., Li, J., … & Bai, Y. (2017). Enhanced cell proliferation and osteogenic differentiation in electrospun PLGA/hydroxyapatite nanofibre scaffolds incorporated with graphene oxide. *PLoS One, 12*(11), e0188352.

Fujihala, M., Morita, K., Mole, R., Mitsuda, S., Tohyama, T., Yano, S. I., … & Nakajima, K. (2020). Gapless spin liquid in a square-kagome lattice antiferromagnet. *Nature Communications, 11*(1), 1–7.

Fukushima, K., Liu, S., Wu, H., Engler, A. C., Coady, D. J., Maune, H., … & Hedrick, J. L. (2013). Supramolecular high-aspect ratio assemblies with strong antifungal activity. *Nature Communications, 4*(1), 1–9.

Geim, A. K. (2009). Graphene: Status and prospects. *Science, 324*(5934), 1530–1534.

Geim, A. K., & Novoselov, K. S. (2010). The rise of graphene. In *Nanoscience and Technology: A Collection of Reviews from Nature Journals* (pp. 11–19), edited by Peter Rodgers, London: Nature publishing group.

Gonzalez-Arraga, L. A., Lado, J. L., Guinea, F., & San-Jose, P. (2017). Electrically controllable magnetism in twisted bilayer graphene. *Physical Review Letters, 119*(10), 107201.

Goodman, R. P., Schaap, I. A., Tardin, C. F., Erben, C. M., Berry, R. M., Schmidt, C. F., & Turberfield, A. J. (2005). Rapid chiral assembly of rigid DNA building blocks for molecular nanofabrication. *Science, 310*(5754), 1661–1665.

Gupta, A. K., & Gupta, M. (2005). Synthesis and surface engineering of iron oxide nanoparticles for biomedical applications. *Biomaterials, 26*(18), 3995–4021.

He, M., Li, Y., Cai, J., Liu, Y., Watanabe, K., Taniguchi, T., … & Yankowitz, M. (2021). Symmetry breaking in twisted double bilayer graphene. *Nature Physics, 17*(1), 26–30.

Hummers Jr, W. S., & Offeman, R. E. (1985). Preparation of graphitic oxide. *Journal of the American Chemical Society, 80*(6), 1339–1339.

Jain, P. K., Huang, X., El-Sayed, I. H., & El-Sayed, M. A. (2007). Review of some interesting surface plasmon resonance-enhanced properties of noble metal nanoparticles and their applications to biosystems. *Plasmonics, 2*(3), 107–118.

Jung, S. M., Lee, E. K., Choi, M., Shin, D., Jeon, I. Y., Seo, J. M., … & Baek, J. B. (2014). Direct Solvothermal Synthesis of B/N-Doped Graphene. *Angewandte Chemie, 126*(9), 2430–2433.

Kannappan, S., Kaliyappan, K., Manian, R. K., Pandian, A. S., Yang, H., Lee, Y. S., … & Lu, W. (2013). Graphene based supercapacitors with improved specific capacitance and fast charging time at high current density. *arXiv preprint arXiv*, 1311.1548.

Kawai, S., Saito, S., Osumi, S., Yamaguchi, S., Foster, A. S., Spijker, P., & Meyer, E. (2015). Atomically controlled substitutional boron-doping of graphene nanoribbons. *Nature Communications, 6*, 80–98.

Lee, C., Wei, X., Kysar, J. W., & Hone, J. (2008). Measurement of the elastic properties and intrinsic strength of monolayer graphene. *Science, 321*(5887), 385–388.

Lee, P. A., Nagaosa, N., & Wen, X. G. (2006). Doping a Mott insulator: Physics of high-temperature superconductivity. *Reviews of Modern Physics, 78*, 17–85.

Li, G., Luican, A., Dos Santos, J. L., Neto, A. C., Reina, A., Kong, J., & Andrei, E. Y. (2010). Observation of Van Hove singularities in twisted graphene layers. *Nature Physics, 6*(2), 109–113.

Li, N., Chen, Z., Ren, W., Li, F., & Cheng, H. (2012). Flexible graphene-based lithium ion batteries with ultrafast charge and discharge rates. *Proceedings of the National Academy of Sciences, 109*(43), 17360–17365.

Li, S., Yang, Z.-D., Zhang, G., & Zeng, X. C. (2015). Electronic and transport properties of porous graphene sheets and nanoribbons: benzo-CMPs and BN codoped derivatives. *Journal of Materials Chemistry C, 3*, 9637–9649.

Liu, K., Zhang, J., Cheng, F., Zheng, T., Wang, C., & Zhu, J. J. (2011). Green and facile synthesis of highly biocompatible graphenenanosheets and its application for cellular imaging and drug delivery. *Journal of Materials Chemistry, 21*(32), 12034–12040.

Liu, L., & Shen, Z. (2009). Bandgap engineering of graphene: A density functional theory study. *Applied Physics Letters, 95*(25), 252104.

Liu, N., Chortos, A., Lei, T., Jin, L., Kim, T. R., Bae, W. G., ... & Bao, Z. (2017). Ultratransparent and stretchable graphene electrodes. *Science Advances, 3*(9), e1700159.

Liu, Z. Y., Xiao, B. L., Wang, W. G., & Ma, Z. Y. (2013). Developing high-performance aluminum matrix composites with directionally aligned carbon nanotubes by combining friction stir processing and subsequent rolling. *Carbon, 62*, 35–42.

Luong, D. X., Bets, K. V., Algozeeb, W. A., Stanford, M. G., Kittrell, C., Chen, W., ... & Tour, J. M. (2020). Gram-scale bottom-up flash graphene synthesis. *Nature, 577*(7792), 647–651.

Mao, J., Milovanović, S. P., Anđelković, M., Lai, X., Cao, Y., Watanabe, K., ... & Andrei, E. Y. (2020). Evidence of flat bands and correlated states in buckled graphene superlattices. *Nature, 584*(7820), 215–220.

Marcano, D. C., Kosynkin, D. V., Berlin, J. M., Sinitskii, A., Sun, Z., Slesarev, A., ... & Tour, J. M. (2010). Improved synthesis of graphene oxide. *ACS Nano, 4*(8), 4806–4814.

Mateo, L. M., Sun, Q., Liu, S. X., Bergkamp, J. J., Eimre, K., Pignedoli, C. A., ... & Torres, T. (2020). On-surface synthesis and characterization of triply fused porphyrin-graphene nanoribbon hybrids. *Angewandte Chemie International Edition, 59*, 1334–1339.

Mayorov, A. S., Gorbachev, R. V., Morozov, S. V., Britnell, L., Jalil, R., Ponomarenko, L. A., ... & Geim, A. K. (2011). Micrometer-scale ballistic transport in encapsulated graphene at room temperature. *Nano Letters, 11*(6), 2396–2399.

Mbam, S. M., Obodo, R. M., Nwanya, A. C., Ekwealor, A. B., Ahmad, I., & Ezema, F. I. (2021a). Research progress in synthesis and electrochemical performance of bismuth oxide. In *Electrode Materials for Energy Storage and Conversion* (Vol. 3, pp. 379–395), edited by Mesfin A. Kebede, Florida: CRC Press.

Mbam, S. M., Obodo, R. M., Nwanya, A. C., Ekwealor, A. B., Ahmad, I., & Ezema, F. I. (2021b). Synthesis and electrochemical properties of graphene. In *Electrode Materials for Energy Storage and Conversion* (pp. 263–277). CRC Press.

Medalia, A. (1978). Effect of carbon black on dynamic properties of rubber vulcanizates. *Rubber Chemistry and Technology, 51*(3), 437–523.

Meng, X., & Xiao, F. S. (2014). Green routes for synthesis of zeolites. *Chemical Reviews, 114*(2), 1521–1543.

Merget, R., Bauer, T., Küpper, H., Philippou, S., Bauer, H., Breitstadt, R., & Bruening, T. (2002). Health hazards due to the inhalation of amorphous silica. *Archives of Toxicology, 75*(11–12), 625–634.

Morell, E. S., Correa, J. D., Vargas, P., Pacheco, M., & Barticevic, Z. (2010). Flat bands in slightly twisted bilayer graphene: Tight-binding calculations. *Physical Review B, 82*(12), 121407.

Moreno, C., Vilas-Varela, M., Kretz, B., Garcia-Lekue, A., Costache, M. V., Paradinas, M., ... & Mugarza, A. (2018). Bottom-up synthesis of multifunctional nanoporous graphene. *Science, 360*, 199–203.

Morozov, S. V., Novoselov, K. S., Katsnelson, M. I., Schedin, F., Elias, D. C., Jaszczak, J. A., & Geim, A. K. (2008). Giant intrinsic carrier mobilities in graphene and its bilayer. *Physical Review Letters, 100*(1), 016602.

Nandkishore, R., Levitov, L. S., & Chubukov, A. V. (2012). Chiral superconductivity from repulsive interactions in doped graphene. *Nature Physics, 8*, 158–163.

Nguyen, G. D., Toma, F. M., Cao, T., Pedramrazi, Z., Chen, C., Rizzo, D. J., ... & Crommie, M. F. (2016). Bottom-up synthesis of $N = 13$ sulfur-doped graphene. *The Journal of Physical Chemistry C, 120*, 2684–2687.

Notarianni, M., Liu, J., Mirri, F., Pasquali, M., & Motta, N. (2014). Graphene-based supercapacitor with carbon nanotube film as highly efficient current collector. *Nanotechnology,* *25*(43), 435405.

Nwanya, A. C., Ndipingwi, M. M., Ikpo, C. O., Obodo, R. M., Nwanya, S. C., Botha, S., ... & Maaza, M. (2020). Zea mays lea silk extract mediated synthesis of nickel oxide nanoparticles as positive electrode material for asymmetric supercabattery. *Journal of Alloys and Compounds, 822,* 153581.

Obodo, R. M., Chibueze, T. C., Ahmad, I., Ekuma, C. E., Raji, A. T., Maaza, M., & Ezema, F. I. (2021a). Effects of copper ion irradiation on $Cu_yZn_{1-2y-x}Mn_y/GO$ supercapacitive electrodes. *Journal of Applied Electrochemistry, 51,* 829–845.

Obodo, R. M., Nwanya, A. C., Arshad, M., Iroegbu, C., Ahmad, I., Osuji, R., ... & Ezema, F. I. (2020a). Conjugated NiO-ZnO/GO nanocomposite powder for applications in supercapacitor electrodes material. *International Journal of Energy Research, 44,* 3192–3202.

Obodo, R. M., Nwanya, A. C., Iroegbu, C., Ezekoye, B. A., Ekwealor, A. B., Ahmad, I., ... & Ezema, F. I. (2020c). Effects of swift copper (Cu^{2+}) ion irradiation on structural, optical and electrochemical properties of Co_3O_4-CuO-MnO_2/GO nanocomposites powder. *Advanced Powder Technology, 1*(4), 1728–1735.

Obodo, R. M., Onah, E. O., Nsude, H. E., Agbogu, A., Nwanya, A. C., Ahmad, I., ... & Ezema, F. I. (2020b). Performance evaluation of graphene oxide based Co_3O_4@GO, MnO_2@GO and Co_3O_4/MnO_2@GO electrodes for supercapacitors. *Electroanalysis, 32,* 2786–2794.

Obodo, R. M., Shinde, N. M., Chime, U. K., Ezugwu, S., Nwanya, A. C., Ahmad, I., ... & Ezema, F. I. (2020d). Recent advances in metal oxide/hydroxide on three-dimensional nickel foam substrate for high performance pseudocapacitive electrodes. *Current Opinion in Electrochemistry, 21,* 242–249.

Ozin, G. A., & Arsenault, A. (2009). *Nanochemistry: A Chemical Approach to Nanomaterials.* London: The Royal Society of Chemistry, 2nd edition.

Pandey, P. A., Wilson, N. R., & Covington, J. A. (2013). Pd-doped reduced graphene oxide sensing films for H_2 detection. *Sensors and Actuators B: Chemical, 183,* 478–487.

Papageorgiou, D. G., Kinloch, I. A., & Young, R. J. (2017). Mechanical properties of graphene and graphene-based nanocomposites. *Progress in Materials Science, 90,* 75–127.

Park, S. Y., Park, J., Sim, S. H., Sung, M. G., Kim, K. S., Hong, B. H., & Hong, S. (2011). Enhanced differentiation of human neural stem cells into neurons on graphene. *Advanced Materials, 23*(36), H263–H267.

Pawlak, R., Liu, X., Ninova, S., D'Astolfo, P., Drechsel, C., Sangtarash, S., ... & Meyer, E. (2020). Bottom-up synthesis of nitrogen-doped porous graphene nanoribbons. *Journal of the American Chemical Society, 142*(29), 12568–12573.

Pedersen, T. G., Flindt, C., Pedersen, J., Jauho, A. P., Mortensen, N. A., & Pedersen, K. (2008). Optical properties of graphene antidot lattices. *Physical Review B, 77*(24), 245431.

Peigney, A., Laurent, C., Flahaut, E., Bacsa, R. R., & Rousset, A. (2001). Specific surface area of carbon nanotubes and bundles of carbon nanotubes. *Carbon, 39*(4), 507–514.

Phong, V. T., Pantaleón, P. A., Cea, T., & Guinea, F. (2021). Band structure and superconductivity in twisted trilayer graphene. *Physical Review B, 104*(12), L121116.

Popov, A. A., Yang, S., & Dunsch, L. (2013). Endohedral fullerenes. *Chemical Reviews, 113*(8), 5989–6113.

Rizzo, D. J., Wu, M., Tsai, H.-Z., Marangoni, T., Durr, R. A., Omrani, A. A., ... & Crommie, M. F. (2019). Length-dependent evolution of type II heterojunctions in bottom-up-synthesized graphene nanoribbons. *Nano Letter, 19,* 3221–3228.

Sanchez-Yamagishi, J. D., Taychatanapat, T., Watanabe, K., Taniguchi, T., Yacoby, A., & Jarillo-Herrero, P. (2012). Quantum Hall effect, screening, and layer-polarized insulating states in twisted bilayer graphene. *Physical Review Letters, 108*(7), 076601.

Shearer, C. J., Slattery, A. D., Stapleton, A. J., Shapter, J. G., & Gibson, C. T. (2016). Accurate thickness measurement of graphene. *Nanotechnology, 27*(12), 125704.

Sheehy, D. E., & Schmalian, J. (2009). Optical transparency of graphene as determined by the fine-structure constant. *Physical Review B, 80*(19), 193411.

Shen, C., Chu, Y., Wu, Q., Li, N., Wang, S., Zhao, Y., ... & Zhang, G. (2020). Correlated states in twisted double bilayer graphene twisted double bilayer graphene. *Nature Physics, 16*(5), 520–525.

Stampfer, C., Schurtenberger, E., Molitor, F., Guttinger, J., Ihn, T., & Ensslin, K. (2008). Tunable graphene single electron transistor. *Nano Letters, 8*(8), 2378–2383.

Stormer, H. L. (1999). Nobel lecture: The fractional quantum Hall effect. *Reviews of Modern Physics, 71*, 875–889.

Sun, B., Wang, B., Su, D., Xiao, L., Ahn, H., & Wang, G. (2012). Graphene nanosheets as cathode catalysts for lithium-air batteries with an enhanced electrochemical performance. *Carbon, 50*, 727–733.

Talirz, L., Söde, H., Dumslaff, T., Wang, S., Sanchez Valencia, J. R., Liu, J., ... & Ruffieux, P. (2017). On-surface synthesis and characterization of 9-atom wide armchair graphene nanoribbons. *ACS Nano, 11*, 1380–1388.

Terrones, M. (2003). Synthesis, properties, and applications of carbon nanotubes. *Annual Review of Materials Research, 33*(1), 419–501.

Tsuei, C. C., & Kirtley, J. R. (2000). Pairing symmetry in cuprate superconductors. *Reviews of Modern Physics, 72*, 969.

Uchoa, B., & Neto, A. C. (2007). Superconducting states of pure and doped graphene. *Physical Review Letters, 98*(14), 146801.

Wang, J., Liu, Y., Chen, M., Zhao, H., Wang, J., Zhao, Z., ... & Wang, J. (2019). Hierarchical porous carbon-graphene-based Lithiume Sulfur batteries. *Electrochimica Acta, 318*, 161–168.

Wang, J., Wang, X., Xu, C., Zhang, M., & Shang, X. (2011). Preparation of graphene/poly(vinyl alcohol) nanocomposites with enhanced mechanical properties and water resistance. *Polymer International, 60*(5), 816–822.

Wang, L., Boutilier, M. S., Kidambi, P. R., Jang, D., Hadjiconstantinou, N. G., & Karnik, R. (2017). Fundamental transport mechanisms, fabrication and potential applications of nanoporous atomically thin membranes. *Nature Nanotechnology, 12*, 509–522.

Wang, X., Li, X., Zhang, L., Yoon, Y., Weber, P. K., Wang, H., ... & Dai, H. (2009). N-doping of graphene through electrothermal reactions with ammonia. *Science, 324*(5928), 768–771.

Wei, D., Liu, Y., Wang, Y., Zhang, H., Huang, L., & Yu, G. (2009). Synthesis of N-doped graphene by chemical vapor deposition and its electrical properties. *Nano Letters, 9*(5), 1752–1758.

Wolf, S. A., Chtchelkanova, A. Y., & Treger, D. M. (2006). Spintronics—A retrospective and perspective. *IBM Journal of Research and Development, 50*(1), 101–110.

Xu, Z., Lu, W., Wang, W., Gu, C., Liu, K., Bai, X., ... & Dai, H. (2008). Converting metallic single-walled carbon nanotubes into semiconductors by boron/nitrogen co-doping. *Advanced Materials, 20*(19), 3615–3619.

Xu, Z., Wang, S., Li, Y., Wang, M., Shi, P., & Huang, X. (2014). Covalent functionalization of graphene oxide with biocompatible poly(ethylene glycol) for delivery of paclitaxel. *Applied Materials and Interfaces, 6*, 17268–17276.

Yamaguchi, J., Hayashi, H., Jippo, H., Shiotari, A., Ohtomo, M., Sakakura, M., ... & Sato, S. (2020). Small bandgap in atomically precise 17-atom-wide armchair-edged graphene nanoribbons. *Communications Materials, 1*(1), 1–9.

Yang, X., Zhang, L., Zhang, F., Huang, Y., & Chen, Y. (2014). Sulfur-infiltrated graphene-based layered porous carbon cathodes for high-performance lithium–sulfur batteries. *ACS Nano, 8*(5), 5208–5215.

Yankowitz, M., Jung, J., Laksono, E., Leconte, N., Chittari, B. L., Watanabe, K., ... & Dean, C. R. (2018). Dynamic band-structure tuning of graphene moiré superlattices with pressure. *Nature, 557*(7705), 404–408.

Yankowitz, M., Wang, J., Birdwell, A. G., Chen, Y. A., Watanabe, K., Taniguchi, T., ... & LeRoy, B. J. (2014). Electric field control of soliton motion and stacking in trilayer graphene. *Nature Materials, 13*(8), 786–789.

Yin, J. X., Ma, W., Cochran, T. A., Xu, X., Zhang, S. S., Tien, H. J., ... & Hasan, M. Z. (2020). Quantum-limit Chern topological magnetism in $TbMn_6Sn_6$. *Nature, 583*(7817), 533–536.

Yoo, E., Kim, J., Hosono, E., Zhou, H., Kudo, Y., & Honma, I. (2008). Large reversible li storage of graphene nanosheet families for use in rechargeable lithium ion batteries. *Nano Letters, 8*(8), 2277–2282.

Zaletel, M. P., & Khoo, J. Y. (2019). The gate-tunable strong and fragile topology of multilayer-graphene on a transition metal dichalcogenide. *arXiv preprint arXiv*, 1901.01294.

Zeng, H., Martella, D., Wasylczyk, P., Cerretti, G., Lavocat, J. C., Ho, C.-H., ... & Wiersma, D. S. (2014). High-resolution 3D direct laser writing for liquid-crystalline elastomer microstructures. *Advanced Materials, 26*(11), 2319–2322.

Zhang, Y., Zhang, Y., Li, G., Lu, J., Lin, X., Du, S., ... & Gao, H. J. (2014). Direct visualization of atomically precise nitrogen-doped graphene nanoribbons. *Applied Physics Letters, 105*, 023101.

Zhu, J., Yang, D., Yin, Z., Yan, Q., & Zhang, H. (2014). Graphene and graphene-based materials for energy storage applications. *Small, 10*(17), 3480–3498.

Zrazhevskiy, P., Sena, M., & Gao, X. (2010). Designing multifunctional quantum dots for bioimaging, detection, and drug delivery. *Chemical Society Reviews, 39*(11), 4326–43254.

2 The Role of Graphene Oxide in Enhancement of Working Principle of Dielectric Capacitors as Energy Storage Device

Hope E. Nsude, Kingsley U. Nsude, and Sabastine E. Ugwuanyi
University of Nigeria

Raphael M. Obodo
University of Nigeria
University of Agriculture and Environmental Sciences
Quaid-i-Azam University
Northwestern Polytechnical University

Malik Maaza
iThemba LABS-National Research Foundation
University of South Africa (UNISA)

F. I. Ezema
University of Nigeria
iThemba LABS-National Research Foundation
University of South Africa (UNISA)

CONTENTS

DOI: 10.1201/9781003215196-2

2.1   INTRODUCTION

Alternative sources of energy are not dependable based on their intermittent nature. To ensure constant use of energy, a cheap and environmental system of energy storage with desirably high-power density becomes indispensable for running everyday used electrical devices. Ideally, the ability to charge fast and also discharge fast is what is required of an efficient power storage system (Wang et al., 2021). Currently, there are batteries, supercapacitors and dielectric capacitors among others in use for energy storage (Tang et al., 2021). Of all these systems of energy storage, the dielectric capacitor has the greatest charging and discharging capability as well as high-power densities and high durability though it suffers low-energy density (Huan et al., 2016). Developing efficient capacitors with high-power density has been a research motivation (Obodo, et al., 2020, 2021).

What prompts the choice of one capacitor over another is the capacitance, emanating from the constituent dielectric material basically if other parameters are the same (Sun et al., 2020). The dielectric material is the substance separating the positive charges from the negative charges. Charges placed on the surface of a dielectric (positive relative permittivity) dissipate slowly unlike charges placed on the surface of a metal (negative relative permittivity) which dissipate faster (Obodo et al., 2020). Therefore, an efficient dielectric for a capacitor should have high permittivity and low losses. The simplest form of dielectric used in capacitors is the air (Obodo et al., 2020). Ceramics, polymer films, paper, mica, etc., have been effectively used as dielectrics for capacitors. It is ideal that the dielectric used in capacitor is thin enough to achieve high capacitance based on the equation:

$$C = \frac{\varepsilon A}{d} \qquad (2.1)$$

where C is the capacitance, A is the area of separation between electrodes, ε is the permittivity, and d is the distance. From equation 2.1, it is evident that to achieve a high capacitance in a capacitor, there should be a larger plate area, higher permittivity, and a smaller gap between the plates. A lot of materials of desirable permittivity have been employed to construct capacitor anode but there is a need for materials of high dielectric value but low-dielectric loss for higher capacitance capacitors as a vital part of electronic gadgets (Obodo et al., 2020, 2021).

Many researchers have done different works on improving the dielectric constants of composite dielectrics using different nanofillers of appreciable dielectric constants such as titanium dioxide (TiO_2), barium titanate (BT), etc. Though their works attempt to improve the energy density, there was the corresponding agglomeration of the nanofillers and conducting channels formation (breakdown strength reduction) leading to low-energy density (Cai et al., 2019).

Most of the current nanocomposites which have been used in a bid to enhance the dielectric permittivity of the capacitor have in one way or other limited the overall performance of the capacitor due to the unavoidable lowering of the breakdown strength accompanying the gains (Bhunia et al., 2021). To attain a high overall dielectric constant, researchers have employed high-dielectric constant nanofillers such as BT, TiO_2, and lead zirconium titanate in composite dielectrics. These also aid in the improvement of polarization and energy density.

It is well known that as the filler concentration increases, the breakdown strength of nanocomposites falls, especially at high-volume fractions (Obodo et al., 2020). The nanocomposites are manufactured at a low-volume fraction to increase their dielectric property while preserving strong breakdown strength (Bhunia et al., 2021). The appropriate dielectric characteristics are also determined by nanofiller concentration and aspect ratios in each layer (Sun et al., 2016). As a result, it's critical to tune the interface by utilizing the optimal combination of nanofillers in each layer to help improve the dielectric constant, breakdown strength, and of course, the energy density (Tang et al., 2021).

In natural materials, achieving a high-dielectric permittivity coupled with breakdown strength in turn makes obtaining high density very difficult (Bhunia et al., 2021; Sun et al., 2020; Tang et al., 2021). Sun et al. (2020) designed P(VDF–HFP) and $BaTiO_3$/P(VDF–HFP) composite and showed that instead of random fillers distribution, fabricating layered nanocomposite could address the problems of coupling between dielectric permittivity and breakdown strength which limits high-energy density in materials

Also, aside from the high-energy density, the suitable capacitors for pulsed-power applications must have a high-discharge speed and discharge periods in the microseconds level (Cai & Song, 2010). Time constant affects the rate of discharging through a load resistor. The time constant is the amount of time it takes for the load's discharge energy to reach 90% of its maximum value (Bhunia et al., 2021). Many researchers have put efforts to improve the discharge time of traditional capacitors.

With $Ba_{0.2}\, r_{0.8}TiO_3$ nanowires/PVDF (Tang & Sodano, 2013), we were able to achieve 2.3 μs discharge time. Pan et al. (2017) got 0.189 μs discharge time with $0.5(Ba_{0.7}Ca_{0.3})$-TiO_3 nanofibers/PVDF. With BNNSs/PVDF, Li et al. achieved 3.4 μs. discharge time. The work of (Bhunia et al., 2021) has shown how to make an inexpensive, simple, and scalable thin-film-based dielectric capacitor. With PVDF-based composite, they reported 0.37 second discharge time. They reported that decreased dielectric losses at lesser frequencies are responsible for the outstanding breakdown strength. Extensive research on polyvinylidene fluoride (PVDF) has shown that its application is limited by hysteresis losses arising from the remnant polarization irrespective of its low cost (Bhunia et al., 2021).

The 2-D nanomaterials made of carbon such as graphene oxides (GOs) and its derivative reduced GOs have been used as nanofillers by researchers of recent because it offers an amalgamation of wonderful physical properties (Chan et al., 2020) (Obodo et al., 2019a). When compared to 0-D nanofillers, 2-D nanofillers have a higher aspect ratio and lesser percolation edges (Chen et al., 2016). Graphite-based 2-D nanomaterials on the other hand, have mostly been investigated for applications requiring a high-dielectric constant. There are lots of reports on the dielectric study of GO or graphene-related material composite as per application in dielectric capacitors (Aliyev et al., 2019; Cai & Song, 2010; Chan et al., 2020; Deshmukh et al., 2015, 2016; Kumar et al., 2015; Singh et al., 2016; Stankovich et al., 2007; Wang et al., 2021).

Discussed in this chapter were the concept of the dielectric capacitor, its working principle, methods of synthesis of GO, the structure of GO, and properties of GO. Finally, the effect of GO on the working principle of the dielectric capacitor as well as the effect of different GO composites with improved dielectric constants in enhancing the capacitance of the dielectric capacitor (Obodo, et al., 2019b).

2.2 CAPACITORS

The capacitor is an electronic device for storing electrical charge. Capacitors are one of the passive components used for storing charges (Liu et al., 2020). Capacitor constitutes part of an important part of almost every advancement in technology (Wang et al., 2021). Technological-wise, there are two types of capacitors which are polarized and non-polarized capacitors for fixed capacitors, and tuning and trimmer capacitors for variable capacitors (Liu et al., 2020). A polarized capacitor works in either positive or negative direction while for non-polarized capacitors, the direction does not matter.

Capacitor variables include dielectric constant or relative permittivity (ε_r) and dielectric loss (Marín-Genescà et al., 2020). The ratio between the capacitance C_x of a capacitor in which the space between and around the electrodes is totally and exclusively filled with the insulator in question and the capacitance C_o measured with the same electrode arrangement in a vacuum is the relative permittivity of an insulator and it is mathematically shown in equation 2.2.

$$\varepsilon_r = \frac{C_x}{C_o} \tag{2.2}$$

When the capacitor's dielectric is entirely made up of insulating material, the phase shift between the applied voltage and the resultant current is the complementary angle of an insulator's loss angle (Marín-Genescà et al., 2020).

In 1900, the first paper capacitor was reported but the use of polymer as dielectrics came into play in the 1950s for polymer capacitor used in flexible electronics (Liu et al., 2020). Of recent, semiconductor/polymer composites became a new discovery because of their inherent outstanding physical and chemical properties. These semiconductors rightly called semiconducting fillers are classified into four. At first, are the 0D fillers with particles having three dimensions below 100 nm; they are mainly nanoparticles. Secondly are the 1D fillers such as nanofibers, nanorods, nanotubes, etc. Thirdly, are the 2D fillers such as transition metal dichalcogenides, graphene, etc. Fourthly, is the 3D fillers such as the graphite, carbon fibers, etc., (Liu et al., 2020). Of recent is the development of multilayered structures resulting from various materials modifications.

2.3 WORKING PRINCIPLE OF A CAPACITOR

A parallel plate capacitor is the simplest of all forms of capacitors. It is simply constructed by placing two conductive plates parallel to each other with a dielectric material in between them as shown in Figure 2.1.

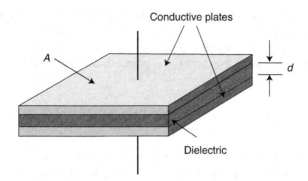

FIGURE 2.1   Structure of a capacitor (Marín-Genescà et al., 2020.)

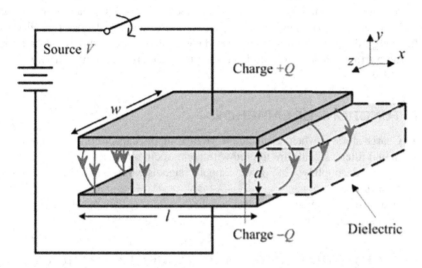

FIGURE 2.2   Parallel plate capacitor with linear dielectric partially inserted (Koh et al., 2014.)

The dielectric could be partially inserted in between the conductive plates as shown in Figure 2.2 where polarization is in the direction of the electric field with force resulting from the x and y components in the fringing field at the electrodes' edge. The force is not dependent on the polarity of the voltage (Koh et al., 2014). The force also tends to pull the dielectric into the electrode.

Capacitors store electrostatic energy in a dielectric between two terminals (Wang et al., 2019). When a voltage source is connected across a capacitor, due to the electric field across the capacitor, positive charges accumulate on the positive plate and negative charges on the negative plate of the capacitor until a maximum charge is stored (Marín-Genescà et al., 2020). The time is referred to as the time of charge of the capacitor. When the source of voltage is removed, the capacitor becomes a source of energy. When a load is connected across the ends of the capacitor, the charge in form of current flows through the load, till the two plates become empty. The time taken to dislodge the charges from the plates is called discharging time.

The quantity of the electrical energy that is stored in a capacitor defines its capacitance. The energy stored in the capacitor is proportional to the voltage applied.

That is,

$$Q = \frac{c}{V}(\text{coulomb/volt})\qquad(2.3)$$

One may say that the capacitance of a capacitor sorely depends on the charge and voltage. The capacitance of a capacitor is determined using the distance of separation of the plates, size of the plate, and the dielectric material is given by equation 2.1. In a circuit, capacitors can be connected in series or in parallel. When the voltage and the charge of a capacitor are known, the capacitance can be measured (Tang & Sodano, 2013).

Three high-level component requirements needed for better capacitor dielectrics include enhanced maximum operating temperature, increased operating electrostatic energy density; meaning increased dielectric constant with low loss without reducing operating field is required and then increased thermal conductivity (Chen et al., 2013).

2.4   FUNCTIONS OF CAPACITORS

- Charge storage; when connected to voltage supply capacitors store charges
- Signal filters; signals are filtered with them in circuits.
- Signal decoupling; in power supply signals, the capacitor subdues high-frequency noise
- energy storage device and power supply filters due to their excellent charge/discharge capabilities (Liu et al., 2020; Tang et al., 2014)

2.5   FACTORS THAT AFFECT CAPACITANCE OF CAPACITOR

The extent to which the strength of a magnetic field is affected by two conductors is defined by capacitance. The capacitance of a capacitor is influenced by three factors which are; the size of conductors, the distance of separation between the plate, and the dielectric material (Marín-Genescà et al., 2020).

- **Size of conductor**: The size of the plate has a great influence on the overall performance of the capacitor. When the area of the plate is larger, greater capacitance is recorded and vice versa. This is because a larger plate permits more charge (field flux) for a particular voltage across the plate.
- **Distance of separation of the plates**: The closer the plates are to each other, the larger the capacitance. This is because the field force becomes greater when the plates are closer, thereby leading to more charge on the plates for a particular voltage across the plates.
- **Dielectric material**: Greater capacitance is gotten for greater permittivity of the dielectric material if all other factors are met. This is because

high-permittivity materials tolerate more field flux; hence more charges are collected for a particular voltage applied.

2.6 GRAPHENE OXIDE (GO) IN BRIEF

GO is simply oxidized graphene. It is made of bonded carbon atoms (Stankovich et al., 2007). Its properties include; easy processing, can be reduced to graphene, dispersible in water and other solvents, non-conductive, high-surface area, high-mechanical strength, fluorescent, hydrophilic (unlike reduced GO) and nice media for surface functionalization finding interesting application in electronics (Chen et al., 2013; Dideikin & Vul', 2019; Hu et al., 2010; Singh et al., 2016).

Graphene was the first discovered 2-D material and became the most prevalent and most effective energy storage material in combination with suitable semiconductors. Brodie in 1859, gave the first report about GO gotten from graphite treated with strong oxidizing blends (Botas et al., 2013; Brodie, 1859). Lots of efforts were put by Staundenmaier and the likes to control the explosive nature of the Brodie method of synthesizing GO by replacing highly fuming HNO_3 (Chen et al., 2013).

This was achieved by Hummer who successively got rid of the HNO_3 and used $KMnO_4$ instead, thereby reducing the explosive nature of GO preparation (Hummers & Offeman, 1958). Apart from the reduction in the explosive nature of GO's preparation, the time taken became highly reduced from days to hours. Residual contaminants like Mn, NO_2, and N_2O_4 from the Hummers method led to new findings on how to eliminate them and this leads to modifications to the method (Chen et al., 2013; Hirata et al., 2004; Marcano et al., 2010). GO finds application in drug delivery (Liu et al., 2008), tribology (Kinoshita et al., 2014; Sarno et al., 2014), antibacterial coating (Akhavan & Ghaderi, 2010; Hu et al., 2010), energy storage (Gao et al., 2015; Wang et al., 2015), etc. GO has been prepared using various techniques such as microwave reduction (Voiry et al., 2016), sonochemical (Kumar et al., 2019), etc.

The number of layers is the primary distinction between graphite oxide and GO. GO is a monolayer or few layered oxygen functionalized (Chen et al., 2013). GO is gotten from graphite oxide by oxidizing with a strong oxidizing agent and exfoliation in water (Kumar et al., 2015). Graphene-based materials, including GO, are good candidates for inclusion into a number of advanced functional materials due to their potential and exceptional characteristics, as well as their simplicity of processing and chemical functionalization.

2.6.1 METHODS OF SYNTHESIS OF GO

Different approaches have been employed by researchers over years in the synthesis of GO. GO can be synthesized using two approaches which are; (i) top-down and (ii) bottom-up approaches. In top-down approach, layers of graphene are separated from graphite. In bottom-up approach, carbon molecules are assembled and formulated into graphene (Wang et al., 2021). Owing to the less yield and time-consuming nature of the bottom up approach, much attention is drawn toward top-down approach (Wang et al., 2015).

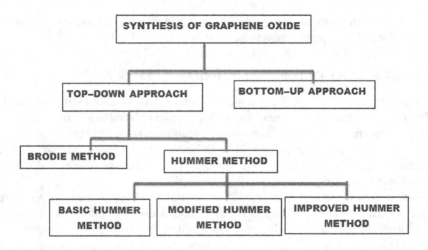

FIGURE 2.3 Synthesis techniques of graphene oxide.

Broadly, there are two methods involved in the synthesis of GO. They are the Brodie method and the Hummer method. The later method is in an attempt to drastically reduce the explosive nature of the former. Hummer method was still modified to contend with some emanating residual contaminants (by oxidation and exfoliation) leading to an improved Hummer method and modified Hummer method as shown in the Figure 2.3.

Small size particles of initiator material are required for the chemical oxidation of graphite. This size determines strongly the properties of the formed GO. It is a technological puzzle so to say how to increase the size of synthesized GO.

By the Hummers method, graphite is oxidized to get GO. The preparation involves;

- Mixing graphite flakes and an oxidizing agent (say $NaNO_3$) in H_2SO_4 and stirred continuously under an ice bath.
- Stirring for 2 hours and addition of potassium permanganate
- Removal of the ice bath and continuously stirring the mixture for hours at a certain temperature little above room temperature to form a brown paste
- Slow dilution of the mixture with water and increase the reaction temperature to the point of effervescence; the color is expected to turn brown
- Further dilution with water and continuous stirring
- Treating with H_2O_2 (reducing residual permanganate and MnO_2 to manganese sulfate); the color is expected to be yellow
- Washing, rinsing, and centrifuging several times
- Filtration and drying to obtain GO powder

The Hummer method is shown schematically in Figure 2.4. The Hummer method described so far may have few changes based on the different synthesis.

Methods employed in the synthesis of GO include mechanical exfoliation (Novoselov et al., 2005), epitaxial growth (Land et al., 1992), chemical reduction of graphite oxide, and liquid phase exfoliation (Stankovich et al., 2007), etc.

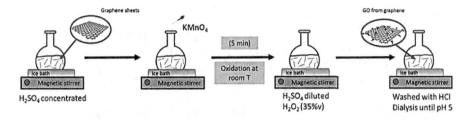

FIGURE 2.4   Schematic representation of Hummer method (Costa et al., 2021.)

Marcano et al., 2010 reported an improved Hummer method in 2010. This new procedure excluded $NaNO_3$, increased KMnO4, and increased the mixing reaction ratio of H_2SO_4 and H_3PO_4. This procedure yielded a larger amount of graphite oxide and lesser toxic residual gases as proposed by Chen et al., 2013. The modified Hummer method involves oxidation as well as exfoliation of graphite sheets. The preparation as reported by Paulchamy et al., 2015 involves;

- Mixing graphite flakes and an oxidizing agent (say $NaNO_3$) in H_2SO_4 and stirred continuously under the ice bath.
- Stirring for 4 hours and addition of potassium permanganate
- Slow dilution of mixture with water and continuous stirring
- Removal of the ice bath and continuously stirring the mixture for hours at a certain temperature little above room temperature to form a brown paste
- Refluxing at the reaction temperature to the point of effervescence for 10–15 minutes, at a lower temperature for about 10 minutes, and much lower temperature like 25°C for 2 hours; the color is expected to turn brown
- Treating with H_2O_2 to stop the reaction; the color is expected to be yellow
- Addition of equal amount into water contained in two separate beakers and stirred for 1 hour and allowed to settle
- Decantation of the remaining water, centrifuging several times, and drying to obtain GO powder

2.6.2   STRUCTURE OF GO

The purification processes, rather than the type of graphite utilized or the oxidation protocol, have a significant impact on the structure of GO (Singh et al., 2016). The precise identity and distribution of oxide functional groups are heavily influenced by the degree of oxidation. Various approaches can be used to identify the chemical makeup of GO, as well as the oxygen-containing functional groups in GO. On the basal plane, hydroxyl and epoxy groups, as well as carboxy, carbonyl, and phenol groups toward the border of GO sheets, have been identified as oxygen functional groups.

The common functional group includes the carboxyl (-COOH), epoxides (C-O-C), carbonyl (C=O) groups, etc. These functional groups account for the exfoliation and hydrophilicity of GO in solutions. These functional groups and defects limit the movement of electrons and hence GO is an electrical insulator. Oxidation is known to appear at the defect sites.

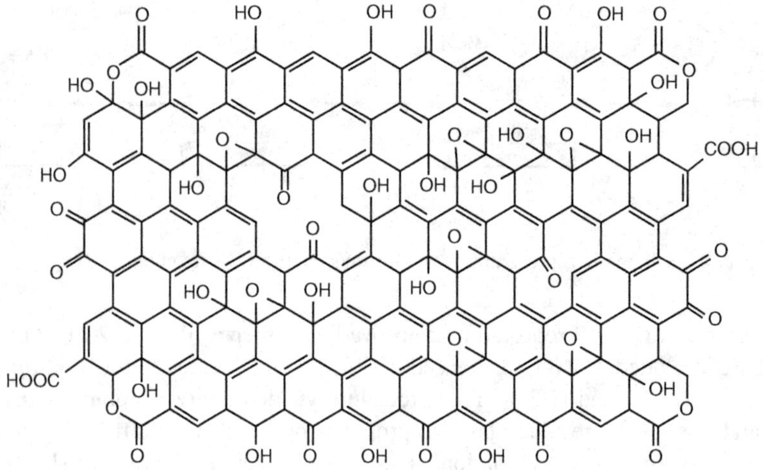

FIGURE 2.5   Structure of Lerf-Klinowski model of GO (Aliyev et al., 2019.)

The stability of GO nanosheets in suspensions is highly influenced by OD and the knowledge of OD helps to understand what the structure of GO is. OD generated along in the synthesis of GO obscures the determination of the functional group of GO.

Many models of GO structures have been in reports both past and recent ones depending on the type of oxygen-containing functional group, the condition of oxidation and the initiator material. Lerf-Klinowski model of GO shows Carboxyl groups at the edges and hydroxyl groups closely located (Siklitskaya et al., 2021). NMR data refutes the carboxyl location (Casabianca et al., 2010), and the closeness of hydroxyl groups to themselves is suspected to cause instability (electrical) in GO structure (Aliyev et al., 2019). Lerf-Klinowski model of GO is taken as the best description of the structure of GO (Dreyer et al., 2010).

Lee model lacks the presence of carboxyl group as well as lactone group. This does not agree with NMR data which has both. (Aliyev et al., 2019) proposed a new structural model of GO which involves OD extraction. It is a modified Lerf-Klinowski model (Figure 2.5).

2.7   EFFECTS OF GO ON THE WORKING PRINCIPLE OF DIELECTRIC CAPACITOR

Enhancement of the working principle of a structural dielectric capacitor (SDC) using GO for various applications including aviation has been reported. Humidity at the level of the atmosphere for aircraft flight is quite low. This causes the generation of static electricity (electrostatics) as a result of the friction between the molecules of the atmosphere and the aircraft body surface. Since the accumulation of these electrostatics negatively affects the aircraft, ways of harnessing it for storage in the conventional capacitor to be used for powering electrical devices on aircraft instead of discharging it, have been recently reported (Xie et al., 2015). To achieve this, a

SDC is used. This conversion involves fabricating carbon fiber reinforced polymer (CFRP) (which competes traditional aluminum alloys) with electrical insulator sandwiched between conductors.

Many research on the effective electrically insulating materials like polymer for use as dielectric layer in carbon based SDC composite have been ongoing (Park et al., 2009). Based on the outstanding properties of GO like its high mechanical properties, it has been considered a promising dielectric for SDC (Nwanya et al, 2020). GO-based SDC has specific capacitance multiple of times greater than polymer-based SDC because of its oxygen functional group which promotes bonding interaction between the GO film and the carbon fiber layer of the SDC (Chan et al., 2020).

Also, the ultra-thin nature of GO is advantageous in term of lesser weight it adds to the composite compared to other dielectric materials. The problem associated with GO-based SDC is the GO film delamination cracking (Carlson et al., 2010). Efforts are made by researchers to improve the interlayer interactions which will improve GO-based SDC delamination resistance capability. Figure 2.6 shows the conversion of conventional CFRP composite to SDC with GO as dielectric, as adapted by (Chan et al., 2020).

Chan et al. (2020) reported on the use of GO on polymer to increase mechanical properties of GO-based SDC composite which can power aircraft navigation light. The GO-based composite showed no redox peaks, which means that it has very fast electrical charging and discharging. Output current and capacitance recorded was reduced when PAA was added probably because of its low dielectric constant (Chan et al., 2020). Outstanding specific capacitance recorded for GO-based SDC unlike PAA-modified GO-based SDC composite could be attributed to wealth of oxygen functional group in GO which are polarizable (Carlson et al., 2010) and this is evident in the CV curve shown in Figure 2.7.

Similar report was given by on GO/PVA composite for different applications including dielectric capacitor. Though GO-based SDCs have much lower specific capacitance as compared to structural supercapacitor or battery, they have appreciably high power density prompting their need in applications requiring fast charge and discharge. Owning to this, GO-based SDCs are very necessary for use in aviation for both energy harvesting and powering electronics.

Kumar et al. (2019) synthesized GO with high dielectric value through modified Hummer's method for various applications including dielectric capacitor. They reported dielectric constant of GO of the order 10^6 with low dielectric loss suitable

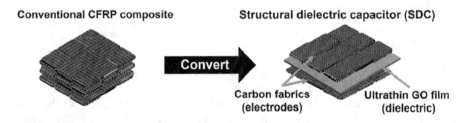

FIGURE 2.6 Use of GO as the dielectric in SDC (Chan et al., 2020). (Reprinted with permission from Elsevier.)

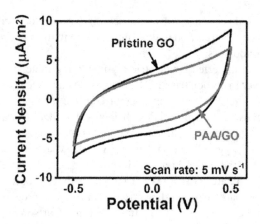

FIGURE 2.7 CV curve of GO-based SDC and PAA-modified GO-based SDC composite at 5 mV s^{-1} (Chan et al., 2020). (Reprinted with permission from Elsevier.)

for application in capacitor. It is important to note that GO aids in the simultaneous improvement of breakdown strength and dielectric constant, as well as the lowering of loss tangent. GO/PVDF interfaces are generally robust due to the growth in functional groups (Paulchamy et al., 2015).

Chan et al. (2018) in their work demonstrated that paper-based dielectric layer in use in structural dielectric capacitor's CFRP; glass fiber reinforced composite dielectrics can be effectively replaced using GO paper. The use of GO paper as a dielectric material in SDC according to their report showed much higher capacitance of 17.13 µF m^{-2} against 2.18 µF m^{-2} reported for printing paper. They recorded that GO has electrical conductivity of the range 5×10^{-6} and 4×10^{-3} S m^{-1} indicating the insulating property of GO. Apart from the electrical conductivity and high dielectric strength of GO, it is known to be highly light-weighted and of excellent mechanical properties (Chan et al., 2020).

Though the energy density of the GO paper based-SDC is low than those of the existing SDCs, the potential application of GO paper in the development of highly efficient SDC cannot be overemphasized considering its working voltage of about 400 V (Chan et al., 2020). Experimental tests were utilized to assess the dielectric, thermal, and microstructural characterization of acceptable materials as dielectrics for capacitors so as to yield data that would increase the knowledge of compounds for ideal use.

2.8 SUMMARY

The research focus for dielectric capacitors is achieving high capacitance using a material of high-dielectric constant and low-dielectric losses without compromising the breakdown strength. Many power electronic systems require capacitors that can store and release huge amounts of electric energy within a short period. GO among other carbon derivatives for its inherent oxygen functional group is famous and finds wider applications, especially in capacitors. GO has an appreciably high-dielectric constant of multiple magnitudes, mechanical properties, and surface area and is tunable. The ease of mass production of graphene oxide-based systems favors its application in material science. Several efforts are being input by different researchers in exploiting the outstanding properties of GO in a dielectric layer of a dielectric capacitor. Experimental tests were utilized to assess the dielectric, thermal, and microstructural characterization of acceptable materials as dielectrics for capacitors in order to produce data that would improve the knowledge of compounds for optimal use. Here in this chapter, we have discussed the basic working principle of a dielectric capacitor, factors that influence the working principle, different methods of synthesizing GO, structure models of GO, properties of GO, and the effect of GO and GO composites on the working principle of a dielectric capacitor.

ACKNOWLEDGMENTS

RMO and IA humbly acknowledge NCP for their PhD fellowship (NCP-CAAD/PhD-132/EPD) award and COMSATS for the travel grant for the fellowship.

FIE graciously acknowledges the grant by TETFUND under contract number TETFUND/DR&D/CE/UNI/NSUKKA/RP/VOL.I and also acknowledge the support received from the Africa Centre of Excellence for Sustainable Power and Energy Development (ACE-SPED), University of Nigeria, Nsukka that enabled the timely completion of this research.

We thank Engr. Emeka Okwuosa for the generous sponsorship of April 2014, July 2016, July 2018, and July 2021 conferences/workshops on applications of nanotechnology to energy, health & Environment and for providing some research facilities.

REFERENCES

Akhavan, O., & Ghaderi, E. (2010). Toxicity of graphene and graphene oxide nanowalls against bacteria. *ACS Nano*, *4*(10), 5731–5736.

Aliyev, E., Filiz, V., Khan, M. M., Lee, Y. J., Abetz, C., & Abetz, V. (2019). Structural characterization of graphene oxide: Surface functional groups and fractionated oxidative debris. *Nanomaterials*, *9*(8), 1180.

Bhunia, R., Sarkar, A., Anand, S., Garg, A., & Gupta, R. K. (2021). Unveiling the role of graphene oxide as an interface interlocking ingredient in polyvinylidene fluoride-based multilayered thin-film capacitors for high energy density and ultrafast discharge applications. *Energy Technology*, *9*(4), 2000905.

Botas, C., Álvarez, P., Blanco, P., Granda, M., Blanco, C., Santamaría, R., Romasanta, L. J., Verdejo, R., López-Manchado, M. A., & Menéndez, R. (2013). Graphene materials with different structures prepared from the same graphite by the Hummers and Brodie methods. *Carbon*, *65*, 156–164.

Brodie, B. C. (1859). XIII. On the atomic weight of graphite. *Philosophical Transactions of the Royal Society of London*, *149*, 249–259.

Cai, D., & Song, M. (2010). Recent advance in functionalized graphene/polymer nanocomposites. *Journal of Materials Chemistry*, *20*(37), 7906.

Cai, Z., Wang, H., Zhao, P., Chen, L., Zhu, C., Hui, K., Li, L., & Wang, X. (2019). Significantly enhanced dielectric breakdown strength and energy density of multilayer ceramic capacitors with high efficiency by electrodes structure design. *Applied Physics Letters*, *115*(2), 023901.

Carlson, T., Ordéus, D., Wysocki, M., & Asp, L. E. (2010). Structural capacitor materials made from carbon fibre epoxy composites. *Composites Science and Technology, 70*(7), 1135–1140.

Casabianca, L. B., Shaibat, M. A., Cai, W. W., Park, S., Piner, R., Ruoff, R. S., & Ishii, Y. (2010). NMR-based structural modeling of graphite oxide using multidimensional 13C solid-state NMR and ab initio chemical shift calculations. *Journal of the American Chemical Society, 132*(16), 5672–5676.

Chan, K.-Y., Jia, B., Lin, H., Zhu, B., & Lau, K.-T. (2018). Design of a structural power composite using graphene oxide as a dielectric material layer. *Materials Letters, 216*, 162–165.

Chan, K.-Y., Pham, D. Q., Demir, B., Yang, D., Mayes, E. L. H., Mouritz, A. P., Ang, A. S. M., Fox, B., Lin, H., Jia, B., & Lau, K.-T. (2020). Graphene oxide thin film structural dielectric capacitors for aviation static electricity harvesting and storage. *Composites Part B: Engineering, 201*, 108375.

Chen, J., Yao, B., Li, C., & Shi, G. (2013). An improved Hummers method for eco-friendly synthesis of graphene oxide. *Carbon, 64*, 225–229.

Chen, Q., Wu, S., & Xin, Y. (2016). Synthesis of Au–CuS–TiO_2 nanobelts photocatalyst for efficient photocatalytic degradation of antibiotic oxytetracycline. *Chemical Engineering Journal, 302*, 377–387.

Costa, M. C. F., Marangoni, V. S., Ng, P. R., Nguyen, H. T. L., Carvalho, A., & Castro Neto, A. H. (2021). Accelerated synthesis of graphene oxide from graphene. *Nanomaterials, 11*(2), 551.

Deshmukh, K., Ahamed, M. B., Sadasivuni, K. K., Ponnamma, D., Deshmukh, R. R., Pasha, S. K. K., AlMaadeed, M. A.-A., & Chidambaram, K. (2016). Graphene oxide reinforced polyvinyl alcohol/polyethylene glycol blend composites as high-performance dielectric material. *Journal of Polymer Research, 23*(8), 159.

Dideikin, A. T., & Vul', A. Y. (2019). Graphene oxide and derivatives: The place in graphene family. *Frontiers in Physics, 6*, 149. https://doi.org/10.3389/fphy.2018.00149

Dreyer, D. R., Park, S., Bielawski, C. W., & Ruoff, R. S. (2010). The chemistry of graphene oxide. *Chemical Society Reviews, 39*(1), 228–240.

Gao, X., Li, J., Xie, Y., Guan, D., & Yuan, C. (2015). A multilayered silicon-reduced graphene oxide electrode for high performance lithium-ion batteries. *ACS Applied Materials & Interfaces, 7*(15), 7855–7862.

Hirata, M., Gotou, T., Horiuchi, S., Fujiwara, M., & Ohba, M. (2004). Thin-film particles of graphite oxide 1: High-yield synthesis and flexibility of the particles. *Carbon, 42*(14), 2929–2937.

Hu, W., Peng, C., Luo, W., Lv, M., Li, X., Li, D., Huang, Q., & Fan, C. (2010). Graphene-based antibacterial paper. *ACS Nano, 4*(7), 4317–4323.

Huan, T. D., Boggs, S., Teyssedre, G., Laurent, C., Cakmak, M., Kumar, S., & Ramprasad, R. (2016). Advanced polymeric dielectrics for high energy density applications. *Progress in Materials Science, 83*, 236–269.

Hummers, W. S., & Offeman, R. E. (1958). Preparation of graphitic oxide. *Journal of the American Chemical Society, 80*(6), 1339–1339.

Kinoshita, H., Nishina, Y., Alias, A. A., & Fujii, M. (2014). Tribological properties of monolayer graphene oxide sheets as water-based lubricant additives. *Carbon, 66*, 720–723.

Koh, K., Sreekumar, M., & Ponnambalam, S. (2014). Experimental investigation of the effect of the driving voltage of an electroadhesion actuator. *Materials, 7*(7), 4963–4981.

Kumar, K. S., Pittala, S., Sanyadanam, S., & Paik, P. (2015). A new single/few-layered graphene oxide with a high dielectric constant of 10^6: Contribution of defects and functional groups. *RSC Advances, 5*(19), 14768–14779.

Kumar, R., Sahoo, S., Joanni, E., Singh, R. K., Yadav, R. M., Verma, R. K., Singh, D. P., Tan, W. K., del Pino, A. P., & Moshkalev, S. A. (2019). A review on synthesis of graphene, h-BN and MoS_2 for energy storage applications: Recent progress and perspectives. *Nano Research, 12*(11), 2655–2694.

Land, T. A., Michely, T., Behm, R. J., Hemminger, J. C., & Comsa, G. (1992). STM investigation of single layer graphite structures produced on Pt (111) by hydrocarbon decomposition. *Surface Science, 264*(3), 261–270.

Liu, L., Qu, J., Gu, A., & Wang, B. (2020). Percolative polymer composites for dielectric capacitors: A brief history, materials, and multilayer interface design. *Journal of Materials Chemistry A, 8*(36), 18515–18537.

Liu, Z., Robinson, J. T., Sun, X., & Dai, H. (2008). PEGylated nanographene oxide for delivery of water-insoluble cancer drugs. *Journal of the American Chemical Society, 130*(-33), 10876–10877.

Marcano, D. C., Kosynkin, D. V., Berlin, J. M., Sinitskii, A., Sun, Z., Slesarev, A., Alemany, L. B., Lu, W., & Tour, J. M. (2010). Improved synthesis of graphene oxide. *ACS Nano, 4*(8), 4806–4814.

Marín-Genescà, M., García-Amorós, J., Mujal-Rosas, R., Massagués Vidal, L., & Colom Fajula, X. (2020). Application properties analysis as a dielectric capacitor of end-of-life tire-reinforced HDPE. *Polymers, 12*(11), 2675.

Novoselov, K. S., Geim, A. K., Morozov, S. V., Jiang, D., Katsnelson, M. I., Grigorieva, Iv., Dubonos, Sv., & Firsov, and A. (2005). Two-dimensional gas of massless Dirac fermions in graphene. *Nature, 438*(7065), 197–200.

Nwanya, A. C., Ndipingwi, M. M., Ikpo, C. O., Obodo, R. M., Nwanya, S. C., Botha, S., . . . Maaza, M. (2020). Zea mays lea silk extract mediated synthesis of nickel oxide nanoparticles as positive electrode material for asymmetric supercabattery. *Journal of Alloys and Compounds, 822*, 153581.

Obodo, R. M., Ahmad, A., Jain, G. H., Ahmad, I., Maaza, M., & Ezema, F. I. (2020a). 8.0 MeV copper ion (Cu^{++}) irradiation-induced effects on structural, electrical, optical and electrochemical properties of Co_3O_4-NiO-ZnO/GO nanowires. *Materials Science for Energy Technologies, 3*, 193–200.

Obodo, R. M., Ahmad, I., & Ezema, F. I. (2019a). Introductory chapter in graphene and its applications. In *Graphene and Its Derivatives-Synthesis and Applications*. Intechopen.

Obodo, R. M., Asjad, M., Nwanya, A. C., Ahmad, I., Zhao, T., Ekwealor, A. B., . . . Ezema, F. I. (2020b). Evaluation of 8.0 MeV carbon (C^{2+}) irradiation effects on hydrothermally synthesized Co_3O_4-CuO-ZnO@GO electrodes for supercapacitor applications. *Electroanalysis, 32*, 2958–2968.

Obodo, R. M., Chibueze, T. C., Ahmad, I., Ekuma, C. E., Raji, A. T., Maaza, M., & Ezema, F. I. (2021a). Effects of copper ion irradiation on $Cu_yZn_{1-2y-x}Mn_y$/GO supercapacitive electrodes. *Journal of Applied Electrochemistry, 51*, 829–845.

Obodo, R. M., Chime, U. K., Nkele, A. C., Nwanya, A. C., Bashir, A. K., Madiba, I. G., . . . Ezema, F. I. (2021b). Effect of annealing on hydrothermally deposited Co_3O_4-ZnO thin films for supercapacitor applications. *Materials Today: Proceedings, 36*, 374–378.

Obodo, R. M., Nwanya, A. C., Arshad, M., Iroegbu, C., Ahmad, I., Osuji, R., . . . Ezema, F. I. (2020c). Conjugated NiO-ZnO/GO nanocomposite powder for applications in supercapacitor electrodes material. *Int J Energy Res, 44*, 3192–3202.

Obodo, R. M., Nwanya, A. C., Hassina, T., Kebede, M. A., Ahmad, I., Maaza, M., & Ezema, F. I. (2019b). Transition metal oxides-based nanomaterials for high energy and power density supercapacitor. In *Electrochemical Devices for Energy storage applications* (p. 7). United Kingdom: CRC.

Obodo, R. M., Nwanya, A. C., Iroegbu, C., Ezekoye, B. A., Ekwealor, A. B., Ahmad, I., . . . Ezema, F. I. (2020d). Effects of swift copper (Cu^{2+}) ion irradiation on structural, optical and electrochemical properties of Co_3O_4-CuO-MnO_2/GO nanocomposites powder. *Advanced Powder Technology.*

Obodo, R. M., Onah, E. O., Nsude, H. E., Agbogu, A., Nwanya, A. C., Ahmad, I., . . . Ezema, F. I. (2020e). Performance Evaluation of Graphene Oxide Based Co3O4@GO, MnO2@GO and Co3O4/MnO2@GO Electrodes for Supercapacitors. *Electroanalysis, 32*, 2786–2794.

Obodo, R. M., Shinde, N. M., Chime, U. K., Ezugwu, S., Nwanya, A. C., Ahmad, I., . . . Ezema, F. I. (2020f). Recent advances in metal oxide/hydroxide on three-dimensional nickel foam substrate for high performance pseudocapacitive electrodes. *Current Opinion in Electrochemistry, 21*, 242–249.

Pan, Z., Yao, L., Zhai, J., Wang, H., & Shen, B. (2017). Ultrafast discharge and enhanced energy density of polymer nanocomposites loaded with $0.5(Ba_{0.7} Ca_{0.3})TiO_3 -0.5Ba(Zr_{0.2} Ti_{0.8})O_3$ one-dimensional nanofibers. *ACS Applied Materials & Interfaces, 9*(16), 14337–14346.

Park, S., Dikin, D. A., Nguyen, S. T., & Ruoff, R. S. (2009). Graphene oxide sheets chemically cross-linked by polyallylamine. *The Journal of Physical Chemistry C, 113*(36), 15801–15804.

Paulchamy, B., Arthi, G., & Lignesh, B. D. (2015). A simple approach to stepwise synthesis of graphene oxide nanomaterial. *Journal of Nanomedicine & Nanotechnology, 6* (1), 1.

Sarno, M., Senatore, A., Cirillo, C., Petrone, V., & Ciambelli, P. (2014). Oil lubricant tribological behaviour improvement through dispersion of few layer graphene oxide. *Journal of Nanoscience and Nanotechnology, 14*(7), 4960–4968.

Siklitskaya, A., Gacka, E., Larowska, D., Mazurkiewicz-Pawlicka, M., Malolepszy, A., Stobiński, L., Marciniak, B., Lewandowska-Andrałojć, A., & Kubas, A. (2021). Lerf–Klinowski-type models of graphene oxide and reduced graphene oxide are robust in analyzing non-covalent functionalization with porphyrins. *Scientific Reports, 11*(1), 1–14.

Singh, R. K., Kumar, R., & Singh, D. P. (2016). Graphene oxide: Strategies for synthesis, reduction and frontier applications. *RSC Advances, 6*(69), 64993–65011.

Stankovich, S., Dikin, D. A., Piner, R. D., Kohlhaas, K. A., Kleinhammes, A., Jia, Y., Wu, Y., Nguyen, S. T., & Ruoff, R. S. (2007). Synthesis of graphene-based nanosheets via chemical reduction of exfoliated graphite oxide. *Carbon, 45*(7), 1558–1565.

Sun, L., Shi, Z., Liang, L., Wei, S., Wang, H., Dastan, D., Sun, K., & Fan, R. (2020). Layer-structured $BaTiO_3/P$ (VDF–HFP) composites with concurrently improved dielectric permittivity and breakdown strength toward capacitive energy-storage applications. *Journal of Materials Chemistry C, 8*(30), 10257–10265.

Sun, S., Li, S., Wang, S., Li, Y., Han, L., Kong, H., & Wang, P. (2016). Fabrication of hollow $NiCo_2O_4$ nanoparticle/graphene composite for supercapacitor electrode. *Materials Letters, 182*, 23–26.

Tang, H., & Sodano, H. A. (2013). Ultra high energy density nanocomposite capacitors with fast discharge using $Ba_{0.2}Sr_{0.8}TiO_3$ nanowires. *Nano Letters, 13*(4), 1373–1379.

Tang, H., Zhou, Z., & Sodano, H. A. (2014). Relationship between $BaTiO_3$ nanowire aspect ratio and the dielectric permittivity of nanocomposites. *ACS Applied Materials & Interfaces, 6*(8), 5450–5455.

Tang, Y., Xu, W., Niu, S., Zhang, Z., Zhang, Y., & Jiang, Z. (2021). Crosslinked dielectric materials for high-temperature capacitive energy storage. *Journal of Materials Chemistry A, 9*(16), 10000–10011.

Voiry, D., Yang, J., Kupferberg, J., Fullon, R., Lee, C., Jeong, H. Y., Shin, H. S., & Chhowalla, M. (2016). High-quality graphene via microwave reduction of solution-exfoliated graphene oxide. *Science, 353*(6306), 1413–1416.

Wang, M., Duong, L. D., Mai, N. T., Kim, S., Kim, Y., Seo, H., Kim, Y. C., Jang, W., Lee, Y., & Suhr, J. (2015). All-solid-state reduced graphene oxide supercapacitor with large volumetric capacitance and ultralong stability prepared by electrophoretic deposition method. *ACS Applied Materials & Interfaces, 7*(2), 1348–1354.

Wang, P., Pan, Z., Wang, W., Hu, J., Liu, J., Yu, J., Zhai, J., Chi, Q., & Shen, Z. (2021). Ultrahigh energy storage performance of a polymer-based nanocomposite via interface engineering. *Journal of Materials Chemistry A*, 9(6), 3530–3539.

Wang, Y., Chen, J., Li, Y., Niu, Y., Wang, Q., & Wang, H. (2019). Multilayered hierarchical polymer composites for high energydensity capacitors. *Journal of Materials Chemistry A*, 7(7), 2965–2980.

Xie, H., Huang, Z., Guo, S., & Torru, E. (2015). Feasibility of an electrostatic energy harvesting device for CFCs aircraft. *Procedia Engineering*, 99, 1213–1222.

3 Graphene – An Energy Storage Material

Venkatesh Koushick
Vel Tech Rangarajan Dr. Sagunthala R&D
Institute of Science and Technology

Jeyachandran Eindhumathy
Saranathan College of Engineering

Chandrasekar Divya
Manonmaniam Sundaranar University

Manickam Anusuya
Indra Ganesan Group of Institutions

Fabian I. Ezema
University of Nigeria

CONTENTS

3.1 INTRODUCTION

Graphene is now being studied extensively across the world because of its unusual qualities such as zero bandgap, extraordinary electron mobility at ambient warmth, elevated thermal conductivity and rigidity, enormous facade region, impermeability to gases, and so on [1]. At normal temperature, graphene charge carriers have core mobility, are massless, and travel a small number of micrometers while retaining their structure. Graphene-based materials have recently acquired prominence in energy storage systems, electronics, chemical sensors, optoelectronics, nanocomposites

DOI: 10.1201/9781003215196-3

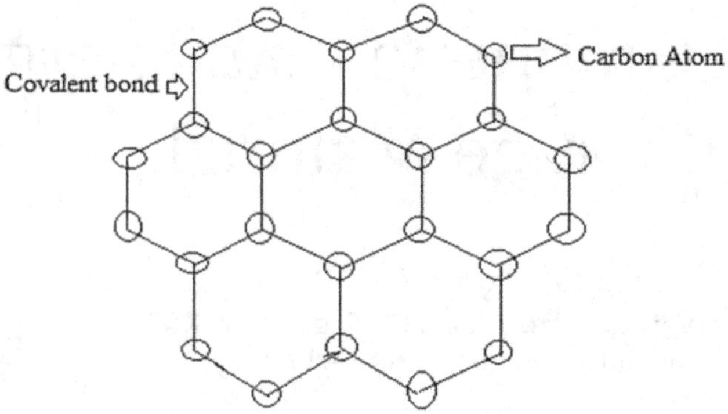

FIGURE 3.1 Structure of graphene.

(NCs), and health applications for instance osteogenic. Graphene is a carbon allotrope in which the carbon atoms are organized in a sole layer [2]. These carbon atoms are set in a honeycomb lattice that is dual-dimensional. In a single Graphene sheet, the carbon–carbon bond distance is around 0.142 nm. One of the unusual and significant aspects of graphene that has piqued the curiosity of researchers is its component electrons, which appear to be massless relativistic particles, resulting in an inconsistent quantum Hall effect and the nonexistence of localization. Graphene has been used in a range of relevance, including supercapacitors and lithium-ion batteries for energy storage, gas detection, and conducting electrodes. Figure 3.1 depicts the structure of graphene, and it may have different arrangements such as heaped, swathe, piece, and rolled.

3.1.1 GRAPHENE OXIDE

Graphene oxide (GO) is a byproduct of graphene oxidation. It is collected from a solitary single molecular layer which had oxygen functions for instance carboxyl, carbonyl, epoxide, or hydroxyl groups. GO is a water-soluble nanomaterial generated by chemically inserting oxygen flaws into a graphite heap [3], followed by thorough exfoliation of the firm into an atomic-thick piece using thermal, substance, or mechanical processes. Reduced GO (rGO) slender films are transparent conductors because they have the same clearness and resistance as carbon nanotube (CNT) networks. The convenient mixture, important solubility and process ability, pale customizable conductivity, outsized surface region, exceptional biocompatibility, and a profusion of low-cost source material are all advantages of GO over other carbon-based products for instance pure graphene, diamond, graphite, CNTs, and fullerene. The degree of oxidation has a significant effect on the electrical arrangement and assets of GO. Figure 3.2 depicts the stages of synthesis from GO to rGO.

At low oxidation, the bandgap is narrow, giving GO semiconductor characteristics. Furthermore, the presence of unpaired electrons results in ferro or paramagnetic characteristics. At saturation oxidation levels, the magnetic characteristics

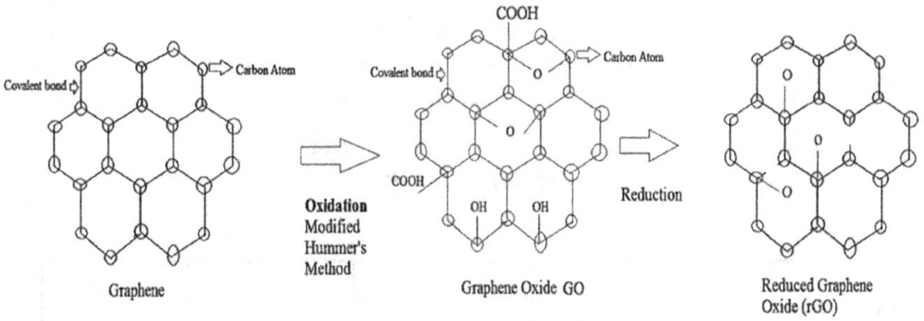

FIGURE 3.2 Stage synthesis of GO to rGO.

decrease and the bandgap approaches insulators. The energy gap was demonstrated to be highly programmable by modifying the quantity of epoxide and hydroxyl on the graphene exterior in configurations with various O_2 concentrations [4]. The improved qualities of GO, together with as perfunctory durability, huge exterior region, changeable ocular and electrical properties, and functionalized Graphene-containing functional groups, have piqued the notice of many researchers. Because of its hydrophilic properties, graphite oxide rapidly diffuses inside H_2O, where it tears into microstructural atomically slim particles, ensuing in mono-layer GO solutions/suspensions. The emergence, transparency, and 'strength' of a conventional GO solution are all dependent on the concentration of the solvent utilized. The modified Hummer's technique is the most well-known method for synthesizing GO today. The major component is the action of graphite powder in a solution of sodium nitrate, potassium permanganate, and sulfuric acid. Whereas GO can be easily isolated in H_2O, it may moreover be isolated in unrefined solvents due to electrostatic repulsion caused by the negative charge of the GO sheets, resulting in a stable colloidal suspension [5]. To discharge the GO product onto any appropriate substrate, drop-casting dip coating, spray coating, spin coating Langmuir–Blodgett, along with void filtering methods can all be employed. Figure 3.3 depicts several GO deposition processes on a suitable basis.

3.1.2 Li-Ion Hybrid Supercapacitors

Li-ion batteries offer an elevated specific energy storage capacity with a charge–discharge cycle life of 1,000 cycles. Li-ion batteries, conversely, have disadvantages such as a combustible electrolyte, a short capacitive current, and a lower exact power. Researchers have experimented with Li-ion hybrid supercapacitors to circumvent these constraints (LIHCs). These LIHCs have elevated specific energy and power. However, because near the unequal kinetics among the anode and cathode, these LIHCs have a significant disadvantage of short precise energy by high-charge/discharge rates [6]. Several researchers have synthesized transition metal organic frameworks (MOFs) (Fe, Co, and Ni) for elevated energy storage density devices. As a result, a detailed sympathetic of the behavior of rechargeable

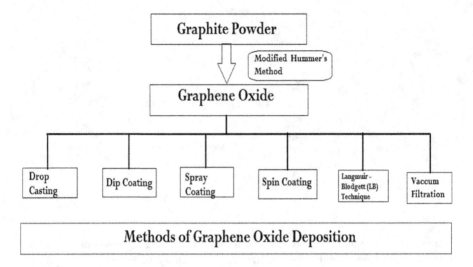

FIGURE 3.3   Various categories of evidence methods of GO.

batteries with supercapacitors is essential within the production furthermore design of complex functional elevated recital energy storage devices such as Li-ion batteries, supercapacitors, and cross-capacitors. If one chooses one of them, one must make a compromise on one or more traits.

As a result, a hybrid technology with exceptionally elevate precise energy, power, current density, and charging instance is required, with these properties not varying at a rapid charge–discharge rate [7]. The focus is on advancements in energy storage devices such as Li-ion and Na-ion rechargeable batteries, pseudo supercapacitors, asymmetric supercapacitors, and cross-supercapacitors with rGO/metal oxides nanocomposites, device of exploit, and increased energy storage capability.

Using exterior augment 3D rGO cross nano-networks, researchers examined high-performance electrode materials for energy storage devices for instance rechargeable batteries, asymmetric, pretend, with mixture supercapacitors [8].

These graphene electrodes have a precise capacitance of 306 F g^{-1} and an energy density of 148.75 Wh kg^{-1}, allowing them to achieve elevated power density and energy density at the same time. Figure 3.4 depicts the construction of batteries and supercapacitors.

3.1.2.1   Batteries

The chemical storage and high energy densities are 100's Wh kg^{-1}. The reactant diffusion, low power density of the battery is 10 W kg^{-1}. It has low cycle life due to degradation.

3.1.2.2   Supercapacitors

The supercapacitor has a surface charge storage capacity of 1–10 Wh kg^{-1} and a low energy density. The supercapacitor has an elevated power density of 1 kW kg^{-1} and a cycle life in the region of 105 cycles.

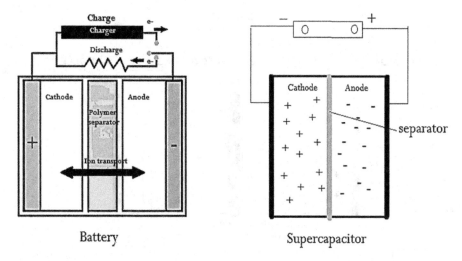

FIGURE 3.4   Structure of battery and super-capacitor.

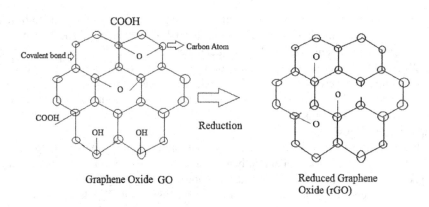

FIGURE 3.5   Structure conversion from GO to rGO.

3.1.3   REDUCED GRAPHENE OXIDE

rGO is a GO whose O_2 concentration has been abridged by thermal, substance, or further means. The use of chemical, thermal, electrochemical, and photo-irradiation approaches to efficiently rGO. The valence structure conversion from GO to rGO is given and shown in Figure 3.5.

Electrochemically synthesized rGO, for example, has substantially superior electrical conductivity and electrochemical performance than chemically or thermally formed rGO materials. After reduction, the specific capacitance of GO rises, with a value of 100–300 F g^{-1} was observed [9].

Because there is an extended term insist for rechargeable devices with elevated specific energy and speedy charging. Surface-enhanced 3D rGO hybrid nano-networks are as well studied as elevated recital electrodes for energy storage devices such as rechargeable batteries, asymmetric supercapacitors, and hybrid with pretend supercapacitors.

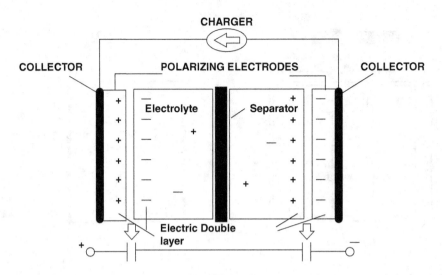

FIGURE 3.6   Electric double-layer capacitance.

To recuperate the electrochemical energy storage efficiency of graphene-based materials, O_2-rich in-plane holes are introduced hooked on graphene backbones using sonochemical engraved and subsequent chemical reduction treatment.

When GO was etched, a huge number of in-plane nano-scale apertures with a high O_2 concentration were created [10]. The etched GO was treated with hydrazine to provide a permeable basal flat with elevated electrical conductivity and electrochemically vigorous O_2 atoms. Figure 3.6 depicts the structure of electric double-layer capacitance (EDLC).

It has been established that high conductivity and in-plane nano-scale apertures of rGO by enhanced electrochemically active O_2 groups increase EDLC with pseudo capacitance recital in excess of an extensive power density range.

An rGO electrode displayed an utmost energy density of 47 W h kg^{-1}, about thrice a time that of rGO without engraving treatment, with an elevated power density able to 100 kW kg^{-1} and a significant energy density [11]. Figure 3.7 depicts the rate of voltage retention versus time. Table 3.1 discusses the comparison of several batteries.

3.1.4   GRAPHENE PROPERTIES AND APPLICATIONS

Magnetic GO, a composite of magnetic nano-particles and GO, has special corporeal and substance features such as nano-size, a huge precise exterior area, and paramagnetic and biocompatible capabilities, making it a potential biomaterial in the pasture of biomedicine. Graphene, on the other hand, is extremely diamagnetic, akin to graphite. Our nanocrystals have just a little paramagnetic contribution visible below 50 K. The findings show that the paramagnetism is caused by a single species of defects, with roughly one magnetic moment per typical graphene crystallite [12].

Graphene is inherently nonmagnetic because all of the outside electrons in carbon hexatomic rings are perfectly connected to create - and - bonds, graphene is

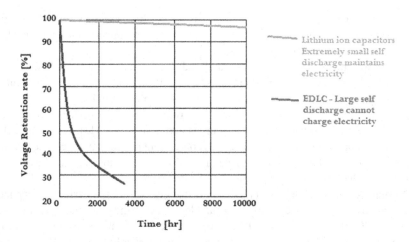

FIGURE 3.7 Voltage retention rate vs. time.

TABLE 3.1

Comparison of Various Batteries (avnet.com, electronicspecifier.com)

Description	EDLC/Polyacene Capacitors	Lithium-Ion Hybrid Capacitors	Lithium-Ion Batteries (LIB)
Electrodes			
Internal resistance	Low	Medium	High
Operating temperature (°C)	−35 to +65	−20 to +80	−15 to +55
Maximum rated voltage	2.3–2.7 V	3.8 V	4.1–4.3 V
Maximum operating voltage	0	2.2	2.5
Capacitance/energy vs device volume	Lowest	2–3 × EDLC	Highest 100 × EDLC
Charge/discharge cycles	100 k	100 k	500–1,000
Self-discharge	Vulnerable	Very low	Very low
Security	Highly stable	Highly stable	Safety consideration required
Voltage monitoring required	No	Yes	Yes

intrinsically nonmagnetic. All the efforts are made to break the symmetric bonds in graphene in order to liberate unpaired electrons and create net spins.

If one of the conductive materials in a supercapacitor has a larger relative surface area than the other, it will better retain an electrostatic charge. Graphene, on the other hand, has far better electrical conductivity than graphite because of the presence of quasi-particles, which are electrons that behave as if they have no mass and may traverse vast distances without scattering [13]. Graphene's excellent electrical conductivity is owing to its zero-overlap semimetal structure, which contains electrons and holes as charge carriers.

These unbound electrons found over and under the graphene page are known as pi () electrons, by the help to strengthen carbon-to-carbon bonding. Graphene is made up of a single layer of graphite. Graphene, such as graphite, conducts electricity well because its electrons are delocalized and free to flow across its surface. Graphene's characteristics make it valuable in electronics and composites [14]. Graphene cells, such as Li-ion batteries, are made up of two conducting plates that are covered with a porous substance and immersed in an electrolyte solution. Li-ion batteries have lesser electrical conductivity than graphene.

This enables faster-charging cells that can also give very large currents. Graphene has enormous promise as a highly sustainable material and for enhancing the sustainability of several businesses. It has recently been shown that graphene can be blended into many materials to make them more ecologically friendly [15].

Because of its extraordinary qualities, graphene is being investigated for application in electronics. It conducts electricity better than any other known material, is extremely durable, does not shatter, and is both flexible and translucent. Transportation, medical, electronics, energy, military, and desalination are just a few of the fields where graphene research is having an influence [16].

Supplementary uses for graphene comprise anti-corrosion covering with paints, competent and distinct sensors, quicker along with more resourceful electronics, flexible displays, proficient solar panels, quicker DNA sequencing, medication liberation, and others.

It is 200 periods stronger than steel and highly lithe; it is the thinnest material conceivable; and it is entirely transparent, transmitting more than 90% of the light. It is the greatest conductor and may serve as a perfect barrier. Because graphene is fabricated on a solitary sheet of carbon atoms tied in a hexagonal prototype, it is exceedingly thin and lightweight, making it an appealing material for nanotechnology applications.

3.2 CONCLUSION

Elevated recital energy storage devices constructed as rGO/Metal oxides NCs are being investigated. The fundamental mechanism of rGO's synergistic action on metal oxide NPs are studied. The Li-ion hybrid supercapacitor's highest power density was 41.25 kWh kg^{-1}. Simultaneously, the highest in progress concentration of the VrG-5 electrode within the rechargeable battery was 574.5 C g^{-1}. Lithe energy storage devices with elevated energy concentration have been devised; however, they are highly expensive.

REFERENCES

1. Kumar H. 2020. Recent advancement made in the field of reduced graphene oxide-based nanocomposites used in the energy storage devices: A review. *J Energy Storage* 33, Oct 2020:102032.
2. Li S, Chen Y, He X, Mao X, Zhou Y, Xu J, Yang Y. 2009. Modifying reduced graphene oxide by conducting polymer through a hydrothermal polymerization method and its application as energy storage electrodes. *Nanoscale Res Lett* 14:226.

3. Diez-Pascual AM, Sanchez JAL, Capilla RP, Diaz PG. 2018. Recent developments in graphene/polymer nano-composites for application in polymer solar cells. *Polymers* 10(2):217.
4. Ramesha GK, Sampath S. 2009. Electrochemical reduction of oriented graphene oxide films: an in situ Raman spectroelectrochemical study. *J Phys Chem C* 113(19):7985–7989.
5. Ramabadran U, Ryan G, Zhou X, Farhat S, Manciu F, Tong Y, Ayler R, Garner G. 2017. Reduced graphene oxide on nickel foam for supercapacitor electrodes. *Materials* 10(11):1295.
6. Yan D, Liu Y, Li YH, Zhuo RF, Wu ZG, Ren PY, Li SK, Wang J, Yan PX, Geng ZR. 2014. Synthesis and electrochemical properties of MnO_2/rGO/PEDOT:PSS ternary composite electrode material for supercapacitors. *Mater Lett* 127:53–55.
7. Zhao C, Shu KW, Wang CY, Gambhir S, Wallace GG. 2015. Reduced graphene oxide and polypyrrole/reduced graphene oxide composite coated stretchable fabric electrodes for supercapacitor application. *Electrochim Acta* 172:12–19.
8. Kumar NA, Choi HJ, Shin YR, Chang DW, Dai LM, Baek JB. 2012. Polyaniline-grafted reduced graphene oxide for efficient electrochemical supercapacitors. *ACS Nano* 6(2):1715–1723.
9. Olabi AG, Abdelkareem MA, Wilberforce T, Sayed ET. 2021. Application of graphene in energy storage device – a review. *Renew Sustain Energy Rev* 135:110026.
10. Kuila T, Mishra AK, Khanra P, Kim NH, Lee JH. 2013. Recent advances in the efficient reduction of graphene oxide and its application as energy storage electrode materials. *Nanoscale* 5, 52–71.
11. Mallard LM, Pimenta MA, Dresselhaus G, Dresselhaus MS. 2009. Raman spectroscopy in graphene. *Phys Rep* 473:51–87.
12. Geim AK, Kim P. 2008. Carbon wonderland. *Sci Am* 298:90.
13. Novoselov KS, Geim AK, Morozov SV, Jiang D, Zhang Y, Dubonos SV, et al. 2004. Electric field effect in atomically thin carbon films. *Science* 306:666.
14. Viculis LM, Mack JJ, Kaner RB. 2003. A chemical route to carbon nanoscrolls. *Science* 299:1361.
15. Berger C, Song Z, Li T, Li X, Ogbazghi AY, Feng R, et al. 2004. Ultrathin epitaxial graphite: 2D electron gas properties and a route toward graphene-based nanoelectronics. *J Phys Chem* 108:19912.
16. Land TA, Michely T, Behm RJ, Hemminger JC, Comsa G. 1992. STM investigation of single layer graphite structures produced on Pt (111) by hydrocarbon decomposition. *Surf Science* 264:261–270.

4 Recent Advances in Graphene Oxide–Based Fuel Cells

Agnes Chinecherem Nkele
University of Nigeria and Colorado State University

Chinedu P. Chime and Fabian I. Ezema
University of Nigeria

CONTENTS

4.1 INTRODUCTION TO FUEL CELLS

The negative effects associated with greenhouse emissions have prompted researchers to delve into the use of fuel cells (Su and Hu, 2021). Fuel cells yield sustainable and clean energy technology with minimal pollution. This is because fuel cells release heat and water vapor as hydrogen and oxygen naturally react. A fuel cell is an electrochemical system that converts the chemical energy of an oxidizing agent like hydrogen and fuel to direct current electrical energy using a pair of redox reactions. Typically, a fuel cell comprises of an electrolyte that sandwiches a positive electrode/cathode and a negative electrode/anode as shown in Figure 4.1. Gas diffusion layers can also be incorporated into fuel cells to minimize oxidation effects. The source of fuel is usually hydrogen. The cathode which is usually nickel is responsible for converting ions to chemical wastes such as water. The anode which could be made of platinum degrades the fuel to ions

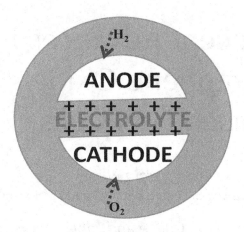

FIGURE 4.1   Schematic of the major fuel cell components.

and electrons. The electrolyte which could be salt carbonates, phosphoric acid, or potassium hydroxide; allows ions to be conducted between the electrodes so as to sustain the reaction chemistry. Fuel such as hydrogen and air is fed to the anode where oxidation occurs to release protons and electrons produced by the catalyst in the anode. Anodic hydrogen ions collect electrons from the oxygen (supplied to the cathode) to release water. The electrons create an electric current as they move through an external circuit to create direct current electricity while the protons combine with oxygen and electrons at the cathode to release heat and water (Fuel Cells, 2020). The voltage flowing through a fuel cell reduces with an increase in current due to activation, mass transport, and ohmic losses. The quantity of heat and water emitted depends on the source of fuel. The choice of components to be incorporated into fuel cells would be based on how efficient, durable, and electrochemically performing (Su and Hu, 2021). The amount of fuel supplied and the chemical activeness determines the quantity of electric current that would flow to the external circuit. Fuel cells do not need to be recharged because they spontaneously generate heat and electricity as long as there is a supply of fuel. They serve as backup for power used in private and commercial residents (don, 2020). Fuel cells yield higher efficiencies with minimal noise, no air pollutants, and incur no wear on moving parts. Fuel cells can be applied in transport systems (Lemons, 1990), utility power stations, commercial and residential buildings, forklifts, etc. However, fuel cells suffer some limitations such as difficulty in storage, heavy-weighted, expensive to manufacture, large size, and poor durability (Kordesch and Simader, 1996). Hydrogen also suffers limitations in its production, delivery, and storage.

Fuel cells are divided into different major classifications namely alkaline fuel cell (AFC), phosphoric acid fuel cell (PAFC), proton exchange membrane fuel cell (PEMFC), solid oxide fuel cell (SOFC), and molten carbonate fuel cell (MCFC). This classification is based on the kind of temperature and electrolyte used by fuel cell (Xiang et al., 2021) as outlined in Table 4.1. Hydrogen fuel cells have low recorded efficiency, heavy in weight, and rigorous mode of production. PEMFC has a proton-conducting membrane that contains the electrolyte where the anode and cathode are immersed. It is made up of bipolar plates, electrodes, catalysts, membranes, and current collectors. PAFC involves

TABLE 4.1

Parameters Showing the Differences between the Classes of Fuel Cells

Fuel Cell	Electrolyte	Power (W)	Working Temperature (°C)	Cell Efficiency	System Efficiency
Metal hydride	Aqueous alkaline solution		≥ 20		
Electro-galvanic	Aqueous alkaline solution		<40		
Direct formic acid	Polymer membrane (ionomer)		<40		
Zinc-air battery	Aqueous alkaline solution		<40		
Microbial	Polymer membrane or humic acid		<40		
Upflow microbial			<40		
Regenerative	Polymer membrane (ionomer)		<50		
Reformed methanol	Polymer membrane (ionomer)		250–300	50%–60%	25%–40%
Direct ethanol	Polymer membrane (ionomer)		>25		
Proton exchange membrane	Polymer membrane (ionomer)	1 W–500 kW	50–100	50%–70%	30%–50%
Redox	Liquid electrolytes with redox shuttle and Polymer membrane (ionomer)	1 kW–10 MW			
Phosphoric acid	Molten phosphoric acid	<10 MW	150–200	55%	40%
Solid acid	Solid acid	10 W–1 kW	200–300	55%–60%	40%–45%
Molten carbonate	Molten alkaline carbonater	100 MW	600–650	55%	45%–55%

a non-conducting electrolyte (phosphoric acid) which allows the movement of hydrogen ions from the anode to the cathode. This fuel cell operates at high temperatures and releases heat that can produce steam for air conditioners and other thermal devices. SAFC has a salt-like structure at reduced temperatures and uses a solid acid material as electrolyte. AFC comprises of two carbon electrodes filled with a catalyst (silver, cobalt oxide, platinum, etc.) while the space separating the electrodes is filled with a concentrated alkaline solution like hydroxides of potassium, sodium, calcium, etc. SOFC uses a solid or ceramic material as the electrolyte, has a slow start-up time, operates at high temperatures, and emits carbon particles. MCFC uses lithium potassium carbonate salt as an electrolyte and converts fossil fuel to gaseous hydrogen in the anode to release hydrogen.

4.2　GRAPHENE OXIDE (GO) AS A FUEL CELL MATERIAL

Graphene is a single-atom layer with a hexagonal crystal structure bonded with sp^2 carbon atoms with high carrier mobility, and is thermally stable (Su and Hu, 2021). It is essential for fuel cells because it is highly conductive (thermally and electrically), resistant to corrosion, relatively affordable, and has a large surface area. Incorporating graphene into PEMFC improves the cell performance and conductive rate of the ions (Yadav et al., 2018). Three-dimensional graphene is porous and conductive with a wide area of the surface. It also serves as an efficient catalyst in fuel cells, especially in creating ultra-durable catalysts. Three-dimensional graphene materials allow more fuel to be diffused and transported into the cells. The mobility of carriers in graphene is not affected by temperature. Electrons are easily transported from the graphene to the substrate due to their efficient electrical performance (Su and Hu, 2021). The strong attribute of ripples in graphene makes it an efficient catalyst that is permeable to hydrogen. Incorporating graphene material into proton exchange membranes would increase the conductivity, current flow, resistance to water, and even distribution of air/fuel (Su and Hu, 2021). The great mechanical feature exhibited by graphene allows it to be used to reinforce composite fuel cell components. The zero band gap of graphene makes the mobile charges continuously adjust and manifest semiconducting features (Yadav et al., 2018).

Graphene oxide (GO) is usually produced when a layer of graphite oxide is chemically oxidized. It is a cheap and active fuel cell that has high-conducting protons at lower temperatures and relative humidity. Incorporating GO would enhance the commercialization of fuel cells with better efficiency. The expensive nature of platinum prompted the development of GO-based nanomaterials as alternative catalysts in fuel cells. This is because GO-based materials have mobile carriers, high stability, increased durability, surface areas that are electrochemically active, and are highly conductive (Graphene Oxide as a Catalytic Agent in Polymer Electrolyte Membrane Fuel Cell | Minnesota State University, Mankato, 2021). They serve as good electrocatalytic materials with more active sites and increased electron mobility. The mechanical and electrochemical features of GO are influenced by defects and structural instability encountered. Oxygen functional groups on the surface of graphene greatly affect its electrochemical performance. It serves as a good electrode material due to its uniform dispersal in water. GO serves as an electrolyte in fuel cells (which operate at low temperature) especially PEFCs because of its increased proton conduction rate. The different characteristics of graphene-based materials are shown in Figure 4.2.

4.3　ENHANCING THE PERFORMANCE OF GRAPHENE OXIDE–BASED MATERIALS

4.3.1　Introducing Dopants

Introducing dopants to GO-based materials alters the density of state and electronic features while increasing the catalytic performance and active sites of graphene. Nitrogen is an n-type electron donor that commonly serves a graphene dopant

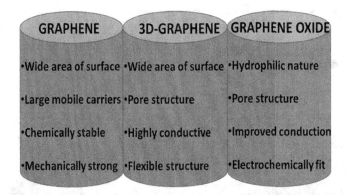

FIGURE 4.2   Different features of graphene-based materials useful in fuel cells.

because it has high electronegativity and exhibits improved stability (Li et al., 2009). Boron is a *p*-type electron acceptor with more carrier concentration and a large density of state, and is highly conductive. Doping graphene with sulfur that is rich in thiophene produces more conductive materials than those rich in sulfonate. Sulfur reduces the affinity of graphene to water and enhances oxygen reduction reactions (Yang et al., 2012). The ferromagnetic property of graphene-based materials is also improved upon doping (Liu et al., 2013).

4.3.2   HYDROGENATION

This process involves converting sp^2 carbon hybrid atoms to sp^3 carbon hybrids. Hydrogenation can also involve the use of hydrogen to reduce GO under ambient conditions (Sofer et al., 2014). Hydrogenated graphene material has more heterogeneous electron transfer which improves the semiconducting and electronic features (Krishna et al., 2012).

4.3.3   SYNTHESIS METHOD INVOLVED

The synthesis method adopted influences the features and structure of graphene-based materials. The bottom-up approach yields graphene of high quality while top-down approach is useful for depositing graphene over a wide substrate surface (Xiang et al., 2021). Other methods such as chemical vapor deposition, hummers method, solvothermal, mechanical exfoliation, one-step synthesis, polymerization, etc. (Allagui et al., 2015; Zhang et al., 2012; Chen et al., 2013; Li and Li, 2013) as shown in Figure 4.3 have also been adopted in fabricating GO-based materials (Yadav et al., 2018). Adopting the right synthesis method makes the fabricated graphene chemically stable and highly conductive.

4.4   RECENT ADVANCES IN GRAPHENE OXIDE–BASED FUEL CELLS

GO-based materials can be usefully applied in different fuel cell components like cathode, anode, electrolyte membrane, etc.

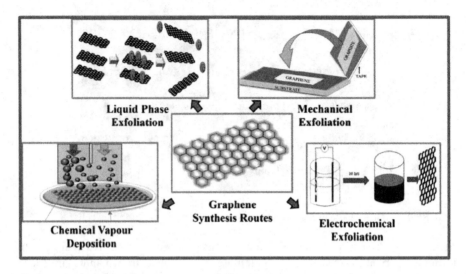

FIGURE 4.3 Different methods applicable in the synthesis of graphene-based materials. (Reprinted with permission from Yadav et al. 2018.)

4.4.1 AS CATHODE MATERIAL

Due to the oxidation-reduction reactions that occur at the cathode, GO-based materials have been incorporated as cathode-catalysts. They can also be used to increase electron mobility when combined with nanoparticles because of the increased number of active sites. Graphene doped with nitrogen via hydrogen exfoliation in proton exchange membrane fuel cell recorded good efficiency, improved stability, and high power density value at 60°C (Vinayan et al. 2012). Formation of agglomerates was minimized because the synthesized materials were durable in acidic media and of high performance. Shu et al. designed a nitrogen-doped graphene that was pre-treated with hydrogen tetraoxosulfate (vi) acid through the hydrothermal method (Shu et al., 2020). The synthesis process has been shown to produce platinum-based electrocatalysts with mobile carriers, high durability, and exhibit high catalytic performance. Zhang et al. dispersed ruthenium on nitrogen-doped graphene after annealing under gaseous ammonia in acidic media (Zhang et al., 2017). The preparation process for the doped graphene material began with dispersing the GO in water, mixing the precursor materials in the solution, and forming the doped composite as shown in Figure 4.4. They obtained a more durable and active fuel cell that was compatible with methanol and more catalytically active. Incorporating ruthenium increased the oxidation-reduction reaction of the formed graphene material and increased the number of active sites.

4.4.2 AS ANODE MATERIAL

GO-based materials are promising tools for improving the transport ability of mobile carriers and making them biocompatible especially in microbial fuel cells. These materials internally reduce the resistance in the anode as they increase the

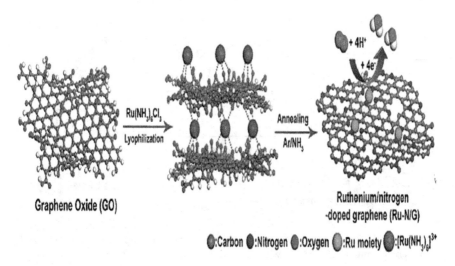

FIGURE 4.4 Description of the stages involved in synthesizing the graphene-doped composite material. (Reprinted with permission from Zhang et al. 2017.)

growth of bacteria in microbial fuel cells (Li et al., 2017). Three-dimensional graphene materials increase the active surface area for bacteria to be attached while lowering the resistance encountered during the diffusion process. Yong et al. produced a biofilm of reduced GO as an anode material through a one-step self-assembly or fishing method (Yong et al., 2014). Excellent electron transfer with increased oxidation and reduction currents was obtained. The increased electrochemical activeness and area of the surface make the films potential microbial fuel cell materials.

A three-dimensional graphene scaffold was produced by Ren et al. to serve as an anode (Ren et al., 2016). A porous morphology that is biocompatible with minimal internal resistance, and increased carrier transport, was obtained. Modifying GO by incorporating a three-dimensional aerogel form of graphene can increase the density of the film and cause the microbial fuel cell to perform better (Li et al., 2020). The hydrophilic nature of the anode is increased and the time for the start-up is shortened. Holmes et al. decorated the anode part of a direct methanol fuel cell with a single graphene layer and noticed that the permeability of methanol reduced as more graphene was introduced (Holmes et al., 2017). The conduction rate of the proton (at reduced temperature) and the cell performance greatly improved.

Luo et al. adopted a capillary compression method to synthesize reduced GO in drops of aerosol so that the adhesive power can be increased as illustrated in Figure 4.5 (Luo et al., 2011). The crumpled nanomaterial was highly compressive with the minimal aggregate formation and improved stability. The roughness encountered due to crumpling enhanced the formation of the film, the performance of the cell, and enhanced aggregate resistance. The quantity of GO concentration greatly affected the size of the crumpled particle.

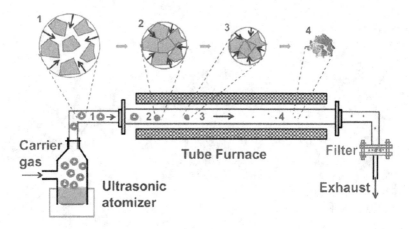

FIGURE 4.5   Synthesis of the crumpled graphene oxide sheets prepared in aerosol drops by compression technique. (Reprinted with permission from Luo et al. 2011.)

4.4.3   As Catalyst in Electrodes

Graphene-based materials serve as good support catalysts for methanol oxidation reactions because they are electroconductive and possess a wide area of the surface (Su and Hu, 2021). To serve as better catalysts in the electrodes, the graphene-based materials would have to be doped with elements like phosphorus, boron, sulfur, etc. These dopants affect the chemical and electronic characteristics, improve oxidation-reduction reactions, and increase the number of active sites (Kong et al., 2013). Two or more elements can be co-doped to make the GO-based materials more conductive, highly efficient, and reduce their energy band gap (Chai et al., 2017), although the mechanism of forming bonds is not easily controlled (Shao et al., 2019).

Graphene doped with boron yields good and highly stable electrocatalyst (Sheng et al., 2012; Kong et al., 2013). Boron is a strong electron acceptor that enables great electrocatalytic activity by generating free electrons. Doping graphene with phosphorus yielded a potential of approximately 0.92 V (Zhang et al., 2013). The thermally annealed doped product exhibited excellent oxidation-reduction reaction and energy storage.

4.4.4   As Electrolyte Membrane

A good electrolyte for fuel cell devices must be very conductive and develop membrane resistance against reactants. Graphene-based materials have been explored as efficient electrolytes to conduct protons in polymer membranes of fuel cells because of their flexible functionalization attribute. Graphene improves the mechanical features of polymer membranes. It also protects the metallic plates because of its strong resistance to corrosion. Intrinsic defects observed on the surface of GO improve its proton conductivity (Lee et al., 2014).

Hatakeyama et al. investigated fuel cell performance using a reduced GO paper membrane (Hatakeyama et al., 2014). They achieved higher proton conductivity as the distance between the layers increased and lower electron conductivity as the quantity of oxygen increased.

Gao et al. studied the transport of protons through a fuel cell membrane made of GO (Gao et al., 2014). They obtained increased oxidation reactions and higher ionic conduction rates.

4.5 CONCLUSIONS AND FUTURE PERSPECTIVES

This book chapter has successfully discussed recent advancements in the use of GO-based materials in fuel cells. The concept of fuel cell as an electrochemical medium for energy production was introduced. The different classifications of fuel cells (based on the kind of fuel used) have also been highlighted. Certain attributes of materials made from graphene and GO like high conductivity, durability, corrosion resistance, etc., which makes them relevant for use in fuel cells have been highlighted. Factors such as the incorporation of dopants, hydrogenation, and the synthesis method adopted have been considered as performance enhancers for GO-based materials. The relevance of GO-based materials as cathode material, anode material, electrode catalyst, and impermeable membrane has also been discussed. The future perspectives on the use of GO-based materials have also been outlined in this concluding section.

Graphene can be better utilized in fuel cells by changing the material structure to a three-dimensional curved shape which would minimize restacking effects. Low-cost and efficient materials can be incorporated into fuel cells to replace platinum. The performance of fuel cell can be improved by using durable electrolytes. Over time, the fuel in cars could be replaced with hydrogen to power vehicles in the form of fuel cell electric vehicles. The vehicles would combine hydrogen stored in a tank with oxygen from the air to generate electricity. This would enhance efficiency and minimize environmental pollution as water and heat would be the by-products. Graphene may also be useful in producing hydrogen fuel cells that are very durable over time.

REFERENCES

Allagui, Anis, Tareq Salameh, and Hussain Alawadhi. 2015. "One-Pot Synthesis of Composite NiO/Graphitic Carbon Flakes with Contact Glow Discharge Electrolysis for Electrochemical Supercapacitors." *International Journal of Energy Research* 39 (12): 1689–97. doi:10.1002/er.3379.

Chai, Guo-Liang, Kaipei Qiu, Mo Qiao, Maria-Magdalena Titirici, Congxiao Shang, and Zhengxiao Guo. 2017. "Active Sites Engineering Leads to Exceptional ORR and OER Bifunctionality in P, N Co-Doped Graphene Frameworks." *Energy & Environmental Science* 10 (5): 1186–95.

Chen, Ji, Bowen Yao, Chun Li, and Gaoquan Shi. 2013. "An Improved Hummers Method for Eco-Friendly Synthesis of Graphene Oxide." *Carbon* 64: 225–29.

don. 2020. "Fuel Cells – Alternate Energy Storage." *Energy Matters.* Accessed October 27. https://www.energymatters.com.auhttps://www.energymatters.com.au/components/fuel-cells/.

Fuel Cells. 2020. *Energy.Gov.* Accessed October 23. https://www.energy.gov/eere/fuelcells/fuel-cells.

Gao, Wei, Gang Wu, Michael T. Janicke, David A. Cullen, Rangachary Mukundan, Jon K. Baldwin, Eric L. Brosha, Charudatta Galande, Pulickel M. Ajayan, and Karren L. More. 2014. "Ozonated Graphene Oxide Film as a Proton-Exchange Membrane." *Angewandte Chemie International Edition* 53 (14): 3588–93.

Graphene Oxide as a Catalytic Agent in Polymer Electrolyte Membrane Fuel Cell | Minnesota State University, Mankato. 2021. Accessed October 22. https://research.mnsu.edu/research-month/research-events/graphene-oxide-as-a-catalytic-agent-in-polymer-electrolyte-membrane-fuel-cell/.

Hatakeyama, Kazuto, Hikaru Tateishi, Takaaki Taniguchi, Michio Koinuma, Tetsuya Kida, Shinya Hayami, Hiroyuki Yokoi, and Yasumichi Matsumoto. 2014. "Tunable Graphene Oxide Proton/Electron Mixed Conductor That Functions at Room Temperature." *Chemistry of Materials* 26 (19): 5598–5604.

Holmes, Stuart M., Prabhuraj Balakrishnan, Vasu S. Kalangi, Xiang Zhang, Marcelo Lozada-Hidalgo, Pulickel M. Ajayan, and Rahul R. Nair. 2017. "2D Crystals Significantly Enhance the Performance of a Working Fuel Cell." *Advanced Energy Materials* 7 (5): 1601216.

Kong, Xiangkai, Qianwang Chen, and Zhiyuan Sun. 2013. "Enhanced Oxygen Reduction Reactions in Fuel Cells on H-decorated and B-Substituted Graphene." *ChemPhysChem* 14 (3): 514–19.

Kordesch, Karl, and Günter Simader. 1996. *Fuel Cells and Their Applications*. Vol. 117. VCh, Weinheim.

Krishna, Rahul, Elby Titus, Luís C. Costa, José C. J. M. D. S. Menezes, Maria R. P. Correia, Sara Pinto, Joao Ventura, J. P. Araújo, José AS Cavaleiro, and José J. A. Gracio. 2012. "Facile Synthesis of Hydrogenated Reduced Graphene Oxide via Hydrogen Spillover Mechanism." *Journal of Materials Chemistry* 22 (21): 10457–59.

Lee, D. C., H. N. Yang, S. H. Park, and W. J. Kim. 2014. "Nafion/Graphene Oxide Composite Membranes for Low Humidifying Polymer Electrolyte Membrane Fuel Cell." *Journal of Membrane Science* 452: 20–28.

Lemons, Ross A. 1990. "Fuel Cells for Transportation." *Journal of Power Sources* 29 (1–2): 251–64.

Li, Jiannan, Yanling Yu, Dahong Chen, Guohong Liu, Dongyi Li, Hyung-Sool Lee, and Yujie Feng. 2020. "Hydrophilic Graphene Aerogel Anodes Enhance the Performance of Microbial Electrochemical Systems." *Bioresource Technology* 304: 122907.

Li, Shuang, Chong Cheng, and Arne Thomas. 2017. "Carbon-Based Microbial-Fuel-Cell Electrodes: From Conductive Supports to Active Catalysts." *Advanced Materials* 29 (8): 1602547.

Li, Xiaolin, Hailiang Wang, Joshua T. Robinson, Hernan Sanchez, Georgi Diankov, and Hongjie Dai. 2009. "Simultaneous Nitrogen Doping and Reduction of Graphene Oxide." *Journal of the American Chemical Society* 131 (43): 15939–44.

Li, Yueming, and Xue-Mei Li. 2013. "Facile Treatment of Wastewater Produced in Hummer's Method to Prepare Mn_3O_4 Nanoparticles and Study Their Electrochemical Performance in an Asymmetric Supercapacitor." *RSC Advances* 3 (7): 2398–2403. doi:10.1039/C2RA22191H.

Liu, Yuan, Nujiang Tang, Xiangang Wan, Qian Feng, Ming Li, Qinghua Xu, Fuchi Liu, and Youwei Du. 2013. "Realization of Ferromagnetic Graphene Oxide with High Magnetization by Doping Graphene Oxide with Nitrogen." *Scientific Reports* 3 (1): 1–5.

Luo, Jiayan, Hee Dong Jang, Tao Sun, Li Xiao, Zhen He, Alexandros P. Katsoulidis, Mercouri G. Kanatzidis, J. Murray Gibson, and Jiaxing Huang. 2011. "Compression and Aggregation-Resistant Particles of Crumpled Soft Sheets." *ACS Nano* 5 (11): 8943–49.

Ren, Hao, He Tian, Cameron L. Gardner, Tian-Ling Ren, and Junseok Chae. 2016. "A Miniaturized Microbial Fuel Cell with Three-Dimensional Graphene Macroporous Scaffold Anode Demonstrating a Record Power Density of over 10000 W m⁻³." *Nanoscale* 8 (6): 3539–47.

Shao, Yanqiu, Zhenshuang Jiang, Qiaoqiao Zhang, and Jingqi Guan. 2019. "Progress in Nonmetal-Doped Graphene Electrocatalysts for the Oxygen Reduction Reaction." *ChemSusChem* 12 (10): 2133–46.

Sheng, Zhen-Huan, Hong-Li Gao, Wen-Jing Bao, Feng-Bin Wang, and Xing-Hua Xia. 2012. "Synthesis of Boron Doped Graphene for Oxygen Reduction Reaction in Fuel Cells." *Journal of Materials Chemistry* 22 (2): 390–95.

Shu, Yasuhiro, Koji Miyake, Javier Quílez-Bermejo, Yexin Zhu, Yuichiro Hirota, Yoshiaki Uchida, Shunsuke Tanaka, Emilia Morallón, Diego Cazorla-Amorós, and Chang Yi Kong. 2020. "Rational Design of Single Atomic Co in CoNx Moieties on Graphene Matrix as an Ultra-Highly Efficient Active Site for Oxygen Reduction Reaction." *ChemNanoMat* 6 (2): 218–22.

Sofer, Zdeněk, Ondřej Jankovský, Petr Šimek, Lýdie Soferová, David Sedmidubský, and Martin Pumera. 2014. "Highly Hydrogenated Graphene via Active Hydrogen Reduction of Graphene Oxide in the Aqueous Phase at Room Temperature." *Nanoscale* 6 (4): 2153–60.

Su, Hanrui, and Yun Hang Hu. 2021. "Recent Advances in Graphene-Based Materials for Fuel Cell Applications." *Energy Science & Engineering* 9 (7): 958–83. doi:10.1002/ese3.833.

Vinayan, Bhaghavathi P., Rupali Nagar, Natarajan Rajalakshmi, and Sundara Ramaprabhu. 2012. "Novel Platinum–Cobalt Alloy Nanoparticles Dispersed on Nitrogen-Doped Graphene as a Cathode Electrocatalyst for PEMFC Applications." *Advanced Functional Materials* 22 (16): 3519–26.

Xiang, Yiqiu, Ling Xin, Jiwei Hu, Caifang Li, Jimei Qi, Yu Hou, and Xionghui Wei. 2021. "Advances in the Applications of Graphene-Based Nanocomposites in Clean Energy Materials." *Crystals* 11 (1): 47.

Yadav, Ramdayal, Akshay Subhash, Nikhil Chemmenchery, and Balasubramanian Kandasubramanian. 2018. "Graphene and Graphene Oxide for Fuel Cell Technology" 57 (29): 9333–50. doi:10.1021/acs.iecr.8b02326.

Yang, Shubin, Linjie Zhi, Kun Tang, Xinliang Feng, Joachim Maier, and Klaus Müllen. 2012. "Efficient Synthesis of Heteroatom (N or S)-doped Graphene Based on Ultrathin Graphene Oxide-porous Silica Sheets for Oxygen Reduction Reactions." *Advanced Functional Materials* 22 (17): 3634–40.

Yong, Yang-Chun, Yang-Yang Yu, Xinhai Zhang, and Hao Song. 2014. "Highly Active Bidirectional Electron Transfer by a Self-assembled Electroactive Reduced-graphene-oxide-hybridized Biofilm." *Angewandte Chemie International Edition* 53 (17): 4480–83.

Zhang, Chenhao, Junwei Sha, Huilong Fei, Mingjie Liu, Sadegh Yazdi, Jibo Zhang, Qifeng Zhong, Xiaolong Zou, Naiqin Zhao, and Haisheng Yu. 2017. "Single-Atomic Ruthenium Catalytic Sites on Nitrogen-Doped Graphene for Oxygen Reduction Reaction in Acidic Medium." *Acs Nano* 11 (7): 6930–41. doi:10.1021/acsnano.7b02148.

Zhang, Chenzhen, Nasir Mahmood, Han Yin, Fei Liu, and Yanglong Hou. 2013. "Synthesis of Phosphorus-Doped Graphene and Its Multifunctional Applications for Oxygen Reduction Reaction and Lithium Ion Batteries." *Advanced Materials* 25 (35): 4932–37.

Zhang, Xiong, Xianzhong Sun, Yao Chen, Dacheng Zhang, and Yanwei Ma. 2012. "One-Step Solvothermal Synthesis of Graphene/Mn₃O₄ Nanocomposites and Their Electrochemical Properties for Supercapacitors." *Materials Letters* 68 (February): 336–39. doi:10.1016/j.matlet.2011.10.092.

5 The Prospects of Graphene Oxide (GO) in Improving Efficiency of Energy Storage Devices

Balasubramaniam Yogeswari
Sri Eshwar College of Engineering (Autonomous)

Manickam Anusuya
Indra Ganesan Group of Institutions

Raju Venkatesh
PSNA College of Engineering &Technology

Fabian I. Ezema
University of Nigeria

CONTENTS

5.1 INTRODUCTION

Carbon is a nonmetallic and tetravalent chemical element. Graphene is considered to be an allotrope of carbon having a monolayer of carbon atoms, tightly packed in a hexagonal honeycomb lattice. Graphene, a "wonder material" [1], possesses many striking features such as the thinnest material having the thickness of one atom, the toughest material which is 100–300 times stronger than steel, best electrical and thermal conductor with tunable electrical and optical properties. Right from the isolation of graphene from graphite in the year 2004, it finds wide application in various scientific disciplines such as batteries, transistors, high-frequency electronics,

DOI: 10.1201/9781003215196-5

biosensors (used to detect glucose, DNA, protein, and bacteria), magnetic and chemical sensors, photodetectors, smart paints in vehicles (alarm will be switched on if someone touches the vehicle), energy generation and storage, etc.

Initially, chemical vapor deposition (CVD) which used toxic materials to grow graphene was the only method used to make large-area graphene. But it was identified as the expensive as well as a complex process also. The problems in this type of production process made graphene initially unavailable. Also, the removal of graphene layers from the metallic substrates without causing any damage to the graphene made the usage of CVD graphene in electronics as one of the challenging tasks. And also, CVD requires high temperature and long deposition times; i.e., the graphene was not cheap and comparatively not easy to fabricate also. Huge attempts were taken to find effective, at the same time economical ways to produce and utilize the various derivatives of graphene. Graphene oxide (GO) was identified as a promising material among its other graphene derivatives.

5.2 GRAPHENE OXIDE (GO)

GO is a yellow solid with a hexagonal lattice structure which is the oxidized structure of graphene having carbon, oxygen, and hydrogen in variable ratios. This gives GO many different, at the same time interesting qualities compared to graphene. Being a monoatomic-layered substance with oxygen, GO is identified as a nonconductive and dispersible in water. GO can be obtained by oxidizing graphite with sturdy oxidizers and acids. Graphite is a cheap and naturally occurring form of crystalline carbon. Figure 5.1 shows the structure of GO.

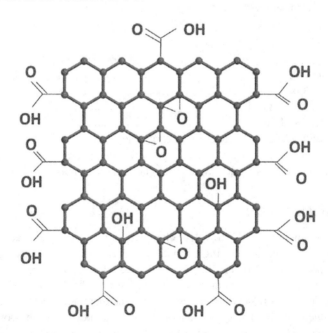

FIGURE 5.1   The structure of GO. (Source: Graphene Oxide I ACS Material, Nov 21, 2017 I ACS MATERIAL LLC.)

5.3 PREPARATION OF GO

Out of different methods available to synthesize GO, modified Hummer's [2] and Staudenmaier's techniques are widely used which involve oxidation of graphite. But the processing techniques and materials used are different [3]. GO has been prepared by using potassium manganate and sodium nitrate in concentrated sulfuric acid in the original Hummer's technique. In modified Hummer's technique, Hummer's reagents have been used with the addition of sodium nitrate [4,5]. In the Staudenmaier method, GO is prepared by using sulfuric acid along with concentrated nitric acid and potassium chlorate [6].

5.4 APPLICATIONS OF GO

With many interesting and remarkable properties, GO finds applications in many fields including research, water treatment, biomedical, solar cells, corrosion resistance, graphene/polymer composite materials, biosensors and multifunctional materials, especially in energy storage applications such as batteries [7,8], supercapacitors [9], fuel cells [10], and many more.

5.5 GO IN ENHANCING PERFORMANCE OF ENERGY STORAGE DEVICES

In addition to $-OH$, $C-O-C$, $C=O$, $-COOH$ groups [11], GO has many oxygen-containing groups which bestows GO many exclusive characteristics for potential relevance's in energy storage devices [10] (Figure 5.2).

Due to the possibility of tailoring its band gap [12] and functionalization, GO presents an efficient mode to alter its physical and chemical properties which improve its

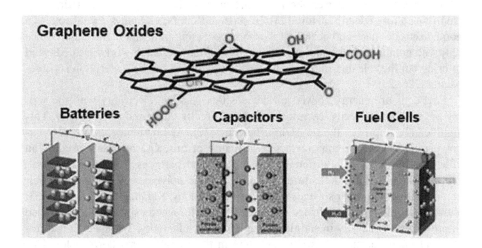

FIGURE 5.2 GO in batteries, supercapacitors, and fuel cells [10].

energy storage applications [13]. Jiang et al. [14] found that by managing the quantity of –O– and –OH functional groups, the band gap of GO can be controlled. Many attempts are being made to synthesize, analyze, and improve the usage of GO in an efficient way to many applications including energy storage. Zheng Bo et al. [15] proposed a method for reducing the GO with caffeic acid which finds applications in sensing and energy storage. In energy storage, hydrogen is considered to be the most promising candidate as a carrier of renewable energy [13]. Here, GO has been identified as suitable for generating hydrogen through a light-assisted method from water splitting [16].

Being rechargeable energy storage device, lithium-ion battery has maximum energy capacitance which holds great scope in meeting the increasing requirement of handy electronic goods. In these versatile batteries, GO is used as electrodes with improved extension cycle life [17] because of its varied chemical and electronic structures which significantly improve their electrochemical performances. The functional groups of GO give the porous structures in anodes of lithium-ion batteries which are considered to be useful for improving their electrochemical performance [18]. Not only in the anode, but using GO in cathode materials demonstrates considerable returns [13] in the form of improved capacity along with good stability [19]. Apart from lithium-ion batteries, GO has been found to be an important material for solving some of the challenges existing in lithium–sulfur batteries. GO is used in lithium–sulfur batteries [20] as a successful electrode material with an improved efficiency. Feiyan Liu [21] used reduced GO in lithium–sulfur batteries as cathode materials with desirable properties and noted a substantial improvement in its performance.

Supercapacitors are energy storage devices that act as a bridge between batteries and usual capacitors. Compared to the normal capacitors, supercapacitors are capable of stocking up extra energy. This makes supercapacitors as attractive devices for energy storage [22]. When lithium-ion batteries are considered, supercapacitors can be taken as a substitute device for stocking up the energy that have exclusive qualities such as high-power densities, bare memory effect, long cycle lifetime and environmentally friendly nature [23]. In supercapacitors, GO plays a vital role as a good electrode material because of its short processing time and higher capacitance [24] with notable power handling capacity [25]. Many research works are dedicated to bring out the role and impact of GO as flourishing electrode material in supercapacitors [26–29].

Fuel cells are energy conversion devices that generate electricity with the supply of fuels. In fuel cells, chemical energy is directly converted to electricity. This significantly improves their system efficiency [30]. At low temperatures, GO has comparatively high proton conduction. Because of this, GO is widely used as an effective electrolyte in a different type of cells and batteries [31]. Farooqui et al. [32] have highlighted the features of GO and some polymer composites and their compatibility with other solvents in fuel cell technology. Ramdayal Yadav et al. [33] showed that the proton exchange membrane fuel cell foisted using graphene yielded improved power density in addition to high ionic conductivity. The specific features of graphene-based materials including GO in the fuel cell are shown below [30] (Figure 5.3).

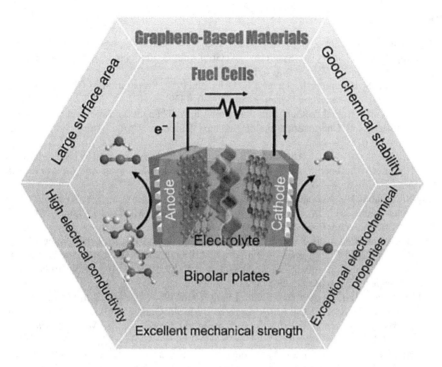

FIGURE 5.3 Fuel cells including graphene-based materials [30] in their components.

Additionally, with a strong hydrophilic nature, GO dissolves in water. This makes GO as a brilliant material used in electrodes [34] and proton electrolytes in fuel cells [35]. Thus graphene-based materials especially GO are being used as various components of fuel cells such as fillers, membranes, and proton-conducting electrolytes.

5.6 CONCLUSION

Due to growing concerns about the environmental safety of fossil fuels and the capacity of energy networks around the world, researchers are progressively more turning their attention to energy storage solutions. As GO is cheaper and easier to manufacture with tunable surface area, excellent electrical conductivity, and good chemical stability, it has captured the attention of many researchers in energy-related applications. Notwithstanding the grand advancement in the graphene materials, notably GO in energy storage and usage, development of established technologies using GO is considered to be a continuing objective in the future and requires more special treatment from diverse phases. In general, incorporating GO-based materials in practical systems for energy applications remains technologically a great tackle. However, by taking the combined better quality and tunable physical qualities of GO itself and adaptable nanomaterials coupled with GO have a brilliant upcoming in the energy. The entire energy storage fraternity is looking for further advancements in handling GO, as a wonder material will continue to grow swiftly, leading eventually to a variety of energy storage devices that would help humanity.

REFERENCES

[1] Balaiah Paulchamy, Gopalakrishnan Arthi, and Durai Lignesh. 2015. A Simple Approach to Stepwise Synthesis of Graphene Oxide Nanomaterial. *Journal of Nanomedicine & Nanotechnology* 6: 253.

[2] William S. Hummers Jr, Richard E. Offeman. 1958. Preparation of Graphitic Oxide. *Journal of the American Chemical Society* 80: 1339.

[3] Cherukutty Ramakrishnan Minitha, Ramasamy Thangavelu Rajendrakumar. 2013. Synthesis and Characterization of Reduced Graphene Oxide. *Advanced Materials Research* 678: 56–60.

[4] Leila Shahriary, Anjali Athawale. 2014. Graphene Oxide Synthesized by Using Modified Hummers Approach. *International Journal of Renewable Energy and Environmental Engineering* 2: 58–63.

[5] N.I. Zaaba, Kai Loong. Foo, Kai Loong Hashim, Soo Jin Tan, Wei-Wen Liu, Chun Hong Voon. 2017. Synthesis of Graphene Oxide using Modified Hummers Method: Solvent Influence. *Procedia Engineering* 184: 469–477.

[6] Hwee Ling Poh, Filip Šaněk, Adriano Ambrosi, Guanjia Zhao, Zdeněk Sofer, Martin Pumera. 2012. Graphenes Prepared by Staudenmaier, Hofmann and Hummers Methods with Consequent Thermal Exfoliation Exhibit Very Different Electrochemical Properties. *Nanoscale* 11: 3515–3522.

[7] Dingchang Lin, Yayuan Liu, Zheng Liang, Hyun-Wook Lee, Jie Sun, Haotian Wang, Kai Yan, Jin Xie, Yi Cui. 2016. Layered Reduced Graphene Oxide with Nanoscale Interlayer Gaps as a Stable Host for Lithium Metal Anodes. *Nature Nanotechnology* 11: 626–632.

[8] Kun Fu, Yibo Wang, Chaoyi Yan, Yonggang Yao, Yanan Chen, Jiaqi Dai, Steven Lacey, Yanbin Wang, Jiayu Wan, Tian Li, Zhengyang Wang, Yue Xu, Liangbing Hu. 2016. Graphene Oxide-Based Electrode Inks for 3D-Printed Lithium-Ion Batteries. *Advanced Materials* 28: 2587–2594.

[9] Wai Kitt Chee, Hong Ngee Lim, Ian Harrison, Kwok Feng Chong, Zulkarnain Zainal, Chi Huey Ng, Nay Ming Huang, 2015. Performance of Flexible and Binderless Polypyrrole/Graphene Oxide/Zinc Oxide Supercapacitor Electrode in a Symmetrical Two-Electrode Configuration. *Electrochimica Acta* 157: 88–94.

[10] Yuheng Tian, Zhichun Yu, Liuyue Cao, Xiao Li Zhang, Chenghua Sun, Da-Wei Wang. 2021. Graphene Oxide: An Emerging Electromaterial for Energy Storage and Conversion. *Journal of Energy Chemistry* 55: 323–344.

[11] Flavio Pendolino, Nerina Armata. 2017. *Graphene Oxide in Environmental Remediation Process.* Springer International Publishing. https://link.springer.com/book/10.1007%2F978-3-319-60429-9

[12] Lizhao Liu, Lu Wang, Junfeng Gao, Jijun Zhao, Xingfa Gao, Zhongfang Chen. 2012. Amorphous Structural Models for Graphene Oxides. *Carbon* 50: 1690–1698.

[13] Fen Li, Xue Jiang, Jijun Zhao, Shengbai Zhang. 2015. Graphene Oxide: A Promising Nanomaterial for Energy and Environmental Applications. *Nano Energy* 16: 488–515.

[14] Xue Jiang, Jawad Nisar, Biswarup Pathak, Jijun Zhao, Rajeev Ahuja. 2013. Graphene Oxide as a Chemically Tunable 2-D Material for Visible-Light Photocatalyst Applications. *Journal of Catalysis* 299: 204–209.

[15] Zheng Bo, Xiaorui Shuai, Shun Mao, Huachao Yang, Jiajing Qian, Junhong Chen, Jianhua Yan, Kefa Cen. 2014. Green Preparation of Reduced Graphene Oxide for Sensing and Energy Storage Applications. *Scientific Reports* 4: 4684.

[16] Te-Fu Yeh, Chiao-Yi Teng, Shean-Jen Chen, Hsisheng Teng. 2014. Nitrogen-Doped Graphene Oxide Quantum Dots as Photocatalysts for Overall Water-Splitting Under Visible Light Illumination. *Advanced Materials* 26: 3297–3303.

[17] Kaikai Lv, Yihe Zhang, Deyang Zhang, Weiwei Ren, Li Sun. 2017. Mn_3O_4 Nanoparticles Embedded in 3D Reduced Graphene Oxide Network as Anode for High-Performance Lithium Ion Batteries. *Journal of Materials Science: Materials in Electronics* 28: 14919–14927.

[18] Shin-Liang Kuo, Wei-Ren Liu, Chia-Pang Kuo, Nae-Lih Wu, Hung-Chun Wu. 2013. Lithium Storage in Reduced Graphene Oxides. *Journal of Power Sources* 244: 552–556.

[19] Sung Hoon Ha, Yo Sub Jeong, Yun Jung Lee. 2013. Free Standing Reduced Graphene Oxide Film Cathods for Lithium Ion Batteries. *ACS Applied Materials & Interfaces* 5: 12295–12303.

[20] Ian V. Lightcap, Prashant V. Kamat. 2013. Graphitic Design: Prospects of Graphene-Based Nanocomposites for Solar Energy Conversion, Storage, and Sensing. *Accounts of Chemical Research* 46: 2235–2243.

[21] Feiyan Liu, Jiyuan Liang, Chang Zhang, Liang Yu, Jinxing Zhao, Chang Liu, Qian Lan, Shengrui Chen, Yuan-Cheng Cao, Guang Zheng. 2017. Reduced Graphene Oxide Encapsulated Sulfur Spheres for the Lithium-Sulfur Battery Cathode. *Results in Physics* 7: 250–255.

[22] Castro-Gutiérrez Jimena, Alain Celzard, Vanessa Fierro. 2020. Energy Storage in Supercapacitors: Focus on Tannin-Derived Carbon Electrodes. *Frontiers in Materials*. https://doi.org/10.3389/fmats.2020.00217

[23] Patrice Simon, Yury Gogotsi. 2008. Materials for Electrochemical Capacitors. *Nature Materials* 7: 845–854.

[24] Bin Xu, Shufang Yue, Zhuyin Sui, Xuetong Zhang, Shanshan Hou, Gaoping Cao, Yusheng Yang. 2011. What is the Choice for Supercapacitors: Graphene or Graphene Oxide? *Energy & Environmental Science* 8: 2826–2830.

[25] Michael P. Down, Samuel J. Rowley-Neale, Graham C. Smith, Craig E. Banks. 2018. Fabrication of Graphene Oxide Supercapacitor Devices. *ACS Applied Energy Materials*. 2: 707–714.

[26] Muhammad Ashraf, Syed Shaheen Shah, Ibrahim Khan, Md. Abdul Aziz, Nisar Ullah, Mujeeb Khan, Syed Farooq Adil, Zainab Liaqat, Muhammad Usman, Wolfgang Tremel, Muhammad Nawaz Tahir. 2021. A High-Performance Asymmetric Supercapacitor Based on Tungsten Oxide Nanoplates and Highly Reduced Graphene Oxide Electrodes. *Chemistry-A European Journal*. https://doi.org/10.1002/chem.202005156

[27] Balasubramaniyan Rajagopalan, Jin Suk Chung. 2014. Reduced Chemically Modified Graphene Oxide for Supercapacitor Electrode. *Nanoscale Research Letters* 9: 535.

[28] Jia-Wei Wang, Wei-Ke Zhang, Chen Jiao, Fang-Yuan Su, Cheng-Meng Chen, Chun-Li Guo, Hui Tong Chua, Sehrina Eahon. 2020. Activated Carbon Based Supercapacitors with a Reduced Graphene Oxide Additive: Preparation and Properties, *Journal of Nanoscience and Nanotechnology* 20: 4073–4083.

[29] Qingqing Ke, John Wang. 2016. Graphene-Based Materials for Supercapacitor Electrodes – A Review. *Journal of Materiomics* 2: 37–54.

[30] Hanrui Su, Yun Hang Hu. 2020. Recent Advances in Graphene-Based Materials for Fuel Cell Applications. *Energy Science and Engineering*. https://doi.org/10.1002/ese3.833

[31] Hikaru Tateishi, Kazuto Hatakeyama, Chikako Ogata, Kengo Gezuhara, Jun Kuroda, Asami Funatsu, Michio Koinuma, Takaaki Taniguchi, Shinya Hayami, Yasumichi Matsumotoa. 2013. Graphene Oxide Fuel Cell. *Journal of The Electrochemical Society* 160: F1175–F1178.

[32] Usaid R. Farooqui, Abdul Latif Ahmad, Noor Ashrina A. Hamid. 2018. Graphene oxide: A Promising Membrane Material for Fuel Cells. *Renewable and Sustainable Energy Reviews* 82: 714–733.

[33] Ramdayal Yadav, Akshay Subhash, Nikhil Chemmenchery, Balasubramanian Kandasubramanian. 2018. Graphene and Graphene Oxide for Fuel Cell Technology. *Industrial & Engineering Chemistry Research* 29: 9333–9350.

[34] Maria Perez-Page, Madhumita Sahoo, and Stuart M. Holmes 2019. Single Layer 2D Crystals for Electrochemical Applications of Ion Exchange Membranes and Hydrogen Evolution Catalysts. *Advanced Materials Interfaces* 6: 1801838.

[35] Wei Gao, Neelam Singh, Li Song, Zheng Liu, Arava Leela Mohana Reddy, Lijie Ci, Robert Vajtai, Qing Zhang, Bingqing Wei, Pulickel M. Ajayan 2011. Direct Laser Writing of Micro-Supercapacitors on Hydrated Graphite Oxide Films. *Nature Nanotechnology* 6: 496–500.

6 Recent Progress in Graphene Water Purification and Recycling Using Energy Storage Devices

Sabastine E. Ugwuanyi and Hope E. Nsude
University of Nigeria

Raphael M. Obodo
University of Nigeria
University of Agriculture and Environmental Sciences
Quaid-i-Azam University
Northwestern Polytechnical University

Ishaq Ahmad
Quaid-i-Azam University
Northwestern Polytechnical University
iThemba LABS-National Research Foundation

Malik Maaza
iThemba LABS-National Research Foundation
University of South Africa (UNISA)

F. I. Ezema
University of Nigeria
iThemba LABS-National Research Foundation
University of South Africa (UNISA)

CONTENTS

DOI: 10.1201/9781003215196-6

6.1   INTRODUCTION

Water is indispensable to all life in the environment as it is one of the most supportive agents of life's existence considered in different areas of life. Water consumption is positively related to the population of an area making use of water as human beings can use water for many purposes – drinking, surface irrigation for crop production, rearing of animals, environmental cleanup, food preparation, industrial production of materials, preparation of science practicals in teaching and learning of physics in the school system in which experimental result is affected by the type of water used in the experiment, washing of materials and bathing(Christ, 2017).

However, water sources are many, but the problem is timely availability of water for a particular purpose in the event of its need (Zeng et al., 2013). For instance, drinking water is not necessarily easily available because of difficulty at a given time to get potable water fit for drinking.

Water pollution is a serious issue in the human environment as it causes so many problems which are viewed from these perspectives (Zeng et al., 2013). It is not fit for drinking, and it is not good for the industrial preparation of substances (Chen et al., 2013), it breads pathogens harmful to the body valuable animals by causing diseases to them (Online Study Materials, 2020). There are many sources of water such as ground water, water vapor, streams, rivers, lakes, and rainfall. However, these water sources can be polluted naturally or by activities of humans (Online Study Material, 2020). Human activity, one of which is agricultural practice, causes a lot of water pollution as a result of the quest for man to meet food needs for the growing human population. This could be done by the use of chemicals for agricultural practices for faster yield and better yield.

Irrigation farming also drags the system of water into environmental pollution as well. Water pollution can be classified into these categories – infectious agents characterized by pathogens and parasitic worms, oxygen demanding waste which comprises organic debris and wastes, inorganic wastes which contain on it acids, metals, salts and radioactive materials, organic chemicals containing petroleum products and other organic wastes and thermal pollutants.

However, the common source of drinking water in the rural areas is rainwater, but for urban areas, it is pipe-borne water. The issue associated with the source in rural areas is the unpredictability of rainfall due to climate change, but for urban areas, the pipe-borne water distribution arrangement is a relatively capital-intensive system which may keep the dependent urban dwellers in this case thirsty for a long time especially when there is a problem in the system (Homaeigohar and Elbahri, 2017). Therefore, it has become imperative to look for a cheaper source of drinking water in case the common sources that are unpredictable and costly in distribution arise to circumvent the problem of water scarcity. This can be done by being innovative, creative and development of quest for researching a recent development for solving water problem. Use of graphene-based devices/materials in water treatment and recycling as a solution to solve water problems is a good recent development in the water supply to a given populace (Yongchen, 2017).

It is known that water may be available, but not for certain uses because of its pollution or contamination. Certain quantity of water is polluted or contaminated with certain dangerous microbes and organic and inorganic pollutants. This water can be treated very well with graphene and graphene-based materials to fit for drinking and even in industries that require pure water for its processes (Chuanand Jing, 2021).

However, environmental pollution has really subjected most of our available water bodies to harmful status consequent upon channeling of industrial products and bye-products to our large water bodies like oceans, seas, rivers and the like. Ocean water is already saline which makes it unfit for drinking and industrial purposes. Microbial contamination and organic and inorganic pollutions of water make the water very unfit for drinking and industrial use, but graphene and graphene-based materials are very good in the purification of such water at a relatively low cost because the treatment process is not capital intensive. In addition, nanotechnology is well in use in the treatment process because of the miniaturization of the treatment system at a relatively low cost (Pei, 2020).

Graphene is a simple layer in a honeycomb cross-section of atomic thickness. Carbon makes up in a plane strata arrangement with the size of atoms giving an idea of graphene (Christ, 2017).

It is considered in 1940 that graphene is taken as a constituent element of graphite (Yan et al., 2019)

Graphene is able to show a very good water treatment material because of its thermodynamic stability in the atmosphere, excellent mechanical properties, high electrical properties, and large surface area (Yongchen, 2017).

However, Homaeigohar and Elbahri (2017) indicated latter list in their industrial waste recycling using nanographeme oxide filters as the derivative technique of graphene for graphene oxide.

6.2   OXIDE OF GRAPHENE SHEET PRODUCTION

One of the derivations of graphene is its oxide, and there are many methods for preparing graphene oxide (Zhang et al., 2016). It is easier to prepare this derivative sheet when the oxide derivative is mixed with water molecules to form a suitable dispersion of graphene oxide using the ultrasonic and storming methods (Avouris and Dimitrakopoulos, 2012). The preparation methods used are explained below (Yongchen, 2017).

6.2.1   Vacuum Filtration Method

This method is easily performed in many places and provides gaseous graphene after preparation of an adequate mixture and its subsequent addition into a sheet for vacuum filtration (Jiang et al., 2016).

The graphene oxide covers the entire membrane equally as a result of the flow-ability of the water on the course of separation.

Oxide derivative of graphene membrane formed was on even dispensability in which the regulation of the mixture determines the density of the graphene sheet (Nupearachchi et al., 2017).

The variables that determine the thickness of oxide of graphene duct and its sheet are pH, quantity of salt in the salt mixture, and pressure. In addition, this makes the graphene oxide membrane prepared in this method comparably flat (Griggs and Medina, 2016).

However, use of vacuum suction method in making oxide graphene sheet makes the membrane size to be very small which makes it difficult to have a large surface area because of the limitation of the equipment used in this method of preparation. Therefore, the membrane prepared using this method needs a supportive material for it to function very well because in addition to the above reason, the oxide of graphene derivative is below 100 nm in thickness.

6.2.2 SPRAY-COATING METHODS

In this method, the desired membrane is formed by spraying a prepared mixture evenly on the base which will be dried of solvent using spray-coating equipment. Agun is used for preheating the absorbent in addition to the liquid miniaturization.

Spray coating is characterized by the advantages of relatively high working efficiently, big working/production area, simpler process and any substrate can be sprayed, but it has relatively poor uniformity of the membrane prepared.

This method is faster foaming and of low cost.

6.2.3 SPIN-COATING METHOD

This method applies a mixture of graphene to the base substance and rotates the substrate material in an increased velocity. With this, there is even distribution of the mixture on the base material which eventually dries to become oxide film. The amount of graphite mixture is regulated during the filming process and the velocity to control the density of the film. This has been carried out through research.

6.2.4 DIP-COATING METHOD

In this method, the substrate is soaked in a mixture of graphite derivative, and after that, removal of the base material and recovery of excess solution are done using an appropriate device. The regulation of the density of the film is done by use of this method by varying adequately the heat level and amount of the mixture with the extraction velocity. The physical-batching process can be used in this method. It was discovered that dip coating in treatment of quartz substrate has been carried out in preparation of graphene-derivative mixture.

6.2.5 STRATA FORMATION METHOD

This is done using polyelectrolyte that forms multilayer film through deposition of charged substrates in the opposite charge. Because of the large oxide group at the oxide derivative of the graphene surface, there is a modification by covalent bonding in the surface structure. It is discovered that trimesic acid chloride can be used to create connectivity in the forming sheet (Weber et al., 2016).

Recent progress in graphene-based devices in water treatment is as follows:

1. The film of oxide of graphene can coexist with water because of the preference bond which is of hydrogen in nature. The formed sheet has wide application in the treatment of water (Yongchen, 2017). Using molecular simulation, Na^+ and Cl^- can be retained in the process. Graphene oxide film can be used for water desalination because of its hydrophobic nature which makes the water flux to be relatively high in comparison to the conventional reverse osmosis membranes. While passing water through graphene sheets, the water molecules experience very little friction in the process (Sun et al., 2014).

2. This has a very great influence in the structuring of the surface quality of the graphene sheet for water treatment for purification, pervaporation and desalination. Improved hydrophobicity can be achieved in the sheet when it is dip-coated with ceramics. For enhancement of the separation of the dye and ions of salt, this method can be encouraged in the preparation (Sun et al., 2015).

3. **Hybrid graphene oxide membrane**: This shows organic films and nanoparticle characteristics with the development of original film material to meet specific needs. In this type of graphene oxide, the membrane permeability is highly improved for organic separation, pervaporation and other fields. Some research has been carried out in this regard in the following ways. Use of a hybrid process produces graphene sheet that decreases the flow rate of water. When made with nanocomposite, the water retention ability of the sheet increases (Surwade et al., 2015).

4. **Graphene-based nanomaterial**: There are many examples in this case. In nanographene oxide-based filter paper, which can be obtained by the Hummers method, it is able to filter out the impurities and reduce the pH level of the industrial wastewater into a reusable form in which the pH is around 7. Graphene-based nanomaterial is a relatively good material for adsorbent in the decontamination of heavy metals from water (Nadar and Mohd, 2020). The applications of graphene are diverse as it can be used in making many nanomaterials for the water treatment process. Graphene being a carbon material with its potential of high applicability in many technical fields can be used in the preparation of nanoforous graphene which is a strange material for making semi-permeable membranes (Gao et al., 2014; Daer et al., 2015; Perreault et al., 2014).

The structure of the graphene is simplified to 2D (Bodzek et al., 2020). The graphene material has many outstanding qualities such as large surface area, high flexibility, relatively low density outstanding heat transfer high conductivity and very good mechanical strength. In addition, it has high charge mobility carrier ability (Cohen-Tanugi et al., 2014). The combination of graphene with other materials broadens its applicability even in water treatment (Pandey et al., 2017).

In construction of fuel cells, it has a very good applicability (Jafri et al., 2010; Wang et al., 2017).

6.3   MODIFIED GRAPHENE AND GRAPHENE BEHAVIOR

Graphene is a single layer of carbon atoms in a two-dimensional hexagonal lattice. It can be imagined as a sheet of paper with a honeycomb structure made of tight carbon atoms, but it is very strong and flexible in better comparison with steel. It is highly conductive. π–π bond, hydrogen bond, hydrophobic effect, covalent and electrostatic interactions are properties that make graphene-based materials to be used as adsorbents in the decontamination of heavy metals from water (Nadar and Mohd, 2020). The conversion of graphene to graphene oxide and subsequent reduction of the graphene oxide to reduced graphene oxide (rGO) makes it highly reactive to molecule interactions, conductive and electrochemically active.

rGO can be used as a supportive network as it can be doped with other atoms such as boron, nitrogen, fluorine or sulfur which grow new specific properties for certain applications. Therefore, it can undergo structural manipulations for ink production, chemical doping and transformation to three-dimensional micro-or macro devices (Baptista-Pires and Radjenovic, 2017).

6.3.1   GRAPHENE MANIPULATIONS IN WATER TREATMENT

Graphene oxide and reduced graphene have been widely harnessed into water treatment through filtration membranes that are uncommon. Reduced graphene oxide and graphene oxide have anti-bacterial and anti-fouling qualities.

RGO has corrosion-free property that makes it a good water treatment and storage material in addition to its impermeability to acid depending on its microstructure. Suspension of graphene oxide in water can be filtered onto a paper and used as a membrane for filtration in which the oxygen atoms present produce a distance between the atomic carbon layers which can let some molecules pass through it and will not allow others to pass through it depending on the size of each.

This has been used in water desalination and contaminant removal.

This interlayed oxygen created distance creates good room for water treatment through the electrical process (Baptista-Pires and Radjenovic, 2017).

Reduced graphene oxide used in water treatment through an electrical process is relatively advantageous because of its chemical-free nature and use of electron as only reagent.

Electrochemical system is also capable of degrading contaminants that are known to be persistent like poly-and perfluoro alkyl substances (Yongchen, 2017).

6.4   CONCLUSION

Graphene is an excellent conductor of electricity. It has high flexibility, its high-strength notwithstanding. It has derivatives for water treatment, and there are many methods used in the formation of the derivatives.

REFERENCES

Avouris, P., & Dimitrakopoulos, C. (2012). Graphene: Synthesis and applications. *Materials Today*, 15, 86–97. http://dx.doi.org/10.1016/S1369-7021(12)70044-5.

Baptista-Pires, L., & Radjenovic, J. (2017). *Graphene and Water Treatment. Consolidated Research group* (ICRA-TECH-2017 SGR 1318).

Bodzek, M., Konieczny, K., & Kwiecińska-Mydlak, A. (2020). Nanotechnology in water and wastewater treatment. Graphene – the nanomaterial for next generation of semipermeable membranes. *Critical Review in Environmental Science and Technology*, 50(15), 1515–1579. http://dx.doi.org/10.1080/10643389.1664258.

Chen, G., Guan, S., Zeng, G., Li, X., Chen, A., Shang, C., Zhou, Y., Li, H., & He, J. (2013). Cadmium removal and 2,4-dichlorophenol degradation by immobilized phanerochaete chrysosporium loaded with nitrogen-doped TiO_2 nanoparticles. *Applied Microbiology and Biotechnology*, 97(7), 3149–3157.

Christ, B. (2017). Enhancing water purification via graphene oxide, holey graphene oxide and lignin membrane architectures. Published senior projects.

Chuan, C. & Jing, R. (2021). *A functionalized grapheme aerogel for efficient water purification.*

Cohen-Tanugi, D., McGovern, R. K., Dave, S. H., Lienhard, J. H., & Grossman, J. C. (2014). Quantifying the potential of ultra-permeable membranes for water desalination. *Energy & Environmental Science*, 7(3), 1134–1141. http://dx.doi.org/10.1039/C3EE43221A.

Daer, S., Kharraz, J., Giwa, A., A., & Hasan, S. W. (2015). Recent applications of nanomaterials in water desalination. A critical review and future opportunities. *Desalination*, 367, 37–489. http://dx.doi.org/10.1016/j.desa.2015.03.030.

Deng, J. H., Zhang, X. R., Zeng, G. M., Gong, J. L., Niu, Q. Y., & Liang, J. (2013). Simultaneous removal of Cd(II) and ionic dyes from aqueous solution using magnetic graphene oxide nanocomposite as an adsorbent. *Chemical Engineering Journal*, 226, 189–200. http://dx.doi.org/10.1016/j.cej.2013.04.045.

Food and Agricultural Organization of the United Nations (2017). Water pollution from Agriculture. A global review. Retrieved fromwww.fao.org/publications.

Gao, Y., Hu, M., & Mi, B. (2014). Membrane surface modification with TiO_2-graphene oxide for enhanced phtocatalytic performance. *Journal of Membrane Science*, 455, 349–356. http://dx.doi.org/10.1016/j.memsci.

Griggs, C. S. & Medina, V. F. (2016). Graphene and graphene oxide membranes for water treatment. In: *Access Science*, New York: McGraw-Hill Education.

Homaeigohar, S., & Elbahri, M. (2017). Graphene membrane for water desalination.

Jafri, R. I., Rajalakshmi, N., & Ramaphrabhu, S. (2010). Nitrogen doped grapheme nanoplatelets as catalyst for oxygen reduction reaction in proton exchange membrane fuel cell. *Journal of Materials Chemistry*, 34, 7114–7117. http://dx.doi.org/10.1039/cojm00467g.

Jiang, Y., Biswas, P., & Fortner, J.D. (2016). A review of recent developments in graphene-enabled membranes for water treatment. *Environmental Science: Water Research & Technology*, 2, 915–922. http://dx.doi.org/10.1039/C6EW00187D

Nupearachchi, C. N., Mahatantila, K., & Vithanage, M. (2017). Application of graphene for decontamination of water; implications for sorptive removal. *Groundwater for Sustainable Development*, 5, 206–215. http://dx.doi.org/10.1016/j.gsd.2017.06.006

Online Study Materials (2020). *Water Pollution, Causes, Effects and Control Measures.* Retrieved fromhttps://www.rpcau.ac.in/wp-content/uploads/2020/03/water-pollution.pdf.

Pandey, R. P., Shukla, G., Manohar, M., & Shahi, V. K. (2017). Graphene oxide based nano-hybrid proton exchange membranes for fuel cell applications: An overview. *Advances in Colloid and Interface Science*, 240, 15–30. http://dx.doi.org/10.1016/j.cis.2016.12.003.

Pei, L. Y. (2020). *Development of Advanced Graphene-based Composites for water purification.* A theses submitted in fulfillment of the requirements for the degree of Doctor of Philosophy. School of Chemical Engineering and Advanced Materials. The University of Adelaide.

Perreault, F., Tousley, M. E., & Elimelech, M. (2014). Thin-film composite polyamide membranes functionalized with biocidalgraphene oxide nanosheets. *Environmental Science &
Technology Letters*, 1(1), 71–76, http://dx.doi.org/10.1021/ez4001356.

Sun, C, Boutilier, M. S. H., Au, H., Poesio, P., Bai, B., Karnik, R., & Hadjiconstanitinou, N. G.
(2014). Mechanism of molecular permeation through nanoporous graphene membranes.
Langmire, 30(2), 675–682. http://dx.doi.org/10.1021/la403969g.

Sun, X.-F., Qin, J., Xia, P.-F., Guo, B.-B., Yang, C.-M., Song, C., & Wang, S.-G (2015).
Graphene oxide-silver nanoparticle membrane for boifouling control and water purification. *Chemical Engineering Journal*, 281, 53–59. http://dx.doi.org/10.1016/j.
cej.2015.060.059.

Surwade, S. O., Sninov, S. N., Vlassiouk, I. V., Unicic, R. R., Veith, G. M., Dai, S., & Malurin,
S. M. (2015). Water desalination using nanoporous single-layer graphene. *Nature
Nanotechnology*, 10(5), 459–464. http://dx.doi.org/10.1038/nnano.2015.37.

Wang, J., Dou, W., Zhang, X., Han, W., Mu, X., Zhang, Y., Zhao, X., Chen, Y., Yang, Z.,
Su, Q., & Xie, E. (2017). Embedded Ag quantum dots into interconnected Co_3O_4
nanosheets grown on 3D graphene networks for high stable and flexible supercapacitors.
Electrochimica Acta, 224, 260–268. http://dx.doi.org/10.1016/j.electacta.2016.12.073.

Weber, J. R., Osuji, C. O., & Elimelech, M. (2016). The critical need for increased selectivity,
not increased water permeability, for desalination membranes. *Environmental Science &
Technology Letters*, 3(4), 112–120. http://dx.doi.org/10.1021/acs.estlett.6b00050.

Yan, W., Lei, G., Pengfei, Q; Xiaomin, L, & Gang, W. (2019). Synthesis of three-dimensional
graphene based hybrid materials for water purification: A Review MDPI.

Yongchen, L (2017). Application of grapheme oxide in water treatment. In *IOP Conference
Series: Earth and Environmental Science*.

Zeng, G., Chen, M., & Zeng, Z (2013). Shale gas: Surface water also at risk. *Nature*, 499(7457),
154. http://dx.doi.org/10.1038/499154c.

Zhang, Y., Wu, B., Xu, H., Liu, H., Wang, M., He, Y., & Pan, B. (2016). Nanomaterials-enabled
water and wastewater treatment. *NanoImpact*, 3, 22–39. http://dx.doi.org/10.1016/j.
impact.2016.09.004.

7 Lithium Ion Battery (LIBs) Performance Optimization using Graphene Oxide

Raphael M. Obodo
University of Nigeria
University of Agriculture and Environmental Sciences
Quaid-i-Azam University
Northwestern Polytechnical University

Hope E. Nsude and Sabastine E. Ugwuanyi
University of Nigeria

David C. Iwueke and Chinedu Iroegbu
Federal University of Technology

Tariq Mehmood
Hohai University

Ishaq Ahmad
Quaid-i-Azam University
Northwestern Polytechnical University
iThemba LABS-National Research Foundation

Malik Maaza
iThemba LABS-National Research Foundation
University of South Africa (UNISA)

Fabian I. Ezema
University of Nigeria
iThemba LABS-National Research Foundation
University of South Africa (UNISA)

DOI: 10.1201/9781003215196-7

CONTENTS

7.1   INTRODUCTION

The recent increase in world population and energy consumption prompted the fabrication of energy storage devices with specific and desirable qualities (Nwanya et al., 2020). The problem of the sustainability of energy production, storage and delivery has received serious attention worldwide due to the rapid depletion of fossil fuel resources as well as problems associated with their environmental pollution (Braithwaile, 1999; Obodo et al., 2021). The high rise in the number of portable electronic and hybrid devices in the world for usage in homes, industries, hospitals, worship centers, automobiles, etc., promoted a tremendous growth in demand for high-capacity electrode materials for batteries (Cao et al., 2015; Obodo et al., 2020). Because of their great energy storage capacity, batteries are now the most often employed energy storage devices used in powering these electronic devices. Chemical energy is converted into electrical energy via electrochemical process in various energy storage systems. Energy storage systems are evaluated using the energy density, usually called specific energy, and the power density, usually called specific power (Obodo et al., 2020). The problem of low battery power density encouraged scientists to look for ways to increase their energy density. The electrodes of lithium-ion batteries (LIBs) have the same electrochemical potential; nonetheless, there is a distinct potential difference between them. When compared with other energy storage devices, LIBs have unique quality in that they function at a constant voltage during the charge or discharge process, whereas the working voltage of other energy storage devices, such as supercapacitors, decreases linearly during the cycling process (Obodo et al., 2020). LIBs storage processes rely on faradic reactions (oxidations/reductions) at the electrode surfaces, which are governed by a sluggish diffusion rate and result in a poor-power density.

The widespread use of batteries appears to be limited due to their poor power density, prompting many studies to improve their energy and power density. Various interest groups have launched massive and zealous research efforts aimed at enhancing LIBs performance with respect to energy and power densities, performance, stability, availability, cost and environmental friendliness, etc. (Dang & Schlçgl, 2010).

Graphene oxide (GO), reduced graphene oxide (rGO), carbon nanotubes (CNTs), activated carbon (AC), carbide-derived carbon (CDC), carbon aerogel (CA) are examples of carbon products possessing desirable qualities that enhance energy storage devices performance (Obodo et al., 2021) and have enticed eccentric attention in major researches and industrial applications, especially in energy storage and conversion devices such as batteries, supercapacitor, etc. Because of their anticipated properties such as low cost, availability, high-surface area and environmental friendliness, these carbon derivatives have gotten a lot of attention in energy storage applications (Adekunle & Ozoemena, 2011). Carbon derivatives have a large specific surface area, which means they accumulate a lot of charge at the electrode–electrolyte interface, making them useful in energy storage device such as LIBs. (Maleki et al, 1999). The electrode designs of GO-based LIBs have a variety of physical, chemical and mechanical properties that improve LIB performance. In this chapter, we looked at the contribution and role of GO in improving the performance of LIBs electrodes performance. The features of GO that led to the enhancement and production of LIBs electrodes that delivered higher energy and power densities were also investigated.

In recent times, $LiCoO_2$, $LiFePO_4$ and $LiMn_2O_4$ cathode materials are most widely used in constructing LIBs electrodes (Nyteń et al., 2005). GO-centered LIBs electrode material progresses in a pseudo tetrahedral feature, which permits two-dimensional lithium-ion diffusion as well as enhancing performance (Nyteń et al., 2005). GO-based LIBs electrodes are widely employed in the fabrication of LIBs because they possess high-theoretical specific heat capacity, great volume capability, little self-discharge, high-discharge voltage and good cycle stability (Obodo et al., 2020, 2021; Zimmerman, 2004. GO not only improves LIBs' reversible specific capacity and specific energy density, but also forms conductive films to replace metal supports in the cathodes and anodes, decreasing their weight. The use of GO also allows for the construction of conformable and transparent batteries, which have immense potential in wearable electronics, smart tissues, and biomedical applications such as long-term implantable prosthetic devices.

7.2 LITHIUM-ION BATTERY (LIBS)

LIBs are rechargeable batteries where lithium ions move from negative electrodes to positive electrodes via an electrolytic solution through an electrochemical process for the period of discharge and back all through charging. LIBs positive electrodes are usually made from interpolated lithium compounds, whereas the negative electrodes are usually comprised of graphite. Table 7.1 is the types of LIBs and their various chemical combination.

Commonly, LIBs negative electrodes are made from carbon material while the positive electrode are usually metal oxides, phosphates, etc. These electrolytes contain lithium salt within an electrolytic solution prospering various electrochemical reactions. Depending on the direction of flow of current in the cell, electrodes electrochemical responsibilities switch between anode and cathode. Figure 7.1 depicts the various types of LIBs with respect to chemical components, mode of operations, shape, size, etc.

TABLE 7.1

Types of LIBs and their Chemical Formular

S/N	Types of LIBs	LIBs Chemical Formular
1	Lithium cobalt oxide	$LiCoO_2$
2	Lithium manganese oxide	$LiMn_2O_4$
3	Lithium nickel manganese cobalt oxide	$LiNiMnCoO_2$
4	Lithium iron phosphate	$LiFePO_4$
5	Lithium nickel cobalt aluminium oxide	$LiNiCoAlO_2$
6	Lithium titanate	Li_2TiO_3

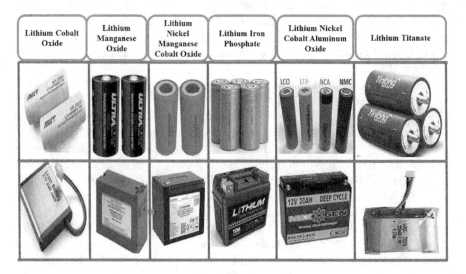

FIGURE 7.1 Images of various types of LIBs.

Lithium cobalt oxide, lithium manganese oxide, lithium nickel manganese cobalt oxide, lithium iron phosphate, lithium nickel cobalt aluminium oxide and lithium titanate are various LIBs discussed in this chapter. Table 7.2 is a summary of the performance evaluation of various types of LIBs.

7.3 WORKING PRINCIPLE LIBS

LIBs are electrochemical cells that operate with two electrodes; a cathode known as the positive electrode and an anode also called the negative electrode disjointed by a separator. Insertion (intercalation) and extraction (deintercalation) are two processes that allow lithium ions to flow into and out of electrode arrangements. To establish interactions between the cathode and anode, the LIB designs are filled with liquid electrolytes. Lithium ions are transferred to and from the positive and negative electrodes in LIBs by oxidizing and reducing the transition metal during charge and

TABLE 7.2

Performance Assessment of Types of LIBs

S/N	Types of LIBs	Specific Power	Specific Energy	Safety/ Durability	Lifespan	Fabrication Cost	LIBs Performance
1	Lithium cobalt oxide	Low	High	Low	Low	Low	Medium
2	Lithium manganese oxide	Medium	Medium	Medium	Low	Low	Low
3	Lithium nickel manganese cobalt oxide	Medium	High	Medium	Medium	Low	Medium
4	Lithium Iron Phosphate	High	Low	High	High	Low	Medium
5	Lithium nickel cobalt aluminium oxide	Medium	High	Low	Medium	Medium	Medium
6	Lithium titanate	Medium	Low	High	High	High	High

discharge (Goodenough & Kim, 2010). LIBs usually use carbon-based anodes and lithium-based decorated or optimised transition metal oxides such as Co, Mn, Ni, etc., or combinations of transition metal oxides and carbon-based materials such as GO, rGO, etc. Liquid electrolytes in LIBs comprise of lithium salts such as lithium hexafluorophosphate ($LiPF_6$), lithium tetrafluoroborate ($LiBF_4$) or lithium perchlorate ($LiClO_4$) within organic solvents such as ethylene carbonate, dimethyl carbonate and diethyl carbonate. The use of a solid as an electrolyte material is a recent advancement in battery technology. Ceramics are the most promising of these because of the intrinsic lithium. Lithium metal oxides are commonly used as solid ceramic electrolytes because they allow lithium ions to move through them more freely. The fundamental advantage of solid electrolytes is that there are no leaks, which is a serious safety concern with liquid electrolyte LIBs. The processes of charge and discharge mechanisms in LIBs are chemical reactions and shown in equation 7.1.

$$LiXXO_2 + C \underset{\text{Discharge}}{\overset{\text{Charge}}{\rightleftarrows}} Li_{1-x}XXO_2 + Li_xC \qquad (7.1)$$

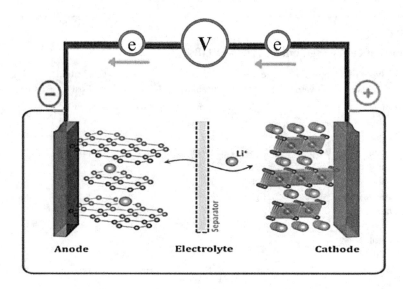

FIGURE 7.2   Schematic diagram of LIBs working principle.

where C is carbon-based material and XX are various combining elements or composites such as cobalt, manganese or cobalt/carbon derivatives, etc.

As soon as LIB has been charged using outer current sources, Li^+ from the cathode migrate to the anode by dispersing in the entire electrolyte and moving across the separator (Pan et al., 2011). The migration of Li^+ makes the cathode Li^+ deficient and the anode lithium sufficient. However, in discharging stage, Li^+ ions vacate the anode, move within the electrolyte to the cathode, whereas the allied electrons are simultaneously accumulated using a current collector to power an outward electrical or electronic device (Ali, 2017) (Figure 7.2).

7.4   LIBS CATHODE MATERIALS

LIBs recent advancement induces and aggravated new research and studies in LIB technologies. As a result, ongoing attempts are made to create efficient LIBs in order to achieve a downsized powerful battery with maximum operating safety. Early studies on lithium-based transition metal oxide materials (LiTMOx) (TM = Co, Ni, Mn, Fe, etc.) started as early as the late 1950s, with a focus on material characteristics and electrochemical performance. Table 7.3 presents various classifications of LIBs cathode materials. New cathode materials have improved the operating voltage and energy density of LIBs, making them ideal for a variety of applications.

To accommodate for charge loss for the period of lithium insertions and extractions from cathode materials in the Generation I LIBs cathode materials, multivalent ion-centered cathode materials are essential. The materials must be structurally stable on the course of lithium movement in the charging/discharging process to provide high-cycle stability. The majority of Generation I LIBs cathode materials are

TABLE 7.3
Classification of LIBs Cathode Materials

S/N	LIBs Cathode Generations	LIBs Cathode Materials	Working Voltage Range (V)	Specific Capacity (mAh g^{-1})	Year
1	Generation I	Mono Li atom-based transition metal (TM) ternary oxides	1.2–3.5	≤ 150	1950–1985
2	Generation II	Mono Li atom-based TM ternary oxides	3.5–4.0	150–200	1985–2000
3	Generation III	Higher Li atom-based TM ternary and quaternary oxides and oxifluorides	3.5–5.0	$200 \geq 330$	2001–2020
4	Generation IV	Li-S or Li-Air systems	1.5–2.5	>1,000	$2020 \geq 2025$

Li, lithium; S, sulphur; TM, transition metals.

ternary oxides based on transition metals, and their electrochemical characteristics are typically investigated in bulk. This explains why bulk materials, despite having lower ionic conductivity and longer diffusion paths, have insufficient capacity when compared to their theoretical capacity.

Professor J. B. Goodenough and his colleagues demonstrated the promise of LIBs in their first-generation cathode materials by hosting TMOs cantered cathodes material. (Goodenough, 2013). Nevertheless, because of the inadequate capacity and poorer operating voltages, it was essential to improve new amendment approaches for 1st generation cathode materials and improve new materials with desirable electrochemical features in order to get the best results. The weak electrical and ionic conductivity of the first-generation cathode materials was a problem and attempts were made to address this in the second generation. In order to change the ionic conductivity as well as the electronic conductivity, doping strategies were employed. Doping high-valency transition metal or non-metallic compounds within the second cation positions improved various materials electronic conductivity and performance.

The third generation focused on developing cathode materials with extremely high-electrochemical capacity (≥ 250 mAh g^{-1}) and good charge/discharge cycle stability. Due to this recent development, cathode materials with improved electrochemical capabilities have exploded. At this phase, increased electrochemical storage capacity was achieved; fabrication of LIBs with higher lithium content, composites of transition metal oxide and various material such as carbon-based materials. Nano structuring and localizing strains caused by doping with other materials afford insight on how to prevent capacity diminishing in LIBs cathode materials, which appear to be a promising contender for high-energy density LIBs cathodes. In this generation of LIBs cathode materials, to compensate for other generations low-electrical conductivity, mostly carbon-based conductive material are used for better performance.

Current generations of rechargeable LIBs (first to third generations) are extensively appropriate, employed for low-power usages in our daily lives and usually constrained by their particular current density ranging from 150 to 200 mAh g^{-1}. The ascending of various generations of LIBs take the form of enormous battery packs for specific applications, such as electric and hybrid automobiles. The high-energy density (>1,000 mAh g^{-1} or higher) is required for power-hungry uses such as electric and hybrid electric vehicles, airlines, and other commuting means for green energy projects. The problematic of limited specific capacity in LIBs can be solved by developing new materials and battery designs with extremely high capacities. Metal air batteries, for example, are showing a lot of promise in overcoming the existing specific capacity limitations, which is an obstacle for contemporary LIBs (Ali, 2017). Lithium-air and lithium-sulphur batteries indicate high dependence with elevated specific energy density. On the other hand, LIBs research and development is still in its early phases and many difficulties like nontoxic and safety (due to lithium's combustible nature) and cyclability must be properly addressed (Pan et al, 2011).

7.5   LIBS ANODE MATERIALS

Negative electrode materials have typically been made of graphite and other carbon-based materials; however, silicon-based alternatives are becoming more popular (Bolotin et al., 2008). These materials are employed because they are plentiful, electrically conductive, and capable of intercalating lithium ions to store electrical charge with minimal volume expansion. Because of its low voltage and good performance, graphite has become the most used substance. Different materials have been used; however, their voltage is high, resulting in a low-energy density. The crucial criterion is that the material has a low voltage; otherwise, the excess capacity is meaningless in terms of energy density.

7.6   GRAPHENE OXIDE (GO)

GO can be synthesized in a variety of ways such as exfoliation of extremely adapted graphite by micromechanical exfoliation, with or without surface processing (Tang & Hu, 2012), epitaxial growth (Yu et al., 2011), chemical vapor deposition (CVD) (An et al., 2011), etc. The most common technique of synthesizing GO using ordinary graphite by the means of modified the Hummers method. Modified Hummers mode of synthesis is always used because its simplicity and large fabrication of GO powder (Novoselov et al, 2005). Figure 7.3 is a sketch of the structural model of GO.

7.7   GRAPHENE IN LIBS

In recent time, scholars and researchers have intensified their studies in GO and other carbon derivatives-based electrode materials for LIBs. Recent findings have shown that carbon derivative can considerably increase LIBs electrodes electrochemical features (Geim & Novoselov, 2007). GO wide surface area as well as enhanced electron

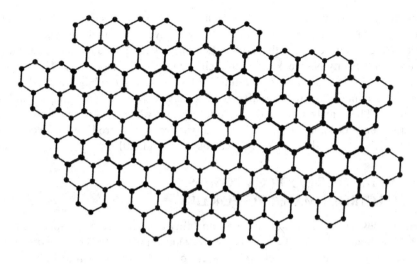

FIGURE 7.3 Schematic diagram of a structural model of GO.

transfer capabilities of graphene significantly enhances electron and ion transmission and diffusion in LIBs electrode materials.

7.7.1 GRAPHENE OXIDE-BASED CATHODE MATERIALS

Metal-based cathode electrode materials such as $LiMn_2O_4$, $LiNiMnCoO_2$, $LiFePO_4$, $LiNiCoAlO_2$, etc., are widely employed as cathode–electrode materials for LIBs due to their availability, low-priced and environmental pleasantness (Manev et al., 1995). On the other hand, their low-electrical conductivities prompted their low-rate capacity, hence, need for augmentation to boost their electrochemical performance. Researchers and scholars have confirmed that addition of GO as a composite is an effective mechanism of improving their conductivities and rate capacities (Zhu et al., 2014). This awesome electrochemical performance of these composites can be credited to higher Li^+ diffusion kinetics and enhanced firmness within a wide voltage range in crystalline metal/GO composites prompted by huge surface area, better electron transfer capabilities of GO, creating significantly improvement in electron and ion transmission and diffusion in cathode materials (Obodo et al., 2020).

7.7.2 GRAPHENE OXIDE-BASED ANODE MATERIALS

GO is extensively used in recent times due to its lightweight, strong electrical conductivity, great mechanical flexibility and chemical stability. GO and other carbon-based materials have opened up innovative opportunities in the realm of LIBs material (Su et al., 2012). When GO-based materials are utilised as LIBs anode material, these features are advantageous. The addition of GO in anodes material resulted in improved electrical conductivities, large surface area ($\geq 2,620\,m^2g^{-1}$), the extraordinary surface-to-volume proportion, thin layers that reduce ion diffusion distances, operational plasticity that allows for creation of flexible electrodes, thermal and chemical firmness that ensures their durability in severe surroundings (Barker et al., 1996). Research have shown that non-carbon-based LIBs anode materials have high theoretical capacity but weaknesses to their usage in LIBs anodes material are volume expansion in the course of lithium/delithiation leading to huge internal stress. Non-carbon anodes decay faster after repeated charging and discharging, according to studies. These materials are disposed to rupture, which brings about poor cycle features, hence GO and other carbon derivatives are used to produce anode composite materials to solve these issues (Goodenough & Kim, 2010).

7.8　METAL OXIDE/GO COMPOSITES ELECTRODE MATERIALS FOR LIBS

Metal oxides with large lithium-storing capability are regarded as prospective alternate anode materials for better capacity LIBs. Due to the volume changes that occur during charge and discharge electrochemical process, as well as metal oxides low conductivities, GO can be employed to improve their electrochemical properties (Obodo et al., 2019).

Metal Oxide/GO composites material possess a unique advantage because oxygen-containing groups in GO can be condensed after the heat treatment and outstanding functional groups including the OH and COOH on the GO leading to strong interaction with the metal ions during the amalgamation procedure (Obodo et al., 2019; Atabaki & Kovacevic, 2013). Because of the small size of GO particles, the electron and lithium-ion traverse a short mean-free path during the lithium-ion insertion/extraction process, resulting in the outstanding rate capability of LIBs composite electrodes.

In addition, GO's high-electron mobility of $15,000\,cm^2\,Vs^{-1}$, the great thermal conductivity of $3,000\,W\,mK^{-1}$ remarkable chemical stability and excellent mechanical properties of GO make it a suitable target for building composite materials employed as the base electrode (Bolotin et al., 2008). Improved electrodes further boost the battery's capacity by allowing more lithium ions to be stored. As a result, the life of graphene-based batteries can be substantially longer than that of traditional batteries (Amin & Maier, 2008).

7.9　CONCLUSION

The growing demand for energy has prompted many academics, individuals, organizations, and countries to invest extensively in developing new energy sources or discovering novel ways/devices to store energy. Lithium-ion batteries are a type of energy storage device that have a number of distinct advantages over traditional batteries. High open-circuit voltage, high-energy density, extended useful life, no memory effect, no pollution, and low-self-discharge rate are among these advantages. Lithium-ion batteries have quickly become the new generation of secondary batteries in recent years due to their beneficial qualities, and they are now widely utilised in mobile phones, laptops, and other portable electronic devices.

Graphene is a graphite monolayer made up of sp^2-hybridised carbon atoms that are organised in a honeycomb crystal structure. Because graphene is a two-dimensional substance, every atom can be regarded a surface atom. Other carbon materials, such as graphite, carbon nanotubes, and fullerenes are made from graphene. Non-carbon anodes decay faster after repeated charging and discharging, according to studies. These materials are prone to rupture, resulting in poor cycle performance, hence, GO and other carbon derivatives are used to produce anode composite materials to resolve these issues.

REFERENCES

Adekunle, A. S., & Ozoemena, K. I. (2011). Electrosynthesised metal (Ni, Fe, Co) oxide films on singlewalled carbon nanotube platforms and their supercapacitance in acidic and neutral pH media. *Electroanalysis, 23*(4), 971–979.

Ali, E. (2017). Low voltage anode materials for lithium-ion batteries. *Energy Storage Materials, 7*, 157–180.

Amin, R., & Maier, J. (2008). Effect of annealing on transport properties of $LiFePO_4$: Towards a defect chemical model. *Solid State Ionics, 178*, 1831–1836.

An, H., Lee, W. J., & Jung, J. (2011). Graphene synthesis on Fe foil using thermal CVD. *Current Applied Physics, 11*, S81–S85.

Atabaki, M. M., & Kovacevic, R. (2013). Graphene composites as anode materials in lithium-ion batteries. *Electronic Materials Letters, 9*, 133–153.

Barker, J., Pynenburg, R., Koksbang, R., & Saidi, M. Y. (1996). An electrochemical investigation into the lithium insertion properties of Li_xCoO_2. *Electrochimica Acta, 41*, 2481–2488.

Bolotin, K. I., Sikes, K. J., Jiang, Z., Klima, M., Fudenberg, G., Hone, J., ... & Stormer, H. L. (2008). Ultrahigh electron mobility in suspended graphene. *Solid State Communications, 146*, 351–355.

Braithwaile, J. W. (1999). Corrosion of lithium-ion battery current collector. *Journal of the Electrochemical Society, 146*, 448–456.

Cao, X., Zheng, X., Shi, B., Yang, W., J, Fan, J., ... & Zhang, H. (2015). Reduced graphene oxide-wrapped MoO_3 composites prepared by using metal-organic frameworks as precursor for all-solid-state flexible supercapacitors. *Advanced Materials, 27*, 4695–4701.

Dang, S. S., & Schlçgl, R. (2010). Nanostructured carbon and carbon nanocomposites for electrochemical energy storage applications. *ChemSusChem, 3*, 136–168.

Geim, A. K., & Novoselov, K. S. (2007). The rise of graphene. *Nature Materials, 6*, 183–191.

Goodenough, J. B. (2013). *Journal of American Chemical Society, 135*, 1167.

Goodenough, J. B., & Kim, Y. (2010). Challenges for rechargeable Li batterie. *Chemistry of Materials, 22*, 587–603.

Maleki, H. et al. (1999). Thermal properties of lithium-ion battery and components. *Journal of the Electrochemical Society, 146*, 947.

Manev, V., Banov, B., Momchilov, B. A., & Nassalevska, A. (1995). *Journal of Power Sources, 57*, 99–103.

Novoselov, K. S. et al. (2005). Two dimensional atomic crystals. *Proceedings of the National Academy of Sciences of the United States of America, 102*, 10451–10453.

Nwanya, A. C., Ndipingwi, M. M., Ikpo, C. O., Obodo, R. M., Nwanya, S. C., Botha, S., ... & Maaza, M. (2020). Zea mays lea silk extract mediated synthesis of nickel oxide nanoparticles as positive electrode material for asymmetric supercabattery. *Journal of Alloys and Compounds, 822*, 153581.

Nyte´n, A., Abouimrane, A., Armand, M., Gustafsson, T., & Thomas, J. O. (2005). *Electrochemistry Communications, 7*, 156.

Obodo, R. M., Ahmad, A., Jain, G. H., Ahmad, I., Maaza, M., & Ezema, F. I. (2020a). 8.0 MeV copper ion (Cu^{++}) irradiation-induced effects on structural, electrical, optical and electrochemical properties of Co_3O_4-NiO-ZnO/GO nanowires. *Materials Science for Energy Technologies, 3*, 193–200.

Obodo, R. M., Ahmad, I. & Ezema, F. I. (2019a). Introductory chapter in graphene and its applications. In *Graphene and Its Derivatives-Synthesis and Applications*. Intechopen.

Obodo, R. M., Asjad, M., Nwanya, A. C., Ahmad, I., Zhao, T., Ekwealor, A. B., ... & Ezema, F. I. (2020b). Evaluation of 8.0 MeV carbon (C^{2+}) irradiation effects on hydrothermally synthesized Co_3O_4–CuO–ZnO@GO electrodes for supercapacitor applications. *Electroanalysis, 32*, 2958–2968.

Obodo, R. M., Chibueze, T. C., Ahmad, I., Ekuma, C. E., Raji, A. T., Maaza, M., & Ezema, F. I. (2021a). Effects of copper ion irradiation on Cu_yZn_{1-2y-x} Mn_y/GO supercapacitive electrodes. *Journal of Applied Electrochemistry, 51*, 829–845.

Obodo, R. M., Chime, U. K., Nkele, A. C., Nwanya, A. C., Bashir, A. K., Madiba, I. G., ... & Ezema, F. I. (2021b). Effect of annealing on hydrothermally deposited Co_3O_4-ZnO thin films for supercapacitor applications. *Materials Today: Proceedings, 36*, 374–378.

Obodo, R. M., Nwanya, A. C., Arshad, M., Iroegbu, C., Ahmad, I., Osuji, R., ... & Ezema, F. I. (2020c). Conjugated NiO-ZnO/GO nanocomposite powder for applications in supercapacitor electrodes material. *International Journal of Energy Research, 44*, 3192–3202.

Obodo, R. M., Nwanya, A. C., Hassina, T., Kebede, M. A., Ahmad, I., Maaza, M., & Ezema, F. I. (2019b). Transition metal oxides-based nanomaterials for high energy and power density supercapacitor. In *Electrochemical Devices for Energy storage Applications* (p. 7). CRC Press.

Obodo, R. M., Nwanya, A. C., Iroegbu, C., Ezekoye, B. A., Ekwealor, A. B., Ahmad, I., ... & Ezema, F. I. (2020d). Effects of swift copper (Cu^{2+}) ion irradiation on structural, optical and electrochemical properties of Co_3O_4-CuO-MnO$_2$/GO nanocomposites powder. *Advanced Powder Technology, 31*, 1728–1735.

Obodo, R. M., Onah, E. O., Nsude, H. E., Agbogu, A., Nwanya, A. C., Ahmad, I., ... & Ezema, F. I. (2020e). Performance evaluation of graphene oxide based Co_3O_4@GO, MnO$_2$@GO and Co_3O_4/MnO$_2$@GO electrodes for supercapacitors. *Electroanalysis, 32*, 2786–2794.

Obodo, R. M., Shinde, N. M., Chime, U. K., Ezugwu, S., Nwanya, A. C., Ahmad, I., ... & Ezema, F. I. (2020f). Recent advances in metal oxide/hydroxide on three-dimensional nickel foam substrate for high performance pseudocapacitive electrodes. *Current Opinion in Electrochemistry, 21*, 242–249.

Pan, A. et al. (2011). High-rate cathodes based on $Li_3V_2(PO_4)_3$ nanobelts prepared via surfactant-assisted fabrication. *Journal of Power Sources, 196*(7), 3646–3649.

Su, C., Bu, X., Xu, L., Liu, J., & Zhang, C. (2012). A novel LiFePO$_4$/graphene/carbon composite as a performance-improved cathode material for lithium-ion batteries. *Electrochimica Acta, 64*, 190–195.

Tang, B., & Hu, G. (2012). Two kinds of graphene-based composites for photoanode applying in dye-sensitized solar cell. *Journal of Power Sources, 220*, 95–98.

Yu, X. Z., Huang, C. G., Jozwiak, C. M., Köhl, A., Schmid, A. K., & Lanzara, A. (2011). New synthesis method for the growth of epitaxial graphene. *Journal of Electron Spectroscopy and Related Phenomena, 184*, 100–103.

Zhu, J., Duan, R., Zhang, S., Jiang, N., Zhang, Y., & Zhu, J. (2014). The application of graphene in lithium ion battery electrode materials. *Springerplus, 3*, 585.

Zimmerman, A. H. (2004). Self-discharge losses in lithium-ion cells. *IEEE Aerospace and Electronic Systems Magazine, 19*(2), 19–24.

8 Effects of Addition of Graphene Oxide on Energy Density of Supercapacitor

Izunna S. Okeke
University of Nigeria

Raphael M. Obodo
University of Nigeria
University of Agriculture and Environmental Sciences
Quaid-i-Azam University
Northwestern Polytechnical University

Eugene O. Echeweozo
Evangel University

Fabian I. Ezema
University of Nigeria
iThemba LABS-National Research Foundation
University of South Africa (UNISA)

CONTENTS

8.1　INTRODUCTION

A supercapacitor (SC), which is also referred to as an ultracapacitor, stores charge by utilizing a huge surface area of conducting material (Majumder and Thakur, 2019). It is one of the remarkable tools that are designed for energy storage. It is viewed as an

DOI: 10.1201/9781003215196-8

electrochemical energy device that has the potential to succeed batteries in storage applications; especially in hybrid vehicles, portable electronics, satellites and heavy machineries (Kaempgen et al., 2009; Saleem et al., 2016). Supercapacitors have configurations that are comparable to that of batteries. It comprises an electrolyte, a separator and a pair of polarizable electrodes (Zhang et al., 2010). The electrodes are the most crucial parts of a SC.

Based on the mechanism through which energy is stored, SCs can be grouped into two categories; (i) the pseudocapacitors (PCs) and (ii) the electric double-layer capacitors (EDLCs) (Yang, 2021). The pseudocapacity storage energy is accomplished via the transfer of faradic charges between the electrode and the electrolyte. This is achieved because of changeable redox faradaic reactions that widely exhibit greater energy density and specific capacitance relative to EDLCs (Xiao et al., 2009). The EDLCs store energy via the interface double layer assembled in between the electrolyte and the electrode (Long et al., 1999). To accomplish high-storage efficiency in EDLCs, the material's pore size and specific surface area should be controlled (Shi et al., 2011).

Generally, SCs are deployed in industries, portable energy sources, transportation, military, memory protection, power electronics and aerospace (Yu et al., 2013) due to their unique features. It exhibits unique characteristics such as long cycle, large power density and current discharge potential, convenient detection, smooth charging and discharging, excellent low-temperature performance and rapid charging speed (Huang et al., 2019). Typical SC charging time is within a couple of minutes, and the life cycle can stretch to 100,000 cycles with little performance degradation (Wu and Cao, 2018). Currently, the material for SC is expected to improve on features such as higher catalytic activities, durability, stability and economical (Balli et al., 2019).

The primary challenge limiting the application of SC is their energy density. Supercapacitors have low energy density; currently, the energy density of a SC is less than 20 Wh kg^{-1} which is still not good enough when compared to that of batteries that is between 30 and 300 Wh kg^{-1} (Simon and Gogotsi, 2008). The outcome of low energy density is bulkier SC devices. The low energy density could be enhanced by growing the effective surface area of electrodes in EDLC or by growing the operation voltage window (Huang et al., 2019). This can be achieved by developing appropriate organic electrolyte, new material with greater surface area, ionic liquid, pore size, surface functional groups and electrical conductivity (Wu and Zhu, 2017, Shi et al., 2011). Metal oxide-based SC devices exhibit high specific capacitance because it permits fast reversible faradic reactions in the interface of electrode and electrolyte. However, they are facing challenges such as inadequate electronic and ionic conductivity (Akhtar, 2021). Conducting polymers (CPs) such as polypyrrole (PPy), polystyrene sulfonate (PEDOT:PSS) and Polyaniline (PANI) have gained recognition as electrode materials for SC applications (Huang et al., 2016; Manjakkal et al., 2020; Gupta and Miura, 2006). This is because CP offers rapid redox reactions which can lead to improved pseudocapacitance. Nevertheless, CPs have a drawback of long-term stability because of mechanical degeneration (Akhtar, 2021). The drawbacks noted in electroactive material for SC applications are being addressed by combining or adding graphene oxide to form hybrid material. It is expected that the addition of graphene oxide in SC material can enhance their energy density.

Graphene oxide (GO) is essential and highly propitious for graphene-related applications in energy storage, optics and electronics. The GOs are integral members in the graphene–graphite family. It can be conveniently made via chemical exfoliation of graphite by the oxidation process (Xu et al., 2008; Faridbod et al., 2016) and is widely considered as a regular two-dimensional oxygenated planar molecular material (Sun, 2019). The GOs are non-stoichiometric chemical compounds of hydrogen, oxygen and carbon in variable proportions that principally yet do not completely rely on synthesis techniques (Marcano et al., 2010). GO has higher concentrations of defects than graphene which can be beneficial in its applications in the semiconductor industry (Park and Ruoff, 2009).

The fundamental concepts discussed in this chapter are the working principle of a SC, techniques deployed in the synthesis of GO, and structure and properties of GO. Furthermore, the effect of GO on the energy density of a SC electrode material is also discussed. These include the modification or tuning of SC electrodes material such as metal oxides, conducting polymers and carbons by addition of GO to improve their energy density.

8.2 SUPERCAPACITOR WORKING PRINCIPLE

Basically, SC technology provides a link between traditional energy storage technologies such as battery and parallel plate capacitors by bringing together parts of their respective principle of operation. Typical SC has its electrodes immersed in the electrolyte just like a battery, while the accumulated electrostatic charges from the electrolyte and electrodes interface are just like a parallel plate capacitor. There is no dielectric present in a SC; nevertheless, high capacitance is achieved by utilizing a monolayer referred to as the Helmholtz layer to isolate the two layers of charges. The monolayer of the solvent molecule is remarkably thin 0.5–1 nm (Halper och and Ellenbogen, 2006).

The electrolyte characterizes the cell voltage and is also the principal source of ions in the SC. It has a peculiar voltage operating window; however, if it is operated outside the window limits gases are created. An electrolyte is expected to have expansive operating windows, low viscosity and chemical stability, high conductivity, economical and availability. The electrolytes deploy in a SC can be categorized into ionic liquid, organic and aqueous (Wang et al., 2012).

As mentioned earlier SCs are of two types: PCs and EDLCs. For PC type, the faradic transfer of charge in the electrode layer develops via thermodynamic and kinetic conditions aided by a redox reaction. To guarantee effective charge and discharge the redox reaction round an electrode should be reversible or semi-reversible electrochemically. The reversibility of redox material employed in a PC simply implies that the redox process adheres to Nerstian behavior (Conway, 1999). Material can be considered a redox material if (i) it is electrochemically active and can adsorbed steadily on a conductive active substrate surface, (ii) solid-state material can intercalate with the substrate and creates a hydride electrode layer. Figure 8.1a and b depicts the working principle of the PCs and EDLCs types of a SC.

In the EDLC type, the electrolyte–electrode interface serves as capacitance by racking up electrostatic charge. The electrolyte can be a solution or a solid. Basically,

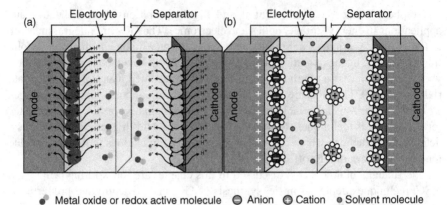

FIGURE 8.1   Diagrammatic depiction of (a) pseudocapacitor and (b) electric double-layer capacitor SC. (Reprinted with permission from Chen, Paul and Dai (2017). Copyright © 2017, Oxford University Press.)

the solid-state electrolyte helps in the conduction of ions and also the separation of the positive electrode from the negative electrode. The accumulations of positive charges on the positive electrode draw an equal number of negative charges about the electrode because of Coulomb's force. The charge equilibrium between the electrolytes and electrodes serves as the EDL.

The capacitance of the interface is defined in equation 8.1:

$$C_{dl} = \frac{\varepsilon A}{4\pi t} \tag{8.1}$$

where ε is the dielectric constant, A is the effective surface area and t is the thickness of the double layer. The energy (E) and power density (P) of SC are described in equations 8.2 and 8.3:

$$E = \frac{CV^2}{2} \tag{8.2}$$

$$P = \frac{V^2}{4R} \tag{8.3}$$

where C is the capacitance, V is the operating voltage window and R is the resistance (Wu and Zhu, 2017). The energy density revolves around C and V, while the power density relies on internal resistance (Miao et al., 2016).

8.3   STRUCTURE AND PROPERTIES OF GO

Several models of GO structures have been proposed by scientists from 1936 to 2018; so far, Lerf–Klinovski's model best describes the GO structure (Dreyer et al., 2014). This model has better interpretability than the majority of the empirical study

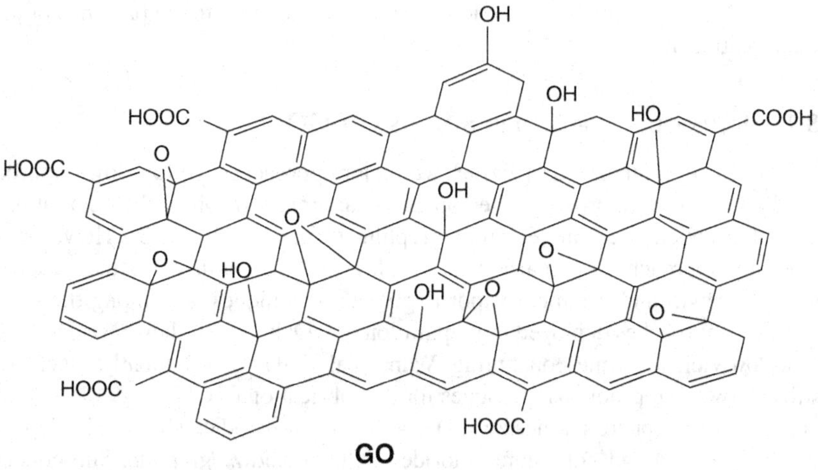

FIGURE 8.2 Depicts the GO structure proposed by Lerf–Klinovski. (Reprinted by permission from Paudics et al. (2021). Copyright © 2021, Elsevier.)

(Rourke et al., 2011). GO consists of sp^2- and sp^3-infused carbon atoms and numerous substituents including carboxylic or epoxy and hydroxyl groups (Siklitskaya et al., 2021). Figure 8.2 illustrates the GO model proposed by Lerf–Klinovsk.

The properties of GO include high mechanical strength, and low electrical and thermal conductivity. Concerning the mechanical properties of GO, the authors has made several efforts to enhance its properties. Dikin et al. (2007) first reported the mechanical properties of GO paper fabricated through individual GO sheets. They reported the strength and stiffness of the GO sample to be 120 MPa and 40 GPa, respectively. Suk et al. (2010) disclosed Young's modulus of a monolayer GO formulated using modified Hummer's techniques to be 207.6 ± 23.4 GPa. This is considerably high but a decrease in order-of-magnitude. Cheng-an et al. (2017) reported an increase in tensile strength (TS) of nanocomposite of GO and polyvinyl alcohol film (PVA) produced via solution casting to be 59.6 MPa. This value is five times more than the TS of PVA film. The significant improvement in the TS of the cement was attributed to oxygen functionalities of GO and OH groups presence in PVA. Abdullah and Ansari (2015) reported GO/epoxy composite prepared at room temperature by casting technique. The addition of 1.5% of GO concentration to epoxy increased the TS from 7 to 13 MPa, modulus rose considerably from 115 to 206 MPa and the load also increased from 126 to 234 N. Babak et al. (2014) reported GO/Cement composite synthesized via ultrasonic technique as well as the effects GO on the TS of cement. They noted a 48% increase in TS of cement with 1.5% GO concentration.

The low electrical conductivity of GO is associated with the interruption in the graphene sp^2-bonding orbitals as well as adding of huge surface groups (Wang et al., 2012). The electrical resistivity of GO is 1.64×10^4 Ω m (Tang et al., 2012). The thermal conductivity of a layer GO film is significantly lower compared to graphene. Naik and Krishnaswamy (2017) reported mean thermal conductivity of GO to be

2.23 W m-K^{-1}. They attributed the low thermal conductivity to the presence of more phonon scattering.

8.3.1 Techniques Deployed in Synthesis of GO

The GO can be synthesized utilizing two major approaches: (i) top-down approach and (ii) bottom-up approach. The top-down approach involves the separation of layers of graphene derivatives from graphite (Wang et al., 2013). However, the bottom-up approach uses carbon molecules in formulating pristine graphene. Epitaxial growth and chemical vapor deposition techniques are among the typical bottom-up approaches deployed in preparation of GO; however, they have drawbacks such as low yield and time-consuming (Wang et al., 2017). Consequently, much effort is shifted toward top-down approaches for the fabrication of GO.

The first attempt to synthesize GO was made by Brodier (1860). He deployed fuming nitric acid and potassium chloride (KCl) to oxidize graphite. Subsequently, Staudenmaier (1898) and Hummers and Offermann (1958) made efforts to improve the process of GO formulations by modifying the Brodier technique. Staudenmaier modified the Brodier technique by replacing the nitric acid with sulfuric acid and slow the addition of KCl in multiple fractions in a single container. Hummers and Offermann improved on the Brodier techniques by deploying a blend of potassium permanganate ($KMnO_4$ as an oxidizer), sulfuric acid (H_2SO_4 as protonated solvent) and sodium nitrate ($NaNO_3$). Hummer's technique involves a mixture of carbon sources (usually graphite flakes) in $NaNO_3$, followed by the introduction of $KMnO_4$. After dilution, the mixture is treated with hydrogen peroxide to eliminate the metal ions from the oxidizer; this yields a yellowish-brown liquid which confirms the formation of GO from pristine graphite. The obtained solids are removed and treated with hydrochloric acid to eliminate metal species. Then, the solution is washed and centrifuged continuously with water till the solution pH becomes neutral. The Hummers techniques are widely used in the preparation of GO because it offers high yield and is also safer. If there is an improvement on this method, it is regarded as a modified Hummers method. Figure 8.3 demonstrates steps taken in converting bulk graphite to GO.

8.4   ADDITION OF GO AND ITS EFFECTS ON THE ENERGY DENSITY OF A SC

The influence of GO on the energy density of a SC has been studied in various ways by researchers. These include modification or tuning of electroactive material utilized in SC such as metal oxides, conducting polymers, metallic sulfide and carbons, by addition of GO to improve the energy density and overall electrochemical properties of these materials. Down, Rowley-Neale, Smith and Banks (2018) formulated GO-based SC through screen-printing techniques. The authors investigated the capacitive characteristics of the GO devices in room temperature ionic liquids and aqueous electrolytes. They reported remarkable power handling capability and energy density of 13.9 kW kg^{-1} and 11.6 Wh kg^{-1}, respectively. Furthermore, they attributed the improved performance noted in the energy of SC to physiochemical properties of GO and not to variations in their morphologies. Figure 8.4 indicates the

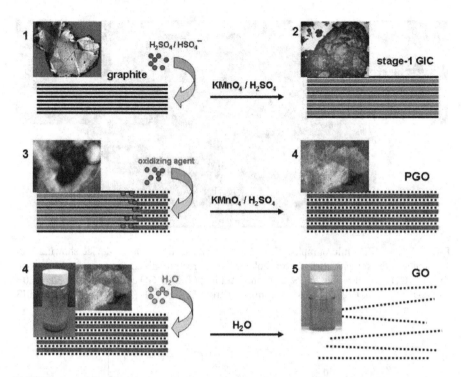

FIGURE 8.3 The three steps taken in converting bulk graphite to graphite intercalation compound (GIC) and final to GO. Solid dark line indicates graphene layers; spotted dark lines indicate single layers of GO; blue lines indicate H_2SO_4/HSO_4^- intercalants; and purple lines indicate a layer of a blend of H_2SO_4/HSO_4^- intercalant along the reduced form of the oxidizing agent. (Reprinted with permission from Dimiev and Tour (2014). Copyright © 2014, American Chemical Society.)

SEM surface micrographs of unmodified screen printed electrodes (SPEs) and different percentages of GO/SPE composite electrodes.

Mane, Malavekar, Ubale, Bulakhe, Insik In and Lokhande (2020) reported synthesized 3% La–MnO$_2$@GO film via binder-free SILAR technique. The results indicated mesoporous sheets of 3%La–MnO$_2$@GO with wide surface area (149 m^{2}g^{-1}), high specific capacitance (C$_s$) of 729 F g^{-1} at the scan rate (SR) of 5 mV s^{-1}) and 94% retention >5,000 CV cycles. At 1.8 V operating windows, the solid-state device, the C$_s$, energy and power density were 140 F g^{-1}, 64 and 1 kW kg^{-1}, respectively. The plot of the energy and power density of the sample is depicted in Figure 8.5. Tanaka et al. (2017) fabricated GO/iron oxide (IO) composite using GO/prussian blue as a precursor at 400°C. The GO was prepared via Hummer's technique. The obtained composite of GO/IO at a ratio of 25:75 indicated a higher surface area (120 m^{2}g^{-1}) relative to IO (93.1 m^{2}g^{-1}) and GO (34.9 m^{2}g^{-1}). In SC, the GO/IO composite with the ratio of 75:25 exhibited greater C$_s$ (91 F g^{-1}) at SR of 20 mV s^{-1}, relative to GO (81 F g^{-1}) and IO (47 F g^{-1}). The improvement in values of C_s was attributed to addition of GO in the material which improved the conductivity by increasing the C_s of the composite and decreasing the resistivity of IO.

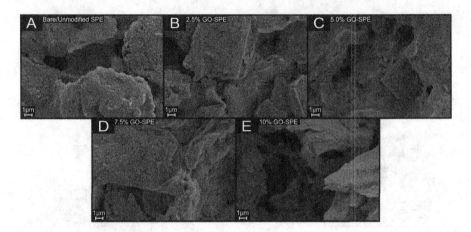

FIGURE 8.4 SEM micrographs of graphite and GO surface in a SC device showing little difference in the samples surface morphologies; (a) Pure unmodified screen printed electrode; (b) 2.5% GO concentration; (c) 5% GO concentration; (d) 7.5% GO concentration; and (e) 10% GO concentration. (Reprinted with permission from Down et al. (2018). Copyright © 2018, American Chemical Society.)

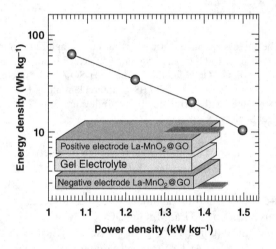

FIGURE 8.5 A plot of energy and power density of 3%La–MnO$_2$@GO. (Reprinted with permission from Mane et al. (2020). Copyright © 2020, Elsevier.)

Iqbal et al. (2017) prepared the aluminum sulfide on a GO film pre-deposited on Ni foam via hydrothermal techniques. The GO-supported Al$_2$S$_3$ exhibited outstanding electrochemical performance which the authors attributed to the excellent electrical conductivity, pore volume and specific surface area of the GO. The specific C_s was 2,178.16 Fg^{-1} at the current density (C_d) of 3 mA cm^{-2}, the energy density was found to be 108.91 Wh kg^{-1} at the C_d of 3 mA cm^{-2}, the power density was 978.9 W kg^{-1} at the C_d of 15 mA cm^{-2} and the electrode stability results indicated 57.84% C_s

FIGURE 8.6 Represents the (a) SEM micrographs of GO-supported Al_2S_3 nanorambutan at various magnification (b) at X3000, (c) at X10,000 and (d) at X30,000. (Reprinted with permission from Ref. Iqbal et al. (2017). Copyright © 2017 Elsevier.)

retention (up to 1,000 cycles). Figure 8.6 reveals SEM micrographs of GO-supported Al_2S_3 nanorambutan at different magnifications. In another study, Wang et al. (2013) formulated nickel sulfide/GO nanocomposite through a hydrothermal technique. The results exhibited NiS/GO nanocomposite with excellent C_s and long cycle life. The C_s and cycle life were 800 F g^{-1} at 1 A g^{-1} and >1,000 cycles, respectively. The authors associated improvement in electrode conductivity and supercapacity behavior with addition of GO to NiS.

Bose, Kim, Kuila, Lau, and Lee (2011) prepared GO/polypyrrole (PPy) nanocomposite through *in situ* polymerization. The estimated energy and power density of the nanocomposite were 94.93 and 3,797.2 W kg^{-1}, respectively. At an SR of 100 mV s^{-1}, high C_s of 267 F g^{-1} was noted. After 500 cycles, the sample retained about 90% of the C_s. The improvement in the electrochemical properties of PPy was attributed to the large surface area of GO. Ajdari et al. (2018) synthesized melamine functionalized GO through a chemical route to enhance electrochemical characteristics of polyorthoaminophenol (POAP). The result indicated POAP/GO-melamine electrodes with high energy and power density of 37.91 and 500 W kg^{-1}, respectively.

The higher efficiency of the composite material was attributed to the large active surface area of GO. Fan et al. (2014) reported an asymmetric SC which was designed by utilizing GO/PPy composite formulated through *in situ* polymerization as the positive electrode and activated carbon as the negative electrode. The results showed that the electrochemical performance of the GO/PPy composite is better compared with individual components. The asymmetric SC maximum energy and power densities were 21.4 and 453.9 W kg^{-1}, respectively. In addition, the SC exhibited outstanding rate capability and cyclic durability. The SEM micrographs showed that PPy nanoparticles were uniformly grown on the surfaces of GO sheets, causing an increase in the specific surface area and electrical conductivity of the composite. According to Österholm et al. (2012), addition of GO into PPy indicates unique and higher porous surface morphology unlike the pure PPy films.

Lei et al. (2013) formulated GO/charged mesoporous carbon (CMK-5 platelets) composite relying on the electrostatic interaction between the two. The composite electrode C_s was 144.4 F g^{-1} in the ionic liquid electrolyte of 1-ethyl-3-methylimidazolium tetrafluoroborate that could be charged/discharged at a voltage of 3.5 V. High energy density (60.7 W h kg^{-1}) and power density (10 kW kg^{-1}) were obtained. Furthermore, the results indicate that straight and short mesochannel of CMK-5 platelets provided a path for swift transport of electrolyte ions while the surface area of the removed GO sheets promoted the formation of EDL capacitance. Liu et al. (2015) formulated a ternary composite of PPy/Bacterial cellulose (BC)/GO through one-step esterification of GO and BC at 70°C and *in situ* polymerization of PPy in GO/BC suspension. The electrochemical performance of the asy-ternary composite films was compared to that of individual PPy films. From the results, the prepared composite exhibited higher C_s (556 F g^{-1}) relative to PPy films (238 Fg^{-1}) in three-electrode asymmetric SC. The observe enhancement in the asy-composite was associated with improved morphology and the combination of PC and EDLC. The composite exhibited excellent capacitance retention (86.2%), while PPy only exhibited 40% retention. The electrical conductivity of the composite was 1,320 S m^{-1}. In addition, the ternary composite exhibited a particularly large energy density of 77.2 Wh kg^{-1} at a power density of 200.1 W kg^{-1}. Figure 8.7 describes the steps taken in the preparation of the BC/GO surface of the samples as well as the SEM micrographs of the samples.

Wang et al. (2015) reported poly(3,4-ethylenedioxythiophene)/4-thienylphenyl functionalized GO (Th-GO) composites synthesized through *in situ* chemical polymerization process. The Th-GO was prepared by Suzuki coupling reaction. From the results, 50 wt% Th-GO with porous structure showed outstanding capacitive behavior compared to other composite materials. It exhibited C_s of 320 F g^{-1} at a current density of 1 A g^{-1}, in addition to excellent capacitance retention (80% at 1 A g^{-1} above 1,000 cycles). Wang et al. (2010) reported GO/PANI composite synthesized through *in situ* polymerization deploying a mild oxidant. Hummer's technique was utilized to prepare the GO from graphite with 12 500 and 500 mesh. High C_s of 746 and 627 F g^{-1} were noted for 12,500 and 500 mesh, respectively, while pure PANI had lower C_s $_{of}$ 216 F g^{-1} at 200 mA g^{-1}. After 500 cycles, capacitance retention of 73% was noted for 12,500 and 64% for 500 mesh relative to 20% indicated by pure PANI. Xu et al. (2016) fabricated a sponge template GO electrode through a combination of GO and polyurethane sponge. In an aqueous electrolyte, the material showed a high energy

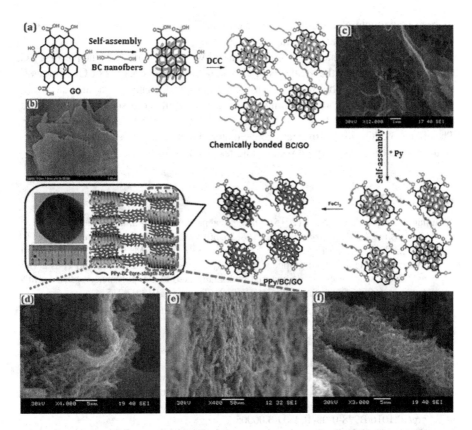

FIGURE 8.7 (a) Preparation steps in preparing BC nanofibers on GO surface and *in situ* polymerization of the composite of PPY/BC/GO. (b) SEM micrograph of pure GO. SEM micrograph of (c) covalent interaction of BC/GO sheets, (d) a single layer of the ternary composite, (e) multilayers of the composite as stacking, and (f) PPy/BC core-sheath blend which served as the link between the stacking. (Reprinted with permission from Liu et al. (2015). Copyright © 2015, American Chemical Society.)

density of 89 Wh kg^{-1} and excellent specific capacity of 401 F g^{-1}. The improved energy density of the sample was attributed to shorter ion transport.

8.5 SUMMARY

One of the drawbacks limiting the application of SCs is their low energy density. Presently, there is a high demand for SC that can turn over greater energy density without relinquishing power density for several applications. In recent years, huge efforts have been made to exploit the unique properties of GO materials in SC applications, especially as an electrode. The electrode is an essential component of a SC that illuminates the electrochemical characteristics of a Sc. Hummer's technique is the most frequent technique often deployed in the preparation of GO. GO has excellent mechanical and optical properties that are tunable as well as a larger surface area. The GO is expected to improve on the limitations of electroactive SC material such as inadequate electronic and ionic conductivity, and mechanical degeneration by controlling the surface area and pore size.

In this chapter, the basic concepts discussed were the working principle of a SC, techniques deployed in the synthesis of GO, structure and properties of GO. In detail, the influence of GO on the energy density of a SC electrode material was also discussed. These include modification or tuning of SC electrodes material such as metal oxides, conducting polymers and carbons by addition of GO to improve their energy density. The effects of GO addition to electroactive material include an increase in active surface area, specific capacitance, pore volume and faster transport of electrolyte ions.

REFERENCES

Abdullah, S. I. & Ansari M.N.M. (2015). Mechanical properties of graphene oxide (GO)/epoxy composites. *HBRC Journal*, *11*(2), 151–156. https://doi.org/10.1016/j.hbrcj.2014.06.001

Ajdari, F. B., Kowsari, E., Ehsani, A., Chepyga, L., Schirowski, M., Jäger, S. et al. (2018). Melamine-functionalized graphene oxide: Synthesis, characterization and considering as pseudocapacitor electrode material with intermixed POAP polymer. *Applied Surface Science*, *459*, 874–883. https://doi.org/10.1016/j.apsusc.2018.07.215

Akhtar, A. J. (2021). *Graphene-Based Materials for Supercapacitor*. IntechOpen. Available from: https://www.intechopen.com/online-first/76804

Babak, F., Abolfazl, H., Alimorad, R. & Parviz, G. (2014). Preparation and Mechanical Properties of Graphene Oxide: Cement Nanocomposites. *The Scientific World Journal*, *2014*, 276323. https://doi.org/10.1155/2014/276323

Balli, B., Şavk, A. & Şen, F. (2019). Graphene and polymer composites for supercapacitor applications. In A. Khan, M. Jawaid, Inamuddin & A. M. Asiri (Eds.), *Nanocarbon and Its Composites*. (pp. 123–151) Woodhead Publishing. https://doi.org/10.1016/B978-0-08-102509-3.00005-5.

Bose, S., Kim, N. H. Kuila, T., Lau, K-T. & Lee J.H. (2011). Electrochemical performance of a graphene–polypyrrole nanocomposite as a supercapacitor electrode. *Nanotechnology*, *22*(29), 295202.

Brodie, B. (1860). Sur le poids atomique du graphite. *Annales de Chimie et de Physique, 59*(466), 472.

Chen, X., Paul, R. & Dai, L. (2017). Carbon-based supercapacitors for efficient energy storage, *National Science Review*, *4*, 453–489. https://doi.org/10.1093/nsr/nwx009

Cheng-an, T., Hao, Z., Fang, W., Hui, Z., Xiaorong, Z., & Jianfang, W. (2017). Mechanical properties of graphene oxide/polyvinyl alcohol composite film. *Polymers and Polymer Composites, 25*(1), 11–16. https://doi.org/10.1177/096739111702500102

Conway, B. E. (1999). *Electrochemical Supercapacitors*. Plenum, New York.

Dikin, D., Stankovich, S., Zimney, E. et al. (2007). Preparation and characterization of graphene oxide paper. *Nature*, *448*, 457–460. https://doi.org/10.1038/nature06016

Dimiev, A. M. & Tour, J. M. (2014). Mechanism of graphene oxide formation. *ACS Nano*, *8*(3), 3060–3068. https://doi.org/10.1021/nn500606a

Down, M. P., Rowley-Neale, S. J., Smith, G. C. & Banks, C E. (2018). Fabrication of graphene oxide supercapacitor devices. *ACS Applied Energy Materials*, *1*, 707–714. https://doi.org/10.1021/acsaem.7b00164

Dreyer, D. R., Todd, A. D. & Bielawski, C. W. (2014). Harnessing the chemistry of graphene oxide. *Chemical Society Reviews*, *43*, 5288–5301. https://doi.org/10.1039/c4cs00060a

Fan, L., Liu, G., Wu, J., Liu, L., Lin, J. & Wei, Y. (2014). Asymmetric supercapacitor based on graphene oxide/polypyrrole composite and activated carbon electrodes. *Electrochimica Acta*, *137*, 26–33. https://doi.org/10.1016/j.electacta.2014.05.137

Faridbod, F., Mohajeri, A., Ganjali, M.R. & Norouzi, P. (2016). Functionalized graphene: synthesis and its applications in electrochemistry. In M. Aliofkhazraei, N. Ali, W.I. Milne, C.S. Ozkan, S. Mitura, J.L. Gervasoni (Eds.), *Graphene Science Handbook: Fabrication Methods* (pp. 149–172) Taylor & Francis, USA.

Gupta, V. & Miura, N. (2006). High performance electrochemical supercapacitor from electrochemically synthesized nanostructured polyaniline. *Materials Letters*, 60, 1466–1469.

Halper och, M.S. & Ellenbogen, J. C. (2006). *Supercapacitors: A Brief Overview*. MITRE Nanosystems Group, McLean, VA.

Huang, S., Zhu, X., Sarkar, S. & Zhao, Y. (2019). Challenges and opportunities for supercapacitors. *APL Materials*, 7, 10090. https://doi.org/10.1063/1.5116146

Huang, Y., Li, H., Wang. Z., Zhu, M., Pei, Z., Xue Q., et al. (2016). Nanostructured polypyrrole as a flexible electrode material of supercapacitor. *Nano Energy*, 22: 422–438. https://doi.org/10.1016/j.nanoen.2016.02.047

Hummers, Jr. W. S. & Offeman, R. E. (1958). Preparation of graphitic oxide. *Journal of the American Chemical Society*, 80(6), 1339–1339.

Iqbal, M. F., Mahmood-Ul-Hassan., Ashiq, M. N., Iqbal, S., Bibi, N. & Parveen, B. (2017). High specific capacitance and energy density of synthesized graphene oxide based hierarchical Al_2S_3 nanorambutan for supercapacitor applications, *Electrochimica Acta, 246*, 1097–1103. https://doi.org/10.1016/j.electacta.2017.06.123

Kaempgen, M., Chan, C.K., Ma, J., Cui, Y. & Gruner G. (2009). Printable thin film supercapacitors using single-walled carbon nanotubes. *Nano Letters*, 9(5), 1872–1876.

Lei, Z., Liu, Z. Wang, H., Sun, X., Lub, L. & Zhao, X. S. (2013). High-energy-density supercapacitor with graphene–CMK-5 as the electrode and ionic liquid as the electrolyte. *Journal of Materials Chemistry A, 1*, 2313–2321. https://doi.org/10.1039/C2TA01040B

Liu, Y., Zhou, J., Tang, J. & Tang, W. (2015).Three-Dimensional, chemically bonded polypyrrole/bacterial cellulose/graphene composites for high-performance supercapacitors *Chemistry of Materials, 27*, 7034–7041. https://doi.org/10.1021/acs.chemmater.5b03060

Long, J.W., Swider, K.E., Merzbacher, C.I. & Rolison D.R. (1999). Voltammetric characterization of ruthenium oxide-based aerogels and other RuO_2 solids: The nature of capacitance in nanostructured materials. *Langmuir, 15*, 780–785. https://doi.org/10.1021/la980785a

Majumder, M. & Thakur, A. K. (2019). Graphene and its modifications for supercapacitor applications. In S. Sahoo., S. Tiwari S. & G. Nayak. (Eds.), *Surface Engineering of Graphene. Carbon Nanostructures* (pp. 113–138). Springer, Cham.

Mane, V. J., Malavekar, D. B., Ubale, S. B Bulakhe, R.N., In, I., & Lokhande, C.D. (2020). Binder free lanthanum doped manganese oxide @ graphene oxide composite as high energy density electrode material for flexible symmetric solid state supercapacitor. *Electrochimica Acta, 335*, 135613. https://doi.org/10.1016/j.electacta.2020.135613.

Manjakkal, L, Pullanchiyodan, A., Yogeswaran, N., Hosseini, E.S. & Dahiya, R. (2020). A wearable supercapacitor based on conductive PEDOT: PSS-coated cloth and a sweat electrolyte. *Advanced Materials, 32*, 1907254. https://doi.org/10.1002/adma.201907254

Marcano, D. C., Kosynkin, D. V., Berlin, J. M., Sinitskii, A., Sun, Z., Slesarev, A., et al. (2010). Improved synthesis of graphene oxide. *ACS Nano, 4*(8), 4806–4814.

Miao, F., Shao, C., Li, X., Wang, K. & Liu, Y. (2016). Flexible solid-state supercapacitors based on freestanding nitrogen doped porous carbon nanofibers derived from electrospun polyacrylonitrile@polyaniline nanofibers. *Journal of Materials Chemistry A, 4*(11), 4180–4187.

Naik, G. and Krishnaswamy, S. (2017). Photoreduction and thermal properties of graphene-based flexible films. *Graphene*, 6, 27–40. https://doi.org/10.4236/graphene.2017.62003

Österholm, A., Lindfors, T., Kauppila, J., Damlin, P. & Kvarnström, C. (2012). Electrochemical incorporation of graphene oxide into conducting polymer films. *Electrochimica Acta, 83*, 463–470.

Park, S. & Ruoff, R. S. (2009). Chemical methods for the production of graphenes, *Nature Nanotechnology, 4*, 217–224. https://doi.org/10.1038/nnano.2009.58.

Paudics, A., Farah, S., Bertóti, I., Farkas, A., László, K., Mohai, M. et al. (2021). Fluorescence probing of binding sites on graphene oxide nanosheets with Oxazine 1 dye. *Applied Surface Science, 541*, 148451. https://doi.org/10.1016/j.apsusc.2020.148451

Rourke, J.P., Pandey, P. A., Moore, J. J., Bates, M., Kinloch I.A., Young, R.J. et al. (2011). The real graphene oxide revealed: Stripping the oxidative debris from the graphene-like sheets. *Angewandte Chemie, 50*(14), 3173–3177.

Saleem, A. M., Desmaris, V. & Enoksson P. (2016). Performance enhancement of carbon nano-materials for supercapacitors. *Journal of Nanomaterials, 2016*, 1537269. http://dx.doi.org/10.1155/2016/1537269

Shi, W., Zhu, J., Sim, D. H., Tay, H. H., Lu, Z., Zhang, X., et al. (2011). Achieving high spe-cific charge capacitances in Fe_3O_4/reduced graphene oxide nanocomposites. *Journal of Materials Chemistry, 21*, 3422–3427. https://doi.org/10.1039/C0JM03175E

Siklitskaya, A., Gacka, E., Larowska, D., Mazurkiewicz-Pawlicka, M. D., Malolepszy, P. A., Stobinski, L. et al. (2021). Lerf–Klinowski-type models of graphene oxide and reduced graphene oxide are robust in analyzing non-covalent functionalization with porphyrins. *Scientific Reports, 11*, 7977. https://doi.org/10.1038/s41598-021-86880-1

Simon, P. & Gogotsi, Y. (2008). Materials for electrochemical capacitors. *Nature Materials, 7*, 845–854. https://doi.org/10.1038/nmat2297

Staudenmaier, L. (1898). Verfahren zur darstellung der graphitsäure. *Berichte der deutschen chemischen Gesellschaft, 31*(2):1481–487.

Suk, J. W., Piner, R. D., An, J. & Ruoff, R. S. (2010). Mechanical properties of monolayer graphene oxide. *ACS Nano, 4*(11), 6557–6564.

Sun, L. (2019). Structure and synthesis of graphene oxide. *Chinese Journal of Chemical Engineering, 27*, 2251–2260. https://doi.org/10.1016/j.cjche.2019.05.003.

Tanaka, S., Salunkhe, R. R., Kaneti, Y. V., Malgras, V., Alshehri, S. M., & Ahamad T. (2017). Prussian blue derived iron oxide nanoparticles wrapped in graphene oxide sheets for electrochemical supercapacitors. *RSC Advances, 7*, 33994–33999. https://doi.org/10.1039/C7RA03179C.

Tang, L., Li, X., Ji, R., Teng, K.S., Tai, G., Ye, J., Wei, C. & Lau, S.P. (2012). Bottom-up syn-thesis of large-scale graphene oxide nanosheets. *Journal of Materials Chemistry, 22*(12), 5676–5683.

Wang, A., Wang, H., Zhang, S., Mao, C., Song, J., Niu, H., et al. (2013). Controlled synthesis of nickel sulfide/graphene oxide nanocomposite for high-performance supercapacitor. *Applied Surface Science, 282*, 704–708. https://doi.org/10.1016/j.apsusc.2013.06.038

Wang, G., Zhang, L. & Zhang, J. (2012). A review of electrode materials for electrochemical supercapacitors. *Chemical Society Reviews, 41*(2), 797–828.

Wang, H., Hao, Q., Yang, X., Lu, L. & Wang, X. (2010). Effect of graphene oxide on the properties of its composite with polyaniline. *ACS Applied Materials & Interfaces, 2*, 821–828. https://doi.org/10.1021/am900815k

Wang, M., Jamal, R., Wang, Y., Yang, L., Liu, F. & Abdiryim. T. (2015). Functionalization of graphene oxide and its composite with poly(3,4-ethylenedioxythiophene) as elec-trode material for supercapacitors. *Nanoscale Research Letters, 10*, 370. https://doi.org/10.1186/s11671-015-1078-x

Wang, X.-Y., Narita, A. & Müllen, K. (2017). Precision synthesis versus bulk-scale fabrication of graphenes, *Nature Reviews Chemistry, 2*(1), 1–10.

Wang, Z., Liu, J., Wang, W., Chen H., Liu, Z., Yu, Q. et al. (2013). Aqueous phase preparation of graphene with low defect density and adjustable layers, *Chemical Communications, 49*(92), 10835–10837.

Wang, Z., Nelson, J. K., Hillborg, H., Zhao, S. & Schadler, L. S. (2012). Graphene oxide filled nanocomposite with novel electrical and dielectric properties, *Advanced Materials, 24*(-23) 3134–3137.

Wu, S. & Zhu, Y. (2017). Highly densified carbon electrode materials towards practical super-capacitor devices. *Science China Materials, 60*, 25–38

Wu Y. & Cao C. (2018). The way to improve the energy density of supercapacitors: Progress and perspective. *Science China Materials, 61*, 1517–1526. https://doi.org/10.1007/s40843-018-9290-y

Xiao, W., Xia, H., Fuh, J. Y. H. & Lu, L. (2009). Growth of single-crystal a-MnO_2 nanotubes prepared by a hydrothermal route and their electrochemical properties. *J Power Sources, 193*, 935–938. https://doi.org/10.1016/j.jpowsour.2009.03.073.

Xu, J., Tan, Z., Zeng, W., Chen, G., Wu, S., Zhao Y. et al. (2016). A hierarchical car-bon derived from sponge-templated activation of graphene oxide for high-perfor-mance supercapacitor electrodes. *Advanced Materials, 28*, 5222–5228. https://doi.org/10.1002/adma.201600586

Xu, Y., Bai, H., Lu, G., Li, C. & Shi, G. (2008). Flexible graphene films via the filtration of water-soluble noncovalent functionalized graphene sheets. *Journal of the American Chemical Society, 130*, 5856–5857. https://doi.org/10.1021/ja800745y

Yang, C. (2021). Review of graphene supercapacitors and different modified graphene elec-trodes. *Smart Grid and Renewable Energy, 12*(1), 1–15.

Yu, A., Chabot, V., & Zhang, J. (2013). *Electrochemical Supercapacitors for Energy Storage and Delivery: Fundamentals and Applications* (1st Edition). CRC Press. https://doi.org/10.1201/b14671

Zhang, L. L., Lei, Z. J., Zhang, J., Tian, X. & Zhao, X. X. (2010). Supercapacitors: Electrode materials aspects. In R.H. Crabtree (Ed.), *Energy Production and Storage: Inorganic Chemical Strategies for a Warming World* (pp. 341–364). John Wiley & Sons, UK.

9 Application of Graphene Oxide in Fuel Cells Fabrication and Performance Optimization

Malachy N Asogwa, Sylvester M. Mbam, and Ada C. Agbogu
University of Nigeria

Raphael. M. Obodo
University of Nigeria
University of Agriculture and Environmental Sciences
Quaid-i-Azam University
Northwestern Polytechnical University

Assumpta C Nwanya
University of Nigeria
University of the Western Cape Sensor Laboratories

Malik Maaza
iThemba LABS-National Research Foundation
University of South Africa (UNISA)

Fabian I. Ezema
University of Nigeria
iThemba LABS-National Research Foundation
University of South Africa (UNISA)

CONTENTS

DOI: 10.1201/9781003215196-9

9.1   INTRODUCTION

The fuel cell is an alternative source of energy just as solar energy, wind, hydrothermal, etc. It is a clean source of energy that poses little or no environmental hazards. With the application of fuel cells as energy alternatives, the environmental concerns due to the poisonous emissions in nature by fossil fuel-based energy sources would be highly minimized. Electric energy is generated in fuel cells when the stored chemical energy undergoes electrochemical reactions using the oxidants and reactants. Fuel cells are of various types such as polymer electrolyte membrane fuel cell (PMFC), alkali fuel cell, phosphoric acid fuel cell, molten carbonate fuel cell and solid oxide fuel cell [1,2]

Currently, polymer electrolyte membrane fuel cell is widely used and has been found very promising and environmentally friendly in its clean energy supplies. It is applied in automobiles, portable power energy devices, etc., and it utilizes mainly hydrogen and oxygen as the reactants and gives out water as byproduct [3–6]. It has a membrane electrode assembly (MEA) which comprises the anode, the cathode and the proton (H^+) conducting membrane [4–6]

These reactions could be depicted by the equations below [5,6]:

$$\text{At the anode}: 2H_2 \rightarrow 4H^+ + 4e^- \tag{9.1}$$

$$\text{At the anode}: O_2 + 4e^- + 4H^+ \rightarrow 2H_2O \tag{9.2}$$

$$\text{Net Reaction}: 2H_2 + O_2 \rightarrow 2H_2O + \text{heat} \tag{9.3}$$

To enhance the efficiency, power output and the overall workability of fuel cells' different catalysts have been advocated; for example, catalysts like platinum and graphene. Platinum catalyst has been widely used because of its efficiency in electro catalysis during the oxidation of hydrogen and the oxygen reduction reactions (ORRs). This, however, causes the emission of carbon monoxide (CO) and other traces of hazardous elements such as CO, sulfur(S), and NH_3 which are unfriendly to the environment. CO poisoning can result due to low-temperature operation (say below 150°C) which results from the high negative change in free energy (ΔG). This low temperature favors CO adsorption on platinum-engineered electrodes because of the affinity of CO on platinum surfaces [1,7].

To curb this challenge, that is, the generation of CO, many improved methods have been brought into place. Methods such as oxidative surface environment and the use of electrical impulse; these methods convert CO to CO_2.

Another challenge in the use of platinum catalysts in fuel cells is its cost. For this reason, different materials, e.g., carbon-based materials, have been used; these are nanotubes, nanofibers, bucky nano balls, graphite particles and nanosheets. These materials can also minimize the traces of CO and the material cost of platinum usage, and they simultaneously improve the efficiency of fuel cells [2–7].

Two-dimensional single-layered 'graphene' has been revolutionary in fuel cell applications, and it is considered as an important material basically because of its steep specific area ($2,630\,m^3g^{-1}$), exceptionally high electrical conductivity (104.63 Scm^{-1}), high flexural strength (44.28 Mpa) and enhanced chemical stability [11].

To produce graphene on large scale, techniques such as hydrogen exfoliation which focuses on solar exfoliation have been used. The use of graphene-based nanomaterial has open new fabrication of low-cost electrocatalyst systems. These are GO/Pt nanoparticles system microwave method along with Nafion/GO membrane; which been engineered with sulphonated graphene oxide (exfoliated) (SGO) through blending [11].

Also, the fabrication of polybenzimidazole (PBI) membrane through the synthesis of 3,3'-diaminobenzidine (GO/PBI), 5-tert-butyl isophthalic acid (GO/BuIPBI), and isocyanate-engineered graphite oxide/BuIPBI (iGO/BuIPBI) membranes as electrolytes, where proton conductivity was achieved using phosphoric acid and the fabrication of phosphoric acid doped PBI/GO electrolyte membranes for high-temperature PEMFCs, was reported in another research finding [12] (Figure 9.1).

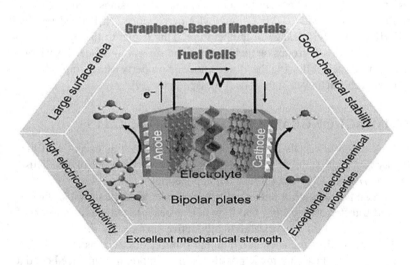

FIGURE 9.1 Schematic of a fuel cell incorporating graphene-based materials in each component [13].

Following from the above, we review the various applications of GO as it is used for fuel cell performance and optimization under the following headings: History, Synthesis, Characterization and Applications.

9.2 GRAPHENE-BASED NANOMATERIALS

Graphene-based nanomaterials include graphene nanoplatelets (GNP), few-layer graphene (FLG), graphene nano-onions (GNO), GO and reduced GO. These are the derivatives of graphene and possess unique properties other than normal graphene; this is because of its excellent mechanical properties.

They possess varying properties such as number of layers, defect density, surface chemistry, quality of graphene sheets, lateral dimension, composition and purity [13–18].

9.3 GRAPHENE OXIDE FOR FUEL CELLS

Graphene is promising for fuel cell applications because it's insulating and hydrophilic. This is due to the presence of various oxygen groups such as epoxide, hydroxide, carbonyls and carboxyls. It has a carbon oxygen ratio of 2:1. Because of these excellent characteristics, properties such as mechanical strength, gas impermeability, surface electrical insulation, gas permeability and proton conductive are retained. This makes GO a good candidate as an electrolyte membrane in fuel cell applications. GO has been massively produced using the Hummers method. This is usually made possible by treatment with $NaNO_2$, $KMnO_4$, H_2SO_4 and DI followed by vacuum drying [19–20].

9.4 PROPERTIES OF GRAPHENE OXIDE

GOs are very hydrophilic forming stable aqueous dispersions such as ethylene glycol, dimethylformamide (DMF), n-methyl-2-pyrrolidone (NMP) and tetrahydrofuran (THF) over a wide range of temperature. This is due to the hydrogen bonding between the hydroxyl groups on their surface and solvent interface. A variety of substrates can be made following the aforementioned property using simple drop-casting method and this achieves high-optical transparency [19].

9.5 IMPACT OF GRAPHENE OXIDE ON FUEL CELLS

Graphene-based materials have excellent properties such as exceptional high conductivity and high stability during the processing of the electrolytes. This makes these materials widely used for fuel cell applications.

Graphene nanomaterials enhance the nano-scaled electrocatalyst surface area for improved transfer of electrons, and then help accelerate the mass transport of reactants, i.e., fuels, to the electrocatalytically active surface. By this, the conductivity is enhanced and then the accumulation and transfer of electrons to the electrocatalytically active surfaces are made possible. Furthermore, the enhanced conductivity helps in accumulation and transport of electrons to the electrocatalytically active surfaces. Functional groups such as carboxyl and epoxies can be employed to modify the surface of graphene, e.g., GO. Sheets of graphene can help hold semiconductor particles like TiO_2. This ability to hold nanoparticles makes it possible for the development of improved electrocatalysts for the fuel cell applications.

Graphite, and its oxides and sub-derivatives when used as source materials in the production of graphene is less cost and produces graphene in bulk quantity.

It also allows the edge planes to connect with catalyst nanoparticles. The catalyst particles can be expedited because of the planar structure of the GO which helps in promoting the active and large surface area.

The enhanced properties of graphene-based materials can be utilized as catalysts for cathodes of PEM fuel cells, due to their catalytic affinity toward ORRs [16–20].

9.6 GRAPHENE AND GRAPHENE OXIDE IN FUEL CELLS

9.6.1 DIRECT METHANOL FUEL CELL (DMFC)

This type of fuel cell has a simple design and very efficient and easy operation. The development of a proton conductive membrane in this type of fuel cell is difficult. Usually, Nafion membranes are widely used. There are other polymeric membranes that could serve as membranes in DMFC application, and these include poly (ether ether ketone) (PEEK), PBI and polysulfone. Nevertheless, for DMFCs, graphene-based membranes receive more attention. It provides a unique two-dimensional structure with flexible functionalization and admirable thermal stability [13–18].

9.6.2 ALKALINE FUEL CELL (AFC)

This involves modification of GO with KOH and it increases the power density, tensile strength and the ionic conductivity of the engineered GO membrane.

Vacancies could exist on graphene surface and could help in metal ion correlation in the GO membrane. This differentiates graphene from other carbon-based materials.

When nitrogen is added at the time of the preparation, it leads to the formation of nitrogen-doped reduced graphene (rGO(N)) which possesses high ORR property.

In AFCs, a Pt/carbon composite material is used as cathode. This enhances the performance of the cell. When added along with cobalt oxide, rGO(N) results in the formation of a newly efficient cathode. This hybrid cathode will upgrade the ORR and electron selectivity of the cell. The effect of doping with nitrogen and the presence of a defect will help the graphene to hold more Co. This would result in the formation of an arrow cathode layer with improved mass transfer [13–18].

9.6.3 DIRECT GLUCOSE FUEL CELL (DGFC)

This type of fuel cell is non-toxic and has simple operation. This is due to the oxidation reaction that takes place in this cell; the oxidation of glucose to carbon dioxide. This produces an open circuit potential of 1.2 V. Graphene serves as the cathode of DGDC to enhance the cell performance. This is achieved by adding GO with 5% polytetrafluoroethylene [13,14].

9.6.4 Direct Ethanol Fuel Cell (DEFC)

This type of fuel cell utilizes ethanol as the fuel. The electrode fabrication is based on Pt catalyst [14].

9.7 GRAPHENE AS ELECTROCATALYST

Graphene is used as electrocatalyst to fasten electrical conductivity. The key component in a fuel cell is the electrode and this is where the gas diffusion occurs. It has three main components; a porous layer of gas, a gas diffusion layer and electrode supporting materials. The gas diffusion layer is capable of conducting electricity. Various materials are employed in the fabrication of fuel cell electrodes; these include graphene, and carbon fiber paper [13–18].

9.7.1 Graphene as an Oxygen Reduction Reaction Catalyst (ORR)

Graphene nanostructured material can be used as catalysts for improved cell operation, durability and catalyst performance. Other nanostructured catalysts include active carbon, platinum (Pt), porous carbon and carbon nanotubes. Similarly, metal carbide, mesoporous silica and conductive polymers have also been researched for application as a nanostructured catalyst in fuel cells.

The overall performance of fuel cells can be enhanced by modifying the anode and cathodes, to improve the oxygen reaction (ORR). This is made possible by incorporating graphene into the electrodes of the cell. This allows the utilization of its macroporosity and the multidimensional electron pathways [13–20].

9.7.2 Graphene Oxide as a Membrane in Fuel Cell

The excessive use of Pt catalyst which can be substituted by GO-based ionic composite at high temperature improves fuel cells performance by increasing the ionic conductivity [13–20].

9.8 CHALLENGES AND OPPORTUNITIES

The major challenge in the fabrication of graphene on large scale is the fabrication processes available are less efficient, which leads to delay in the bulk supply of graphene which is source material for the production of GO. The inefficiency of the processes involved in the production of graphene on large scale generates an excessive number of irregularities in the graphene layers. Another challenge in the production of graphene is that the processes involved are of high cost. Also, electric current discharge occurs in graphene when charged and it is very difficult to prevent the occurrence.

On the other hand, high conductivity of graphene is an advantage and this provides great opportunities in its applications in solar cells.

Nanohybrids of GO are utilized in the production of high-performance technological materials of great importance.

The properties of graphene nanocomposites exhibited by layered GO-based membranes are affected by morphology control, good conductivity, thermal and mechanical stability and water retention ability [15–18].

9.9 SYNTHESIS OF GRAPHENE

The graphene can be synthesized via various methods such as chemical vapor deposition (CVD), liquid phase exfoliation (LPE), mechanical exfoliation, electrochemical exfoliation and bottom-up synthesis [14–18].

9.9.1 SYNTHESIS OF GRAPHENE OXIDE

GO has been synthesized using the Hummers method, and this is an improved synthesis method. It has been found to produce GO with higher sheets with a better structural formation and a higher monolayer yield. It uses potassium permanganate ($KMnO_4$) and sodium nitrate ($NaNO_3$).

The improved Hummers' method excludes $NaNO_3$ to prevent the generation of toxic gases, using ice instead of liquid water to prevent the high temperature rise. This increases the yield and the degree of oxidation.

Another method that uses potassium ferrate (K_2FeO_4), instead of $KMnO_4$, as an oxidant has been used to produce GO. This is a fast, scalable method of GO synthesis. This method is called the green method.

The reliability of this method has been questioned as almost no researchers have used it [13–17].

9.10 CHARACTERIZATION TECHNIQUES FOR GRAPHENE OXIDE

The morphologies (microscopic) of GO can be characterized using transmission electron microscopy (TEM) and dynamic force microscope (DFM).

The structures of GO can be found through characterization with Fourier transform infrared spectra analyzer (FT–IR) and X-ray diffraction (XRD). The optical properties of GO can be studied using UV spectrometer (UV–VIS). The thermal stability of GO is investigated by employing thermo gravimetry analysis [12–18].

9.11 FUTURE APPLICATIONS

When graphene is engineered, the materials would contain transparent flexible electrodes, graphene/polymer electrolyte composites. And these properties are suitable for energy storage, organic electronics and sensors. This enhances the performance of fuel cells. They are widely used for other energy applications in supercapacitors, sensors, satellites, automobiles, submarines, etc.

With improved techniques, graphene or GO is engineered to have controlled size, high quality thereby outperforming the graphene-based composites.

Graphene-based composites, on the other hand, are capable of solving various energy-related and environmental concerns and hence are being utilized in the conversion of solar to chemical energy in photocatalysis [9–13].

REFERENCES

[1] J. L. Hall, "Cell Components," *Phytochemistry*, vol. 26, no. 4, pp. 1235–1236, 1987, doi: 10.1016/s0031-9422(00)82398-5.

[2] N. Mahato, H. Jang, A. Dhyani, and S. Cho, "Recent Progress in Conducting Polymers for Hydrogen Storage and Fuel Cell Applications," *Polymers*, vol. 12, no. 11, p. 2480, 2020.

[3] A. Kulikovsky, "Impedance and Resistivity of Low–Pt Cathode in a PEM Fuel Cell," *J. Electrochem. Soc.*, vol. 168, no. 4, p. 044512, 2021, doi: 10.1149/1945-7111/abf508.

[4] A. Baroutaji, J. G. Carton, M. Sajjia, and A. G. Olabi, "Materials in PEM Fuel Cells," *Ref. Modul. Mater. Sci. Mater. Eng.*, pp. 1–11, 2016, doi: 10.1016/b978-0-12-803581-8.04006-6.

[5] A. Baroutaji, J. G. Carton, M. Sajjia, M. Ramadan, and A. G. Olabi, "Materials in PEM Fuel Cells," *Ref. Modul. Mater. Sci. Mater. Eng.*, pp. 1–30, 2021, doi: 10.1016/b978-0-12-815732-9.00134-0.

[6] J. A. Shahrukh Shamim, K. Sudhakar, and B. Choudhary, "A Review on Recent Advances in Proton Exchange Membrane Fuel Cells : Materials, Technology and Applications," *Adv. Appl. Sci. Res.*, vol. 6, no. March 2016, pp. 89–100, 2015.

[7] J. Tjønnås, F. Zenith, I. J. Halvorsen, M. Klages, and J. Scholta, "Control of Reversible Degradation Mechanisms in Fuel Cells: Mitigation of CO Contamination," *IFAC-PapersOnLine*, vol. 49, no. 7, pp. 302–307, 2016, doi: 10.1016/j.ifacol.2016.07.309.

[8] C. Trellu, B. P. Chaplin, C. Coetsier, and R. Esmilaire, "Electro-oxidation of Organic Pollutants by Reactive Electrochemical Membranes Manuscript submitted to Chemosphere – Special Issue ' Electrochemical Advanced Oxidation Processes for the Abatement of Persistent Organic Pollutants ' – for consideration," 2018.

[9] L. P. L. Carrette, K. A. Friedrich, M. Huber, and U. Stimming, "Improvement of CO Tolerance of Proton Exchange Membrane (PEM) Fuel Cells by a Pulsing Technique," *Phys. Chem. Chem. Phys.*, vol. 3, no. 3, pp. 320–324, 2001, doi: 10.1039/b005843m.

[10] P. Divya and S. Ramaprabhu, "Platinum-Graphene Hybrid Nanostructure as Anode and Cathode Electrocatalysts in Proton Exchange membrane Fuel Cells," *J. Mater. Chem. A*, vol. 2, no. 14, pp. 4912–4918, 2014, doi: 10.1039/c3ta15181f.

[11] J. Ortiz Balbuena, P. Tutor De Ureta, E. Rivera Ruiz, and S. Mellor Pita, "Enfermedad de Vogt-Koyanagi-Harada," *Med. Clin. (Barc)*, vol. 146, no. 2, pp. 93–94, 2016, doi: 10.1016/j.medcli.2015.04.005.

[12] N. Üregen, K. Pehlivanoğlu, Y. Özdemir, and Y. Devrim, "Development of Polybenzimidazole/Graphene Oxide Composite Membranes for High Temperature PEM Fuel Cells," *Int. J. Hydrogen Energy*, vol. 42, no. 4, pp. 2636–2647, 2017, doi: 10.1016/j.ijhydene.2016.07.009.

[13] H. Su and Y. H. Hu, "Recent Advances in Graphene-Based Materials for Fuel Cell Applications," *Energy Sci. Eng.*, vol. 9, no. 7, pp. 958–983, 2021, doi: 10.1002/ese3.833.

[14] C. Xue, J. Zou, Z. Sun, F. Wang, K. Han, and H. Zhu, "Graphite Oxide/Functionalized Graphene Oxide and Polybenzimidazole Composite Membranes for High Temperature Proton Exchange Membrane Fuel Cells," *Int. J. Hydrogen Energy*, vol. 39, no. 15, pp. 7931–7939, 2014.

[15] Uregen, N., Pehlivanoğlu, K., Ozdemir, Y., and Devrim, Y. "Development of Polybenzimidazole/Graphene Oxide Composite Membranes for High Temperature PEM Fuel Cells," *Int. J. Hydrogen Energy*, vol. 42, no. 4, pp. 2636–2647, 2017.

[16] Antolini, E. "Graphene as a New Carbon Support for Low-Temperature Fuel Cell Catalysts," *Appl. Catal. B: Environmental, 123–124*, 52–68, 2012.

[17] Zhang, L.-S., Liang, X.-Q., Song, W.-G., and Wu, Z.-Y. "Identification of the Nitrogen Species on N-Doped Graphene Layers and Pt/NG Composite Catalyst for Direct Methanol Fuel Cell," *Phys. Chem. Chem. Phys.*, vol. 12, no. 38, p. 12055, 2010.

[18] Zhang, J., Yang, H., Shen, G., Cheng, P., Zhang, J., and Guo, S. "Reduction of Graphene Oxide Ascorbic Acid," *Chem. Commun.*, vol. 46, no. 7, 1112–1114, 2010.

[19] Ambrosi, A., Chua, C. K., Bonanni, A., and Pumera, M. "Electrochemistry of Graphene and Related Materials," *Chem. Rev.*, vol. 114, no. 14, 7150–7188, 2014.

[20] Scott, K. "Freestanding Sulfonated Graphene Oxide Paper: A New Polymer Electrolyte for Polymer Electrolyte Fuel Cells," *Chem. Commun.*, vol. 48, no. 45, 5584, 2012.

10 Role of Graphene Oxide (GO) in Enhancing Performance of Energy Storage Devices

S. Meyvel
Chikkaiah Naicker College

M. Malarvizhi
K.S. Rangasamy College of Technology

M. Dakshana
Sathyabama Institute of Science and Technology

P. Sathya
Salem Sowdeswari College

CONTENTS

DOI: 10.1201/9781003215196-10

10.1 INTRODUCTION

Energy has usually played an essential and fundamental role in human survival and civilization. Inside the uncertain level of childhood, guys discovered to apply hearth from volcanoes, lightning, and other resources [1]. With the potential to obtain, keep, and care for the fireplace, guys started to live in safe shelters and eat cooked meals. Rapid ahead, beginning inside the 1st-century advert, steam engines had been designed, evolved, and remained the dominant electricity source properly into the 20th century. They were the riding pressure behind the commercial revolution, and that they took advantage of commercial-use driving machines that have been broadly used in factories, generators, and mines. To this point, one-of-a-kind electricity assets had been made available to us: coal, nuclear, hydro, sun, and so on. We enjoy our abilities to make things larger, quicker, extra cozy, and less expensive. With the rapid economic increase in high-populace nations like China and India and our desire to live and work in what we want (we call this a catchy "globalization"), our unexpected thirst for power appears to create some issues. It becomes now not expected [2].

Although power storage technology no longer constitutes electricity sources or electricity manufacturing, it may be becoming a vital part of electricity supply and use infrastructures; they also provide additional benefits for improving the steadiness, strength quality, and reliability of energy technology and intake structures. There is a splendid need to shop for electric energy, no longer best for digital cellular gadgets such as mobile telephones and computer systems, but additionally for transportation, load-leveling of electrical grids, and the efficient commercialization of renewable sources (along with the sun and wind energy). The traits of numerous energy garage gadgets are stored energy (Wh), most power (W), length, weight, preliminary cost, and lifelong [3].

10.2 TRENDY INTRODUCTION TO ENERGY GARAGE

Electricity production has made notable development by using renewable and smooth energy sources, including solar power and wind energy. But all such electricity resources have a common problem with saving and releasing energy for better performance and efficiency. For this cause, battery, gas cell, flywheel battery, and top-notch capacitor may be used for high strength density, excessive rate, and discharge cutting-edge ranges, and an excellent capacitor may be favored for lots of programs with extensive running advantage, temperature variety for big no and high sturdiness, and OFF/ON cycles. Therefore, they may be favored for lots of programs.

According to the working precept of the remarkable capacitor, it can be labeled as double-layer first-rate capacitor, electrochemical outstanding capacitor, and hybrid exceptional capacitor. The hybrid supercapacitor (SC) device has two distinct forms of electrodes. One of them is a double-layer SC fabric such as activated carbon; the other one is electrochemical-SC cloth together with ruthenium dioxide. This kind of SC includes the benefits of double-layer SC and electrochemical SC, which could be utilized in high strength and strength density calls for structures. The general needs of SCs are excessive operating voltage, massive capacitance, and occasional resistance for pulse energy delivery programs, particularly pulsed power delivery of electromagnetic release device programs. However, the operating voltage of contemporary SCs is very low (<3.5 V), which has limited their different packages. To satisfy the requirements of excessive voltage, engineers historically connect multiple SCs in the collection. But the entire capacitance of the electricity storage tool decreases and the internal resistance will increase. Based on the high operating voltage of the electrolytic capacitor, a hybrid SC concept was proposed [4].

To reap both high power and electricity densities, numerous hybrid capacitors, including activated carbon and pseudocapacitance electrodes, have been proposed and investigated currently. If a pseudocapacitance electrode with very small polarization is used in preference to one of the AC electrodes in EDLCs, the voltage variation of the electrode throughout charging and discharging hardly ever takes place, resulting in a first-rate boom in the capacitance and electricity density of the cellular [5].

The hybrid SC has fast charging and discharging capabilities. Also, its excessive voltage functionality makes it useful for high-voltage packages, which include electric-powered cars [6].

10.3 ELECTROCHEMICAL ENERGY GARAGE

Electrochemical strength garage is a way used to store power in chemical shape. This storage technique takes advantage of the fact that both electrical and chemical

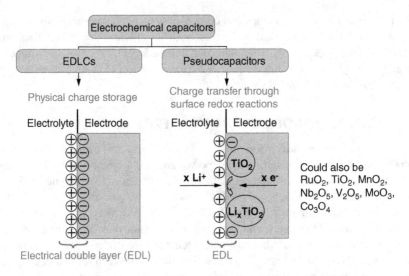

FIGURE 10.1 Classification of electrochemical capacitors as electric double-layer capacitors and pseudocapacitors [7]. (Reproduced by permission of Elsevier.)

electricity proportion the identical carrier, the electron. This commonality lets in for restricting losses from conversion from a shape to someone else.

Electrochemical capacitors (ECs), also known as supercapacitors or ultracapacitors, are typically divided into categories based on their distinct power garage mechanisms. Particularly, Figure 10.1 shows electric double-layer capacitors (EDLCs) and pseudocapacitors. First, EDLCs physically keep charges in electric bilayers that shape close to electrode/electrolyte interfaces. Therefore, the manner is entirely reversible, and the cycle of lifestyles is endless. Alternatively, pseudocapacitors keep power no longer best thru the electric double layer, as in EDLCs, but also thru rapid surface oxidation–discount (redox) reactions and feasible ion intercalation at the electrode.

The performance of ECs is determined by way of the aggregate of electrode fabric and electrolyte used. There are three most important categories of electrode materials used for ECs: (i) Carbon-based substances, (ii) Transition steel oxides, and (iii) Conductive polymers. Furthermore, three kinds of electrolyte materials are used for ECs, which include (i) aqueous electrolytes, (ii) natural electrolytes, and (iii) ionic liquids.

The performances of the ECs can be compared on the Ragone graph displaying the respective electricity and strength densities as proven in Figure 10.2 for distinctive electrical energy garage devices. Because of their bodily rate garage, capacitors have enormous electricity densities but low energy densities than batteries and fuel cells. Then again, batteries and gasoline cells have huge strength densities; however, low power densities because of their slow response kinetics. ECs bridge the space between capacitors and batteries/fuel cells. They offer the opportunity of retaining the high electricity density of batteries without sacrificing the high strength density of capacitors.

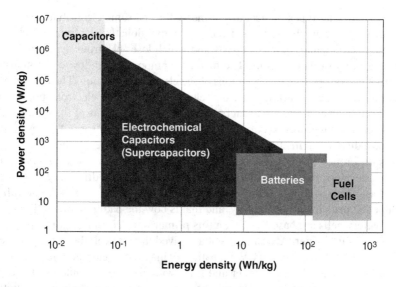

FIGURE 10.2 Ragone plot [7]. (Reproduced by permission of Elsevier.)

10.4 BATTERIES AND CAPACITORS

A battery is a tool that converts the chemical energy contained in its energetic materials without delay into electric strength thru an electrochemical oxidation discount (redox) reaction. This sort of response involves the transfer of electrons from one fabric to another via an electrical circuit. While the battery is regularly used, cellular is the electrochemical unit used to generate or shop electrical power. To recognize the differences between a cellular and a battery, a battery has to be considered with one or extra of these cells related in collection or parallel or each, depending on the favored output voltage and potential.

A battery works at the oxidation and discount reaction of an electrolyte with metals. While two distinctive metallic substances, known as electrodes, are placed in a diluted electrolyte, oxidation and reduction reactions take place on the electrodes, respectively, relying on the electron affinity of the steel of the electrodes. Due to the oxidation response, one electrode is negatively charged, referred to as the cathode, and due to the discount reaction, another electrode, known as the anode, is positively charged. The cathode bureaucracy is a terrible battery terminal, while the anode paperwork is a fantastic battery terminal. To properly understand the fundamental principle of the battery, we first know a few basic concepts concerning electrolyte and electron affinity. In truth, when two different metals are immersed in an electrolyte, there will be a capacity dissimilarity produced between metals.

It has been observed that once some unique compounds are introduced to water, they dissolve and produce negative and high-quality ions. This kind of compound is referred to as an electrolyte. Famous examples of electrolytes are nearly all styles of salts, acids, and bases. The strength released by way of a neutral atom when accepting an electron is called electron affinity. Because the atomic structure is distinctive for one-of-a-kind substances, the electron affinity of different materials may also be unique.

If one-of-a-kind types of steel are immersed in the same electrolyte answer, one in all of them gains electrons, and the alternative donates electrons which metal (or metallic compound) gains electrons and which loses electrons depending on the electron affinity of those metals. The metallic compound with low electron affinity will benefit electrons from the poor ions of the electrolyte solution. Then again, metal with excessive electron affinity releases electrons, and those electrons pop out into the electrolyte answer and are brought to the advantageous ions of the solution. As a result, such a metal profits electrons, while the other loses electrons. As a result, there might be a distinction in electron concentration between those two metals.

This difference in electron concentration causes an electrical capability difference that develops among the metals. This electric potential difference or emf may be used as a voltage supply in any digital or electrical circuit. This is the overall and fundamental principle of the battery, and that is how the battery works [8].

All battery cells are based solely on this primary principle. Let's discuss one at a time. As we said earlier, Alessandro Volta evolved the first cellular battery, which is popularly referred to as the easy Volta battery. This kind of easy cell may be created very quickly. First, take a container and fill it with electrolyte diluted sulfuric acid. Next, we dip zinc and a copper rod into the solution and externally join them with an electric charge. Now your simple voltaic cellular is entire. Current will begin flowing through the external load.

10.5   CATEGORY OF CELLS OR BATTERIES

Electrochemical batteries are classified into four vast classes. A primary cell or battery cannot be easily recharged after one use and discarded after discharge. Maximum primary cells use electrolytes in absorbent fabric or a separator (i.e., loose or liquid electrolyte loose) and are consequently called dry cells.

A secondary cellular, or battery, is a cell that, after use, maybe electrically charged to its original pre-discharge nation through passing modern thru the circuit within the opposite course of cutting-edge for the duration of discharge. Figure 10.3 represents the technique of recharging.

Secondary batteries are divided into subcategories totally based on their meant use. Figure 10.4 illustrates the classification of battery.

- Cells that provide power on demand are used as energy storage devices. Such cells are generally linked to primary strength sources to be charged on demand. Examples of this kind of secondary switchgear encompass emergency fail-safe and backup energy resources, aircraft structures, and desk-bound electricity storage structures for load balancing.
- Cells that are frequently used as number one cells might be recharged after use instead of discarded. Examples of such secondary cells include, in general, transportable patron electronics and electric-powered automobiles.

The third class of batteries is usually referred to as a backup battery. What distinguishes the backup cellular from number one and secondary cells is that an essential aspect of the cell is separated from the ultimate components till just earlier than activation. The maximum regularly isolated factor is the electrolyte. This battery

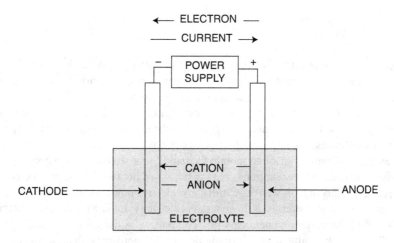

FIGURE 10.3 Recharging a cell. (Reproduced with the reference [8].)

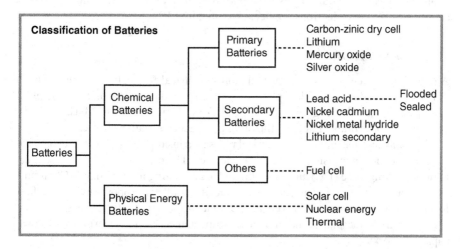

FIGURE 10.4 Classification of battery. (Reproduced with the reference [9].)

structure is usually determined in thermal batteries, where the electrolyte stays inactive within the strongnation until the melting factor of the electrolyte is reached, permitting ionic conduction, thereby activating the battery. Backup batteries efficaciously dispose of the opportunity of self-discharge and minimize chemical degradation. Most spare batteries are used best as soon as possible and then discarded. Reserve batteries are utilized in timing, temperature, and stress touchy blasting gadgets in missiles, torpedoes, and different weapon systems.

Spare cells are generally categorized into the following four categories:

- Water-powered batteries
- Electrolyte active batteries

- Gas-powered batteries
- Heat-activated batteries

The gasoline cell represents the fourth category of batteries. Gas cells are just like batteries; besides, all lively substances are no longer an integral part of the device (as in a battery). In gasoline cells, the energetic components are fed into the batteries from an outside source. A gasoline mobile differs from a battery. It can generate electrical electricity if the active materials are provided to the electrodes but stop working without such substances. An application of gas cells has been in cryogenic fuels utilized in spacecraft. However, gasoline mobile technology for terrestrial packages has been sluggish to use, despite recent trends having rekindled hobbies in systems such as grid energy, load balancing, on-web page mills, and electric-powered vehicles.

Electrochemical batteries will generate electricity by liberating the electricity saved within the battery's chemicals [9]. The battery is used for electrical applications. Rechargeable batteries generally undertake extraordinary shapes, consisting of button cells, cylindrical, prismatic, and sachet batteries. Lead-acid and Ni-Cd batteries have been used for the long term. Ni-MH and Li-ion batteries (LIBs) have been relatively new. Ni-MH and LIBs have played vital roles in determining the sizable adoption of portable electronic devices, specifically LIBs [10].

10.6 CAPACITOR

A capacitor is an electronic factor that stores electrical fees. The capacitor is made of two close conductors (typical plates) separated by a dielectric material. The plates gather an electrical charge when linked to the electricity source. Thus, one plate accumulates a favorable rate, and the other plate gets a negative charge.

Capacitance is the amount of electrical rate saved inside the capacitor at a voltage of 1 V. Capacitance is measured in farad (F) devices. The capacitor cuts the present day in direct cutting-edge (DC) and short circuits in alternating modern-day (AC) circuits.

The capacitance of a capacitor, measured in farads, is proportional to the floor location of the two plates, as well as the dielectric permittivity ε, the smaller the distance among the plates, the greater the capacitance. So first, permits test how a capacitor works.

10.7 CAPACITOR STRUCTURE

A capacitor is a product of two metal plates separated by an insulating fabric called a dielectric. Plates are conductive and are commonly made of aluminum, tantalum, or different metals. At the same time, dielectrics can be made of paper, glass, ceramics, or any insulating cloth that blocks the waft of cutting-edge.

10.8 HOW CAPACITOR WORKS

First, we will word that a metallic compound commonly has an identical quantity of undoubtedly and negatively charged particles, which means that it is electrically neutral.

Suppose we connect a power source or battery to the metal plates of the capacitor. The electrons from the plate related to the positive terminal of the battery will start to pass toward the plate connected to the negative terminal of the battery. But, due to the dielectric among the plates, the electrons will not bypass the capacitor and will start to build upon the plate.

After a wide variety of electronics have been collected at the plate, the battery will not have enough electricity to push the brand new electronics into the plate due to repulsion from the electronics already there.

At this factor, the capacitor is genuinely fully charged. Unfortunately, the primary plate advanced a terrible internet charge. The second one moved an equal net fantastic fee, growing an electric area with an attractive force keeping the capacitor's charge.

10.9   CAPACITOR DIELECTRIC RUNNING PRECEPT

A dielectric contains polar molecules; this means that they could exchange their orientation relying on the loads on the two plates. Hence, molecules align themselves with the electric discipline to draw more electrons; that is, while the capacitor is fully charged, if we put off the battery, it is going to preserve the electric rate for a long time, appearing as a strength garage.

Now, if we brief the two ends of the capacitor thru a load, a modern-day will begin flowing via the pack. Electrons accumulated from the first plate will pass to the second plate until each plate becomes electrically impartial once more. So that is the primary working principle of a capacitor and now permits a look at some software examples.

10.10   APPLICATIONS

Even though capacitor can work as a decoupling capacitor, we will see how it may work for AC to DC converter, sign filtering, and energy storage.

10.11   CAPACITORS AS ELECTRICITY STORAGE

Some other pretty obvious use of capacitors is for energy storage and supply. Even though they can shop much decrease energy compared to a battery of equal length, they have a far better lifespan. They may transmit electricity an awful lot quicker, making them greater appropriate for packages with excessive. A burst of energy is required.

Batteries and capacitors appear similar in that they each keep and launch electric strength. However, the critical difference between them is how they paint in different ways in setting-up programs. One kind variety of batteries is prominent with their chemical structure. The chemical unit, referred to as the cell, consists of three primary components; an effective terminal called a cathode, a negative terminal called the anode, and an electrolyte—the battery fees and discharges thru a chemical reaction that creates voltage. The battery can offer a constant DC voltage. Rechargeable batteries' chemical electricity transformed into energy may be reversed using outside electric electricity to restore the rate. However, batteries, in trendy, offer better strength density for storage, while capacitors have faster rate and discharge competencies.

10.12 SUPERCAPACITOR

10.12.1 FUNDAMENTALS OF SUPERCAPACITORS

The clinical community is one that specializes in power due to the changing global landscape. In this context, there are efforts to expand and refine power garage gadgets. The SC has recently obtained attention in layout and production as an energy storage tool like a battery. An SC, also called an electric-powered double-layer capacitor (EDLC) or ultra capacitor, is a tool whose capacitance values are plenty higher than ordinary capacitors. Compared to electrolytic capacitors, they can keep 10–100 times extra energy in keeping with the unit extent. Moreover, they could take delivery of and deliver fees a whole lot faster than conventional batteries, and their life cycle is an awful lot longer than rechargeable batteries. In widespread, because the call indicates, "top-notch" capacitors have excessive electricity density and power density and the capacity to keep charge with long lifestyles cycles [11]. Specific capacitance is used to reflect the belongings of the energetic material on a single electrode. It is most reliably measured by way of the three-electrode cell and can be derived from the two-electrode cell. It includes electrolyte-containing electrodes separated via a separator.

10.12.2 TYPES OF SUPERCAPACITORS

Relying on the electrode substances, SCs may be divided into three classes:

  i. Electrochemical double-layer SCs,
 ii. Pseudo-SCs, and
iii. Hybrid SCs.

The SCs can be divided into two in line with the storage mechanism which is shown in Figure 10.5.

- **Electrochemical double-layer capacitor (EDLC)**: EDLCs keep charge (electrostatically) within the Helmholtz bilayer on the interface among an electrode surface and an electrolyte. It consists of electrodes (anode, cathode), separator, and electrolyte. The strength density is 100 instances that of traditional capacitors. When the electrode is charged, the ions move closer to the floor to stabilize the system. In the case of EDLC, the particular capacitance of every electrode is identical to that of the parallel plate capacitor [12].
- **Pseudocapacitor**: Faradaic garage (electrochemical) in the dummy capacitor is performed through redox reactions, electron charge transfers, and possible ion intercalation at the electrode. Reversible redox reactions occur by using a potential difference to a faradaic capacitor, permitting charges to bypass the double layer, similar to batteries.

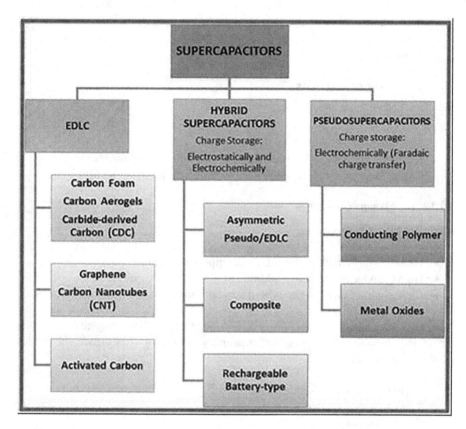

FIGURE 10.5 Overview of the types and classification of supercapacitors [13]. (Reproduced by permission of RSC.)

- **Hybrid capacitors**: In those capacitors, one electrode is often double-layered, and the other is broadly speaking pseudocapacitance, e.g., lithium-ion capacitor (LIC) due to their excessive power capacitance and long-cycle existence (more than a hundred times battery lifecycles), those have obtained a lot of interest and deliver a superb danger to construct more.

Electrochemical double-layer capacitors can be divided into activated carbon, carbon aerogel, and carbon nanotube (CNT). Similarly, pseudocapacitors can be divided into two classes: conductive polymers and steel oxides. Finally, hybrid capacitors can also be divided into asymmetric hybrid, battery-type hybrid, and composite hybrid.

10.12.3 Electrostatic Double-Layer Capacitors (EDLCs)

Materials used and Properties:

- Graphene oxide, CNT, and so forth consisting of carbon substances or derivatives.

- Better double-layer capacitance than faux capacitance.
- Charge separation of some angstroms (0.3–0.8 nm), electrode/electrolyte interface.

10.12.4 PSEUDOCAPACITORS

Materials used and Properties:

- Steel oxide or a conductive polymer.
- Excessive pseudocapacitance further to double-layer capacitance.
- Charge storage, faradaic redox reactions, intercalation, or electrosorption.

10.12.5 HYBRID CAPACITOR

Materials used and Properties:

- Uneven hybrid, battery hybrid, and composite hybrid.
- High pseudocapacitance in addition to double-layer capacitance.
- Charge garage takes place via both faradaic redox reactions and intercalation.

The performance of ultracapacitors is determined through the aggregate of electrode fabric and electrolyte. There are three primary categories of electrode substances used for ultracapacitors.

1. Carbon-primarily-based substances
2. Transition steel oxides
3. Conductive polymers

In addition, there are three types of electrolyte materials used for us:

1. Aqueous electrolytes
2. Natural electrolytes
3. Ionic beverages.

Three varieties of faradaic approaches arise at the electrodes in the dummy capacitor, particularly reversible adsorption, redox reactions of transition metal oxides, and reversible electrochemical doping-dedoping in conductive polymer-based electrodes. Blessings of SC over other storage devices are long charge/discharge cycles and extensive operating temperature range. To increase the SC electricity density, techniques had been explored with the aid of researchers recently: (i) developing excessive-capacitance electrode substances and (ii) developing electrolyte substances with massive capability windows [14].

10.13 ELECTRIC DOUBLE-LAYER CAPACITORS

The electric double-layer capacitor (EDLC) has been an excellent excessive-energy electricity supply for digital conversation devices and electric-powered automobiles.

The advantages of EDLC are its cost-effectiveness and longer lifecycle compared to trendy secondary batteries. EDLC uses a double layer fashioned on the electrode/electrolyte interface, wherein electrical costs collect on the electrode surfaces, and oppositely charged ions are organized on the electrolyte side. EDLC electrode materials must have a prominent floor place for rate deposition and a pore structure appropriate for electrolyte wetting and rapid ionic movement. Presently, activated carbons or molecular sieving carbons are used as EDLC electrode materials. Even though these conventional carbons have a large floor area, EDLC applications are pretty constrained as they include pores starting from micropores (<2 nm diameter) to macropores, and the pores are randomly connected [15]. Electrolytes do not easily wet micropores, and the exposed surface in micropores cannot be used for charge storage. Moreover, even when the electrolyte wets the micropores, the ionic movement in such tiny pores can be so slow that the high-velocity capability, which is one of the advantages of EDLCs, may not be realized [16]. If the pores are randomly connected, both charge storage and speed capacity are even more limited. Electrolytes may not wet blind or isolated pores, and irregular pore connectivity makes ionic movement difficult [17,18]. Therefore, high surface area carbon materials containing regularly interconnected mesopores (>2 nm) are pretty suited for the EDLC electrode [19].

Double-layer capacitance—electrostatic garage of electrical power obtained by using separation of charge in a Helmholtz bilayer at the interface among a conductive electrode surface and an electrolytic solution electrolyte which is configured in Figure 10.6. Fee distance separation in a bilayer is within the order of some angstroms (0.3–0.8 nm) and is static at origin [20]. Helmholtz laid the theoretical

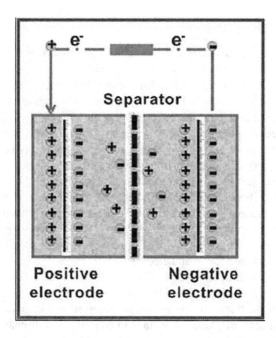

FIGURE 10.6 Configuration of electric double-layer capacitor (EDLC) [22]. (Reproduced by permission of RSC.)

foundations for the double-layer phenomenon. It is far utilized in every EC to keep electric energy. While Helmholtz first used the time period "double layer" in 1853, he predicted charge layers on the interface between multiple metals. He then compared this steel/metallic interface with the metal/aqueous solution interface in 1879 [21].

In this model, the interface consisted of a layer of electrons at the electrode surface and an unmarried layer of ions within the electrolyte. Figure 10.6 suggests the polarization phenomenon and the formation of double-layer capacitance C. More excellent specifically, business EDLCs, where electricity garage is dominant, acquired by double-layer capacitance, save energy by forming an electrical double layer of electrolyte ions on the conductor floor electrodes. Because EDLCs are not constrained using batteries' electrochemical fee switch kinetics, they could rate and discharge at a much better fee with lifetimes of more than 1 million cycles [22]. Business EDLCs are based on two symmetrical electrodes impregnated with electrolytes containing tetraethyl ammonium tetrafluoroborate salts in natural solvents. Contemporary EDLC with organic electrolyte operates at 2.7 V, attaining strength densities of around 5–8 and 7–10 Wh kg^{-1}. EDLC power density is decided using the running voltage and unique capacitance (farad/gram or farad cm^{-3}) of the electrode/electrolyte system. The specific capacitance is related to the particular surface place (specific surface area or SSA) handy with the aid of the electrolyte, the interfacial bilayer capacitance, and the electrode fabric density. The usage of high specific floor vicinity blocking off and electronically conductive electrode permits high capacitance by charging the double layer—an appealing candidate who meets this criterion.

10.14 PSEUDOCAPACITOR

The second one, elegance, represents so-called pseudo-SCs or faradaic SCs. This SC form is used much less regularly than EDLC SCs and is commercially furnished most successfully via some groups. With the use of operation ethics, they will be closer to batteries than to capacitors. Pseudocapacitance is an issue wherein electrode materials intermediate electron transfer and is going via redox reactions. The pseudocapacitance seems at the electrode surfaces, in which faradaic reactions originate. The responses are associated with the passage of energy at some stage in the double layer, just like battery charging or discharging. Figure 10.7 illustrates the schematic example of the mechanism of a primary pseudocapacitor.

But, capacitance breeds because of the specific relation that may be considered as the amount of energy established (Δq) and the exchange of potential (ΔV). So that the $d(\Delta q)/d(\Delta V)$ is equivalent to the capacitance C. All through charge and discharge, there arises a redox reaction (happening from the bonds within the compounds) and power switch among electrolyte and electrode.

Power is not stored inside the "dielectric" layer; however, it is represented with the aid of the electricity of molecule bonds. The drawback of these systems is the basis itself. Because of the truth, the electrodes are stressed and degrade faster than the electrostatic storage principle at some stage in charging and discharging. It is miles connected with growing the internal resistances of SCs. Pseudocapacitors have each electrode shaped thru pseudocapacitive materials like ruthenium oxide (RuO_2) or

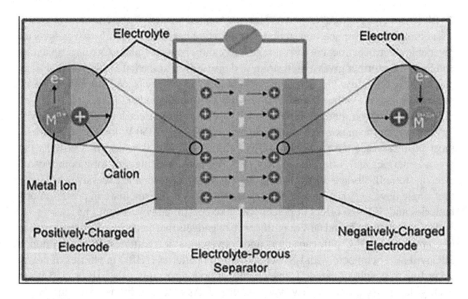

FIGURE 10.7   Schematic illustration of the mechanism of a basic pseudocapacitor [23]. (Reproduced by permission of RSC.)

manganese dioxide (MnO_2). The stableness and cyclability are decreased than in EDLC SCs, at the side of reduced charging performance and more extended time reaction (lower discharge rate).

10.15   HYBRID CAPACITOR

This text presents a short introduction to the hybrid SC for familiarity. It has released a brand new hybrid SC device on the lookout for a strength garage tool with higher-acting scientists. It is a mixture of electrochemical and double-layer excellent capacitors. It has the gain of high electricity density and excessive strength density. Its working voltage is better than the splendid present-day capacitor (<3.5 V), which makes it unique and facilitates fulfilling the requirements of high pulse power gadgets and high electric density, such as electric cars. The equivalent circuit is studied, analyzed, and simulated. Similarly, various experimental tests have been accomplished at the hybrid SC, and the results have been discussed.

  A hybridized tool idea is based totally on an aqueous asymmetric ultracapacitor. The bad electrode is an inert activated carbon, and the fantastic electrode is a floor redox-activated oxide. Such systems are designed to compete with electric double-layer (EDLC) ultracapacitors and are not intended to replace secondary batteries with a lot better strength but decrease energy. We strongly argue that this ancient view is indeed accurate. Given the overall performance benefits and limitations of hybrid-ion capacitors (HICs), their specialized niche lies in ultracapacitor-like excessive-energy high-recyclability programs, but wherein three to four times higher electricity than an ultra cap guarantees extra protection, device complexity, and reduced cycle existence.

Electrical energy storage systems play an essential position in the customer electronics, automobile, aerospace, and stationary markets. There are frequently two gadgets for reversible electrochemical power storage, secondary batteries, and ECs (ultracapacitors and SCs). The former gives a high strength density, even as the latter offers high energy and high recyclability. The dominant present-day technology is a LIB primarily based on a ceramic oxide cathode containing Li and a graphite anode [24].

For instance, commercial Panasonic LIBs offer specific electricity of up to 200 Wh kg^{-1}. However, the maximum unique electricity is beneath 350 W kg^{-1}. In assessment, most business ECs have notable energy rankings as excessive as 10 KW kg^{-1}, but precise energies are within the variety of 5 Wh kg^{-1}. A rising aim for a complicated electric strength storage device is to provide both high power and excessive electricity in a single device. A HIC is a notably new tool. This is intermediate in power among batteries and SCs and offers in precept SC-like control and cycle rankings.

HICs represent a brand new magnificence of gadgets that can bridge the performance of commercial EDLC ultracapacitors and conventional ion batteries. In our opinion, a HIC will never compete with LIBs or sodium-ion batteries (NIBs) in phrases of power according to weight or extent. Consequently, it no longer replaces a LIB or NIB when energy density or extraordinary power is prime. As noted, sodium- or lithium-based ion capacitors (HICs and LICs) can turn in four or five times better power values than EDLC ultracapacitors. LIBs can offer a high electricity density. However, their packages are regularly confined through their low strength densities. The latest technological innovation has been the development of LICs—they integrate the benefits of the high power density of EDLCs and the high energy density of LIBs using an EDLC electrode and a LIB electrode in the production of the LIC device. For a LIC cell, faradaic methods arise at the anode (LIB electrode) and usually have far more ability than the cathode (EDLC electrode) present process non-faradaic reactions. Therefore, the capacitance of a LIC device is ready twice that of an asymmetrically built EDLC. Furthermore, a carbon electrode without pre-lithiation generally has a better electrochemical potential at approximately 3 V than a Li/Li+ electrode. Consequently, carbon materials inclusive of activated carbon (AC) and graphite are regularly used as excellent electrodes (cathode). However, the electrochemical potential of the doped carbon fabric can also decrease significantly after pre-lithiation. As a result, the open-circuit voltage for a LIC-stuffed cellular will boom significantly when a pre-ground carbon electrode is used as the anode to enhance its energy garage ability [25].

SCs can also be categorized into subsequent lessons according to the rate garage phenomenon. SCs can be made in one-of-a-kind geometric shapes, including thin movie and sandwich kind as flexible SCs and micro-SCs as planar SCs. Bendy SC is an outstanding fast rechargeable electrochemical electricity storage tool that mixes the benefits of high storage potential and electricity output, in addition to excessive malleability without a sizable lack of overall performance. Figure 10.8 gives a meaningful description of the classification of capacitor along with the materials used for respective capacitors.

10.16   HYBRID CAPACITORS—LIC, NIC, KIC

A hybrid system aggregates capacitors and batteries to fill the distance between strength and power densities. The LIC is the most commonplace hybrid machine.

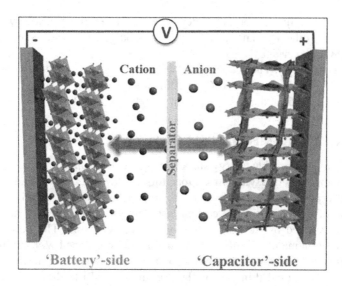

FIGURE 10.8   Mechanism of hybrid capacitors [10]. (Reproduced by permission of ACS.)

But the lithium supply is much less, so scientists are looking ahead to the sodium-ion capacitor (NIC), which is more brilliant than LICs. And the next technology hybrid machine is the potassium-ion capacitor (KIC).

The hybrid capacitor is an aggregate of EDLC and pseudocapacitors. It is shaped by combining a capacitive and battery-type cathode, which gives high electricity and strength density traits. Those capacitors consist of particular high capacitance with progressed power density in comparison to EDLCs and faradaic capacitors. Electricity garage in EDLC is completed via repeatable redox reactions between the lively electrode cloth and the lively devices present in the electrolyte solution. Those garage mechanisms are mixed collectively, forming the power storage mechanism of hybrid SCs. One-half of a hybrid SC acts as an EDLC, and the alternative 1/2 acts as a pseudocapacitor. Enormously hybrid SCs have excessive strength densities in addition to extreme strength densities [26].

Significant research has centered on new substances to grow the energy density of ECs. In reality, since each parameter is proportional to the energy density, increasing the operating voltage and capacitance is essential: (i) replacement of electrolytes with new durable electrolytes, (ii) alternative of existing materials with high capacitance pseudo substances, and (iii) evolution of hybrid mobile configurations. Lithium, sodium, and potassium capacitors fall into this class. HICs were evolved to effectively aggregate the long life cycle, excessive electricity density, and excessive electricity density.

10.17 LITHIUM-ION CAPACITOR (LIC)

High-overall performance strength garage devices are incredibly beneficial in sustainable transportation systems. LIBs and SCs are electricity storage technology for their unique roles in patron electronics and grid power storage. However, in the current country of the artwork, each device is used as hybrid electric-powered vehicles and many others. Insufficient for plenty packages. LICs are combinations of LIBs and SCs that bridge the gap between those two devices, increasing overall performance exceedingly. The primary HIC was made by Amatucci et al. It turned into a named lithium-ion capacitor (LIC) in 2001, wherein nanostructured $Li_4Ti_5O_{12}$ acts because of the anode and activated carbon (AC) because of the cathode.

10.18   RUNNING MECHANISM OF LIC

The LIC is a rechargeable energy storage gadget that belongs to the magnificence of hybrid capacitors or asymmetric capacitors. It may be categorized among LIBs and electric-powered double-layer capacitors (EDLCs). The positive electrode makes use of porous activated carbon like conventional EDLCs. The electrode becomes organized by the carbonization of the precursors. The poor electrode also uses carbon material with an extensive Li-ion pre-doped with lithium ion in its poor electrode. The electrolyte used is an organic-primarily-based carbonate aggregate. The critical thing era is the pre-doping of Li to the anode carbon to boom the electricity density. The Li foil is ready near the coupled electrodes supported by the porous cutting-edge creditors and connected to the advanced porous collector of the anode. After impregnation of the electrolyte, pre-doping proceeds by using dissolving Li inside the electrolyte and moving to the anode to decrease the ability of the anode carbon. The anode potential is diminished via the charging technique of the LIC. Since the capacitance of the anode is appreciably extra than that of the cathode, the possibility of the poor electrode can be stored low at some point of cell discharge [10]. The parent shows the simple shape of lithium ion, EDLC, and LICs.

It can be visible that the negative LIC electrode includes Li-doped carbon. The LIC equivalent capacitance is formed via the acceptable electrode capacitance C^+ in collection with the poor C^-.

Because C^- is an awful lot better than C^+, the capacitance of the LIC cellular is sort of equal to the capacitance of the activated carbon-electrolyte C^+. It should be stated that thermal runaway technology is confined because the cathode material includes: activated carbon as in conventional EDLCs. In this newsletter, prismatic LIC cells of 3,300 F have been researched via JM energy.

The primary functions of this LIC kind are summarized under:

- Nominal capacitance: 3,300 F
- Weight: 350 g
- Maximum voltage: 3.8 V
- Minimum voltage: 2.2 V

10.19   APPLICATIONS AND NECESSITIES

Over the last decade, EDLCs have been implemented in many applications wherein the highest power is wanted. But, the energy content of EDLCs is not enough, so several parallel stacks are required. In this context, LICs may be an exciting answer to the in-vehicle package where better supreme powers with higher strength content than EDLCs are still desired. Moreover, the choice of the investigated prismatic LIC has benefits in tight integration in a package. Similarly, the colorful shape gives a critical area benefit, which may be vital in cellular applications such as buses, trams, and subways [27].

10.20   SODIUM-ION CAPACITORS (NICS)

Ample sodium assets and affordable redox capacity ($Na/Na^+ = -2.7$ V) permit the creation of sodium-ion-primarily-based HICs [28]. The primary NIC was designed in 2012, where V_2O_5/CNT served as a composite anode and AC as a cathode [29]. The combined NIC with an excessive cell voltage of 2.8 V exhibited the most power density of ~40 Wh kg^{-1}.

Sodium-ion hybrid capacitors (NHCs) are promising for big-scale electrical strength storage, taking advantage of the high abundance and occasional cost of sodium resources. NHCs commonly consist of electrodes for redox reactions on the battery anode and ion sorption at the EDL cathode. However, the sluggish redox reaction kinetics of the battery-kind electrode is not matched by the fast capacitive absorption within the capacitor-kind electrode. As a result, NHCs face a monetary undertaking for each high power and strength density, specifically lacking suitable battery-kind electrodes with speedy redox reactions.

10.21   CHARGE GARAGE MECHANISM
AND EQUATIONS OF NICS

NICs can be classified into three sorts consistent with whether or not electrolyte is consumed in the electrochemical system: electrolyte-ingesting mechanism, Na-ion trade mechanism, and hybrid mechanism. It is miles in short defined as follows:

A. **Electrolyte intake mechanism**: In this system, battery materials which include Na+ de/intercalation compounds, steel oxides, and carbonaceous substances, act as anodes. In contrast, capacitor-kind substances with AC derived from biomass and metallic-natural framework, graphene, and CNTs regularly act as cathodes. During the charging method, similar to the mechanism of SCs, under the action of a voltage, cations and anions separate to move to the anode and cathode, respectively, but Na+ enters the Na+-containing compounds, or a discount reaction takes vicinity at the anode. In preference to simple bodily adsorption of Na+. At some point in the release technique, the Na ions depart the anode and go back to the electrolyte with desorption of the anions from the cathode to restore fee balance. The conventional example is the $Na_2Ti_3O_7$//AC device and so forth [30].

B. **Na-ion exchange mechanism**: According to this machine, the cathode is the battery substance that offers Na ions, and the anode is the capacitor-kind substance. For the duration of rate-discharge techniques, the electrolyte awareness stays regular and most effective, the "Rocking

Chair" plays a role inside the transfer of nations, much like NIBs in contrast to batteries, Na ions are separated from the cathode and adsorbed on the anode surface even as the NIC is charging, and vice versa. A successful design can assist $MXene//Na_2Fe_2(SO_4)_3$, $AC//NVOPF@PEDOT$ systems, etc. includes [31].

C. **Hybrid mechanism**: This type of NIC highlights that one or both electrodes incorporate each battery material and capacitor material. Throughout charging, Na ions go away from the cathode and input into the electrolyte. The AC present at the cathode absorbs unfastened anions in the electrolyte, even as all the Na ions are de-intercalated from the cathode and added to the anode via the electrolyte throughout the release system, AC releases the absorbed anions into the electrolyte to stability a number of the intervening Na ions from the anode, every other part of the Na ions, $NTP@rGO/Na_3V_2(PO_4)_3/C$ gadget, etc. [32,33].

10.22 POTASSIUM-ION CAPACITORS (KICS)

KICs also began in 2017. LIBs, portable electronics, electric automobiles (electric vehicles or EVs), and many others are energy garage gadgets. Lately, naturally ample together with NIBs, potassium-ion batteries (PIBs), calcium-ion batteries (improvement of opportunity energy garage gadgets with cation resources), and zinc-ion batteries are receiving growing attention. Among them, PIBs show promising packages because of the following advantages. (i) Potassium has a reasonably excessive energy density (-3.04 V vs. SHE (Standard Hydrogen Electrode)) due to its low general reduction potential (-2.93 V vs. SHE), which is near lithium. (ii) The fee is low because of the abundance of okay (1.5% via weight within the earth's crust), and the manufacturing value of KPF_6 is likewise plenty lower than its Li or Na analogs. (iii) K^- ion shipping kinetics in electrolytes are quicker than Li^+ and Na^+ than Li^- ions and Na^- ions attributed to the smallest Stokes K^+ radius. However, outsized K+ (1.38 Å ionic radius) regularly triggers heavy diffusion kinetics and dramatic extent growth in solids, primary to unsatisfactory fee overall performance and cycle stability. To cope with this trouble, various structural change tactics for electrode substances (canonization, designing open-body systems, the use of alloy-type substances, and electrolyte change approach) had been advanced with improved electrochemical performances. Effective for optimizing the electrochemical performances of Li-ion and Na-ion garage structures. Usually, the hybrid energy garage device consists of a capacitor-kind cathode and a battery-kind anode, thereby attaining the incorporated benefits of excessive strength density capacitors and the excessive electricity density of batteries. But, this hybrid idea has hardly ever been stated in potassium-ion power storage gadgets [34].

10.23 HYBRID MATERIALS FOR SC

The available hybrid component materials for SC are listed below. Figure 10.9 reveals the chart of materials for SC applications.

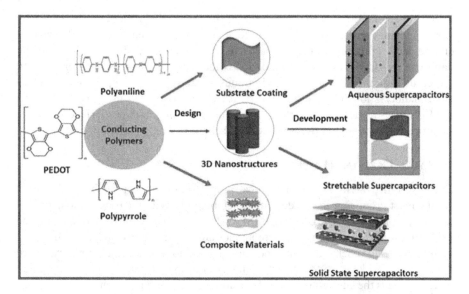

FIGURE 10.9 Types of conducting polymers [47]. (Reproduced by permission of Wiley.)

10.23.1 NANOPOROUS CARBON

Carbon is a simple and powerful cloth within the circle of relatives of energy storage materials. It suggests low value, outstanding loop balance, and a vast operating voltage window [35]. Fullerenes, CNTs, graphene, and carbon nanohorns are well-known nanocarbon substances in power storage, strength conversion, sensing, separation, purification, and the biomedical era. The porosity of the cloth will vary depending on the synthesis condition and practice method. Coconut shell, corn cob, bamboo, pitchstone, and so forth. In contrast to conventional agricultural waste, nanoporous carbon will be of different types.

1. Nanoporous carbons from self-assembling fullerene crystals
2. Nanoporous carbons from metallic-natural frames
3. Nanoporous carbons from tough and tender templates
4. Nanoporous carbons from natural biomass

It includes a pore length of 100 nm or less, with common organic and inorganic frameworks helping everyday framework operation. In recent years, NPC has received more attention due to its terrific porosity houses. Relying on the pore size, NPC is divided into three classes, notably [36].

- Large pores (>50 nm)
- Mesoporous (2–50 nm)
- Microporous (≤2 nm)

The capacitance of SCs mainly relies upon the surface region of the carbon fabric used to make the electrode because the physical adsorption of electrolyte ions primarily refers to the strength storage of the material. Capacitance is driven by various factors with solvent, electrolyte, electrolyte concentration, fee/discharge charge, and temperature taken about carbon substances. Single-walled carbon nanotube (SWCNT) turned into the most widely used carbon nanomaterials with a specific region of $1,300\,m^2g^{-1}$ and famous accurate, unique capacitance as electrode material for SC [37]. Symmetrical SCs are reviewed using carbon-based total electrodes, activated carbons, template carbons, carbide-derived carbons, exfoliated carbon fibers, overseas atoms carbon, CNTs, and nanofibers [38].

10.23.2 GRAPHENE HYBRID

The electronic, thermal, and mechanical properties of graphene hybrid substances are the most promising in this developing global [39]. With the usage of a macroporous three-dimensional (3D) sponge (SP) substrate lined with decreased graphene oxide (rGO), we deposited two different materials $Ni(OH)_2$ and $Co(OH)_2$ with the aid of an inexpensive chemical bath deposition method to produce open-pore, excessive floor place hybrid electrodes of EDLC and pseudocapacitive materials at low value.

Graphene is the emerging second generation of SCs with high conductivity, chemical balance, high ability, and excellent flexibility. The specific floor area of graphene is theoretically $2,630\,m^2g^{-1}$. Graphene synthesized via mechanical exfoliation and chemical vapor deposition (CVD) is not appropriate for the practice of SCs because of its extreme manufacturing value and occasional manufacturing efficiency. The production of the fabric using moist chemical methods, particularly graphene oxide and reduced graphene oxide (RGO), is very reasonably priced and reasonable [40]. Current traits are graphene-based totally SC electrode material with macrostructural complexity, precisely [41].

- Zero-dimensional (0D) (unfastened-standing graphene dots and particles)
- One-dimensional (1D) (fiber and thread kind systems)
- Two-dimensional (2D) (graphene and graphene-primarily-based nanocomposite films)
- Three-dimensional (3D) (graphene foam and hydrogel-based nanocomposites)

3D primarily based graphene material with macroporous shape suggests higher electrochemical overall performance than mono structure and graphene energetic nanomaterials. Recent paintings advanced a flexible and cost-effective approach to increasing recent honeycomb $CoMoO_4$ blended with 3D graphene foam. This coupled hybrid electrode material showed outcomes with excellent unique capacitance together with 2,741, 2,329, 2,098, 1,882, 1,746, 1,585, 1,488, 1,348, and 1,101 Fg^{-1} at a current density of 1.43, 2.43, 5.00, 9.28, 12.8, 15.71, 22.85, 35.71, and 85.71 A g^{-1} (aqueous solution) inside the potential variety of 0–0.9 V, respectively. Asymmetric capacitor (AC as the cathode) showed higher electricity density, and aqueous symmetric capacitor confirmed better strength density. It has been observed that the combination of 3D graphene and NHC-like $CoMoO_4$ can be used as a solid electrode for SCs [42].

10.23.3 METAL OXIDES

$LiMn_2O_4$ and manganese oxide (MnO_2)/CNT nanocomposites are fine and negative electrode materials typically utilized in metallic oxide electrodes. The asymmetric hybrid capacitor changed into capable of supplying an exceptionally high power of as much as 56 Wh kg^{-1} at a selected power of 300 W kg^{-1} depending on the total weight of the $LiMn_2O_4$ and MnO_2/CNT nanocomposite at both electrodes [43]. Thus, it appears to be an appealing electrode material for the century. Factors affecting the metallic oxide-based electrode materials overall performances are: (i) different shapes due to different preparation techniques, (ii) use of numerous experimental situations, and (iii) contribution of synergistic impact with exclusive material combos. Transition metal oxides (TMOs) MnO_2, V_2O_5, SnO_2, Fe_2O_3, Co_3O_4, etc., are more environmentally pleasant and have occasional costs.

Ruthenium oxide (RuO_2) is the maximum researched electrode cloth due to its high conductivity, incredible reversibility, and splendid pseudocapacitive behavior. Low-cost substances have been investigated for SC electrode substances in aqueous solutions. The take a look at RuO_2/AC nanocomposites with 1 M H_2SO_4 discovered that once the Ru content increases, the floor vicinity of the composite decreases uniformly because of the blockading of the AC with the aid of sizeable RuO_2 debris in its mesopores. Every other look with 40 wt% RuO_2 showed the very best strength density (17.6 Wh kg^{-1}), and 80 wt% RuO_2 confirmed inadequate ratio ability mentioned by using Kim and Popov, respectively.

MnO_2 is the material that has been appreciably researched after RuO_2 because of its low price and environmental friendliness. Studies specializing in MnO_2/CNT composites confirmed proper capacitive conduct with the highest particular capacitance of 356 F g^{-1} at 0.5 M Na_2SO_4. Furthermore, the combination of other metals with Pb, Fe, and graphene on MnO_2-based totally electrodes was observed to be with maximum conductivity and charge garage ability of the composite electrodes by using price vendors via the doping technique [44].

Another look at steel oxide discovered the simpler TMO, namely, Cu_2O and CuO with hole systems, valid for each EC and battery. CuO is commonly used in those electrode materials, and although Cu_2O has an excessive precise capacitance of 2,250 F g^{-1} [45], it is far insignificant.

10.23.4 CONDUCTING POLYMER

Polydithienothiophene, poly(3-p-fluorophenyl thiophene), and poly(3-methyl thiophene) (PMeT) as high-quality electrodes and activated carbon as a bad electrode are broadly used as conductive polymers. Composite electrodes, active materials (PMeT or activated carbon AC1 and AC3), conductive additive (graphite, SFG44, Timcal or acetylene black, AB, Hoechst), binder (carboxymethylcellulose, CMC, Aldrich), polytetrafluoroethylene (PTFE) (Du Pont), and SCs have been blended with propylene carbonate (computer, Fluka) 1 M Et_4NBF_4 as electrolyte. Electrochemically conductive polymers will give high ability. That is also a recreation-converting era in the area of hybrid SCs.

Conductive polymers can be taken as SC electrode material in various paper-work because of the doping styles of conductive polymers and one-of-a-kind types of doped conductive polymers. There are three kinds of conductive polymers which are shown in Figure 10.10, i.e., (i) precisely same p-kind doped conductive polymer, (ii) exceptional styles of p-kind conductive polymer, (iii) one conductive polymer is p-type, and the opposite is n-type doped. The capacitor has an excessive voltage distinction in the charged state due to the one-of-a-kind conductive polymer electrode cloth and the difference in capability doping variety [46].

For the SC electrode material, there are exclusive conductive polymers together with polypyrrole (PPy), poly(3,4-ethylene dioxythiophene) (PEDOT), polyaniline (PANI), and poly(3-hexylthiophene) and their corresponding derivatives. Spurious capacitive behaviors thru the redox process offer high precise/volumetric energy and power densities for the SC (Simotwo and Kalra, 2016; Zubair et al., 2016). With the aggregate of electrospinning without polymeric additives and precipitation polymer-ization without surfactant, electroactive polytriphenylamine nanofiber cathodes have been produced with an electrochemical capacitance of 200 F g^{-1} (Ni et al., 2015). An excessive cut-off voltage of 4.2 V, an energy density of 370 W h kg^{-1}, and an electricity density of 34 kW kg^{-1} have been finished in a natural electrolyte gadget when installed in a hybrid EC.

Maximum conductive polymers are hard to dissolve in common solvents to affect the spinnability of electrospinning strategies. To boom, the rotatability of conductive polymers, and easily reversible polymer precursors need to be brought as solvents [47]. Polymer composites are correct with carbon-derived substances and showcase proper electrochemical, electric, and mechanical houses with stepped-forward particular capacitances. The diameter of the carbon fibers covered with PPy nanowires

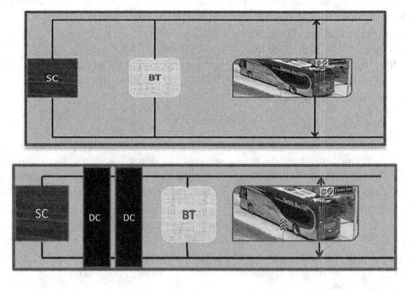

FIGURE 10.10 HESS (a): Passive configuration [59]. (Reproduced by permission of Elsevier). HESS (b) Semi-passive configuration [59]. (Reproduced by permission of Elsevier.)

changed to about 7.4 μm, while the duration of the PPy nanowires was about 0.5–1.1 μm, and their diameter became approximately 85–128 nm.

Although conductive polymer nanofiber electrodes show high capacitance and excessive electricity density compared to CNF-based electrodes, so-called capacitors have low electricity density due to the very gradual reactions between conductive polymers and electrolytes. Moreover, carrying out polymers showcase poor cyclability and conductivity in comparison to carbon-based total electrodes. Therefore, the venture stays to improve the electricity density and power density of conductive polymer nanofiber electrodes [48].

10.23.5 Metallic-Organic Framework (MOFs)

Metallic-organic frameworks (MOFs) are porous materials first described using Yaghi et al. [15]. The synergistic impact between nickel metallic-natural framework (Ni-MOF) and $Fe(CN)^{64-}/Fe(CN)^{63-}$. The precise second layered crystal structure of Ni-MOF can provide an area for Fe $(CN)^{64-}$/Fe $(CN)^{63-}$ garage and diffusion, and $Fe(CN)^{64-}/Fe(CN)^{63-}$ in a Ni-MOF electrode. Electron relay all through charge-discharge tactics through coupling Ni(II)/Ni(III). MnO/RGO, V_2O, Fe_2O_3, and ZnS are other MOFs that provide a reversible capability of over 1,000 mAh g^{-1} for 100 cycles at 50 mA g^{-1} and might withstand a current density of 3,000 mA g^{-1} for 400 cycles. Noticeable capability reduction [49].

Metal-natural frameworks (MOFs) are a new magnificence of porous crystalline substances, including a 3D community of steel ions (Ni, Co, Cu, Mn, Cr, Zr, Fe, and Zn) or their clusters associated with multifunctional natural systems. Ligands (BDC, amino-BDC, BTC, and imidazole). This material consists of excessive pore extent, huge surface regions, multiple topologies, and adjustable pore length. MOFs include the insulating nature aspect in increasing their packages, particularly within the electronics fields. Various conductive materials had been introduced with MOFs to enhance their capability for programs (e.g., rechargeable batteries, optoelectronics, and SCs). The supply of multiple steel ions and organic binders facilitated the advent of several MOFs with one-of-a-kind capabilities. Those MOFs are utilized in high-tech packages, which include drug delivery, sensors, catalysis, and storage/separation. Latest research developments have located that MOFs and their hybridized nanostructures (i.e., metal oxides and porous carbons) are attracting attention to improve electrochemical and digital devices.

10.24   ESSENTIAL OF GRAPHENE OXIDE IN SUPERCAPACITOR APPLICATIONS

SCs have received quite a little attention because of their better electricity densities, better efficiency, and more extended staying power as compared to rechargeable batteries. Industrial gadgets are primarily based on symmetrical activated carbon electrodes and organic electrolytes and feature higher voltage stability than aqueous media. Technological and financial challenges consisting of reliability, protection, lengthy life cycle, strength density, and high power are frequently met through electric automobiles and transportable electronic devices. To meet those specs for LIBs and SCs, it is necessary to understand the role of the active cloth for these gadgets. The principle of operation of those devices offers two electrodes immersed in an electrolyte and separated by using a membrane film. While the potential is created, cations or anions from the electrolyte or electrode pass thru the electrolyte to another electrode. Electrodes primarily based on the adsorption method have to provide a better floor place to promote most adsorption regions, so we observed carbon materials along with graphene, graphite, and CNTs as a lively cloth with electrodes based totally on progressed adsorption manner. In this manner, choosing the suitable electrode fabric with extra structural stability turned into a principal undertaking. All through the loading and unloading system, the addition/removal of ionic species, together with segment modifications of the host material, consequences in loss of specific potential after numerous cycles [50]. Both battery and SC structures have their high-quality and poor sides, so there may be a concept that we can both use and create a gadget with better overall performance.

Consequently, flexible SCs require electrode materials that are no longer handiest, have the suitable electrochemical homes, and have excessive mechanical integrity in case of bending or folding, compact layout, and lightweight residences. The planar configuration layout enables fast ion shipping inside the dimensional course using presenting planar channels for electrolyte ions. These varieties of SCs take the performance of flexible SCs to a higher degree. Out of all electrode substances, the role of graphene oxide is explored in this chapter.

10.25   FUNCTION OF GRAPHENE OXIDE

Graphene is the cloth used for the SC. Graphene oxide (cross) is the oxidized shape of graphene. It miles a monatomic layered cloth formed with the aid of the oxidation of cheaper and easier-to-be-had graphite. Graphene oxide is accessible to the procedure because it is far dispersible in water and other solvents. Graphene oxide is not conducive due to its lattice oxygen; however, it can be reduced to graphene via chemical techniques.

It consists

1. Unheard of electric conductivity (calculated over $15,000 \, cm^2 \, v^{-1} s^{-1}$)
2. Splendid thermal conductivity (between 1,500 and 2,500 $Wm^{-1} \, K^{-1}$)
3. Massive floor place ($2,630 \, m^2 g^{-1}$, a great deal larger than approximately $900 \, m^2 g^{-1}$ carbon black, and about $100–1,000 \, m^2 g^{-1}$ CNT)
4. Excessive mechanical energy (internal tensile power of 130 GPa and Younger's modulus of 1 Pa)
5. Low density

2D graphene is widely used in many fields, including biomedical applications (imaging, sensor, and many others) and electricity applications (gasoline mobile, secondary battery, capacitor, and so forth). While the second graphene has shown remarkable performance in lots of packages, crucial problems need to be addressed:

- Graphene sheets (GSs) may be easily re-stacked, which blocks the energetic sites of electrocatalysts and will increase mass switch resistance, central to harmful electrocatalytic properties
- Graphene sheet lacks micro to macropores to help speedy mass transfer and make bigger the surface vicinity. Therefore, a 3D graphene-primarily-based structure with high porosity is one way to resolve these problems. Micro (<2 nm) and meso (2–50 nm) sized pores can significantly expand their active surface area. In comparison, macro (>50 nm) sized pores act as open channels, which is essential in accelerating the mass transfer price. And assisting surface accessibility [51].

10.26 GRAPHENE OXIDE TECHNIQUE WITH SOLUTION-BASED TECHNIQUES

One of the fundamental advantages of graphene oxide is that it is dispersible in water. This makes it feasible to apply answer-based totally strategies. The primary approach for the production of graphene films is CVD. However, this approach calls for excessive temperatures and relatively lengthy deposition instances, making it high-priced. It also limits aggregation with substrates that can tolerate high temperatures making it hard to build upon polymers.

Answer-based totally strategies encompass spray, spin, and dip coating, in addition to Langmuir–Blodgett deposits. An outline of those strategies may be downloaded underneath.

10.27 REDUCTION OF GRAPHENE OXIDE TO GRAPHENE

Another advantage of graphene oxide is that it may be decreased to graphene by using chemical, thermal, or electrochemical methods. The material produced is called RGO. RGO is one of the most obvious answers to use while massive portions of graphene are required for business packages, including electricity storage. The discount system is essential as it has a significant impact on the first rate of the rGO produced.

Reducing cross the usage of chemical discount is very scalable; however, unluckily, the high quality of RGO produced is commonly poor. Thermally reducing move calls for temperatures of 1,000°C or better, which harm the shape of graphene platelets. However, the overall best of the RGO produced is pretty suitable. Electrochemical techniques have been shown to produce very excessive exceptional RGO, almost identical to pristine graphene. But, the approach nevertheless suffers from scalability issues. As soon as the RGO has been produced, numerous methods can be used to functionalize it. This will enhance the homes of RGO film for use in an expansion of programs.

10.28 POSITION OF GRAPHENE OXIDE IN ALL FORMS OF SUPERCAPACITOR

The critical issue of this bankruptcy is graphene oxides role in all the types of SCs for high-energy garage packages. Graphene is a unique and attractive electrode fabric because of its atom-thick-dimensional (second) shape and beautiful houses. First, graphene theoretically has a particular floor region as excessive as ~2,600 m^2g^{-1}, twice that of SWCNTs and lots better than most carbon blacks and activated carbons. This structure-property makes graphene materials exceptionally acceptable for the formation of electrochemical bilayers (electric double layers). The interfacial bilayer capacitance on one side of a graphene sheet has been measured to be about 21 μF cm^{-2}; consequently, the graphene electrodes theoretical maximum gravimetric precise bilayer capacitance became calculated to be approximately 550 F g^{-1}. 2D, graphene substances, especially chemically changed graphene (CMG), can be received on a large scale at a deficient value using graphite, graphite oxide, and derivatives as precursors. Third, graphene advanced electron mobility, electron emission throughout charge/discharge processes, improves the overall performance of ECs. Fourth, graphene sheets are particular second constructing blocks for self-meeting into 3D macroscopic materials with controlled microstructures that can manufacture EC electrodes without the want to mix conductive components or binders. Mechanical, chemical, thermal, and electrochemical stabilities. But, maximum graphene-based total ECs have low volumetric power densities due to the low viscosity of their electrodes. Moreover, the strength densities of graphene-based ECs are also restrained with the aid of much fewer mass loadings of graphene substances than activated carbons in industrial ECs. But, graphene is a promising electrode fabric for generating high-rate, excessive-power, bendy, or small ECs. When you consider that in 2008, graphene substances had been explored to provide EDLCs; Ruoff and coworkers are pioneers in this subject [52] of their early work, a CMG material became synthesized by discounting graphene oxide (cross) in an aqueous dispersion using hydrazine hydrate. Discount for the duration of chemical remedy, CMG sheets, and abnormal debris with diameters in the variety of 15–25 μm, they accrued in. The unique floor area (SSA) of this CMG agglomerate turned into determined as 705 m^2g^{-1} with the aid of an N_2 absorption Brunauer–Emmett–Teller (wager) take a look at. Powdered CMG changed into processed into the lively electrodes of the ECs with the use of PTFE as a polymeric binder. The SCs of these CMG-based EDLCs have been measured at about 135 and 99 Fg^{-1} in aqueous and organic electrolytes, respectively. Those values are tons decrease than the theoretical most calculated for single-layer graphene of 550 F g^{-1}. The distinctly small SSAs and the out-of-control microstructure of these CMG electrodes substantially decreased the SCs of CMG-based ECs.

Numerous effective strategies, self-assembly, chemical activation, and mild remedy have been developed to improve the SSA of graphene substances without degrading their electric properties. Inside the early degrees, a lot of attempts turned into positioned into developing 3D graphene architectures with open-pore structure, minimizing the re-stacking of graphene sheets and maximizing their SSA exposure to electrolytes [53]. In terms of polymer and colloid technology, CMG sheets are 2D amphiphilic polyelectrolyte. With hydrophobic conjugated basal planes and hydrophilic oxygenated agencies, the hydrophilic-hydrophobic stability among the interplanar van der Waals force and electrostatic repulsion of the CMG sheets dominates their solution behavior. It regulates their self-meeting properties in aqueous environments. In 2010, we pronounced a handy one-step technique to assemble a self-assembled decreased graphene oxide (rGO) hydrogel through a hydrothermal manner [54]. The ensuing rGO hydrogel has an interconnected 3D porous structure with pore sizes ranging from micrometers to several micrometers. And a huge SSA of $964\,m^2g^{-1}$ helps the accessibility of the graphene sheets to the electrolyte to generate EDL costs. EDLC, primarily based on this rGO hydrogel, has a yield of 22,152 $F\,g^{-1}$, which is about 50% better than EDLC with rGO agglomerate particles as electrodes tested below the same situations (100 F g^{-1} at a scanning rate of 20 mV) SC exhibited. The SC of the rGO hydrogel turned into similarly optimized to 222 F g^{-1} (at a discharge charge of 1 A g^{-1}) by again a chemical reduction of the rGO hydrogel with hydrazine. Overall performance keeps excessive SC of 165 F g^{-1} at a fast discharge fee of 100 A g^{-1}, 74% (222 F g^{-1}) of that measured at a present-day discharge density of 1 A g^{-1}. Chemical reduction of pass sheets in their aqueous dispersions can also produce rGO hydrogels, which give an alternative way to put together 3D rGO architectures. Graphene substances with huge SSAs and excessive conductivity are required to make high-performance EDLCs. Oriented microstructures in graphene-based electrodes can enhance the fee competencies of EDLCs. Graphene composites ought to provide ECs with progressed EDLC performances or the advantages of batteries.

Carbon-primarily-based "spacers" can often contribute to small amounts of SC for the corresponding composites. To further boom the power densities of graphene-based ECs, including electroactive components inclusive of conductive polymers (CPs) or inorganic compounds is necessary to provide additional redox (or pseudo-) capacitances. In truth, programs of redox materials to fabricate pseudo-capacitances had been substantially explored. However, these electrode substances have numerous negative aspects.

First, the low conductivity of neutral CPs and steel oxides or hydroxides limits fast fee transport inside the electrodes. Therefore, they frequently need to be mixed with conductive additives to improve their conductive residences.

Second, redox reactions that offer false capacitance frequently occur at the surfaces of redox materials; as a result, these materials need to be formed into nanostructures to grow their SSA. However, nanostructures are often discontinuous, which reduces the conductivity of their macroscopic electrodes and complicates the electrode fabrication approaches. Third, the extent of trade as a result of redox reactions can ruin nanostructures. Consequently, the cycling stability of p.c. primarily based on pure redox materials is often unsatisfactory. All of these problems may be addressed by mixing redox substances with graphene or its derivatives to create uniform composites with controlled micro and nanostructures. Graphene additives can offer composites with stepped-forward conductivities, robust substrates for attaching nanostructures, and strong scaffolds to sustain and buffer volume modifications. Graphene composites with CP, along with PANI, PPy, and PEDOT, had been investigated for programs in ECs. Advances in this location are summarized in a recent evaluation [55]. Among CPs, PANI has usually been studied for this cause due to its low value, sturdy electrochemical activity, and best electrochemical balance. Consequently, we will awareness our dialogue on graphene/PANI composites. Graphene/PANI composites can be produced thru electrochemical polymerization, self-assembly, and chemical polymerization [56]; as an instance, we organized a sandwich-like composite paper in which graphene sheets and PANI nanofibers (PANI-NFs) are evenly dispersed. This sandwich structure could make the maximum of the coolest conductivity of rGO sheets and substantially boom the interfacial area of each component. Also, rGO layers can restrict the extent of exchange of PANI-NF in the course of the charge/discharge process. As a result, AT-based rGO/PANI-NFs composite film exhibited higher volumetric SC, decreased inner resistance, and higher cycle balance in comparison to a pure PANI-NF film. Lately, Li's organization organized rGO/PANI composite movies by using oxidative chemical polymerization of aniline within rGO hydrogels with laterally orientated microstructures.

In this composite, the spaces between graphene sheets firstly occupied via water have been changed with the PANI to create a perfect sandwich-like structure. Composite movie containing 48% (wt.) PANI showed an excessive SC of 530 F g^{-1} at a current density of 10 A g^{-1}. With the aid of doing away with the contribution of the rGO issue, the SC of PANI on my own was calculated to be around 938–1,104 F g^{-1}. Extra importantly, while the modern-day density accelerated as much as 100 A g^{-1}, this EC retained 96% of its SC measured at 10 A g^{-1}. After repeated price/discharge for 10,000 cycles, the capacitance retention of this percent is measured at 93%, indicating a perfect cycle balance. This painting demonstrates the significance of engineering nanoarchitecture of graphene scaffolds to increase excessive-performance graphene-based composite electrodes.

Graphene composites with inorganic compounds, including steel oxides and hydroxides, have also been implemented in graphene-primarily-based percent. But, RuO_2 is luxurious, which severely limits its practical applications. Various metallic compounds, which include nanostructured $Ni(OH)_2$ and MnO_2, are promising substitutes. As an instance, it has been suggested that single-crystalline $Ni(OH)_2$ hexagonal nanoplates develop in situ on CMG sheets and shape a $Ni(OH)_2$/CMG composite [57] $Ni(OH)_2$/CMG composite 2, with excellent cyclic stability. It showed a high SC round 1,335 or 935 Fg^{-1} (based totally on $Ni(OH)_2$ or mixed mass) at a discharge modern of 8 A g^{-1}. The manipulate experiment based totally on the physical combination of $Ni(OH)_2$ and CMG exhibited a decrease SC with the aid of a component of about 1.8, suggesting the significance of direct growth of the nanostructure onto the CMG sheets to optimize the rate switch between the CMG sheets and the nanostructures. Similarly, the SC of the composite prepared by growing $Ni(OH)_2$ on cross sheets became tested to be decreased than bodily combined $Ni(OH)_2$ and CMG, and the importance of using conductive CMG changed into emphasized. In some other examine, $Ni(OH)_2$ changed into plated onto columnar graphene frames with vertically aligned CNTs (VACNTs) organized via growing VACNTs on thermally multiplied exceptionally ordered pyrolytic graphite. The graphene electrodes confirmed a high unique capacitance (1,065 Fg^{-1}) with an excellent rate capacitance and exquisite cycling potential. In this composite, 3D VACNT-graphene served as well-accomplishing pathways for electrons and mechanical scaffolds to remove cracking due to the extended exchange of metal oxides. Also, the gaps between graphene sheets facilitate ion diffusion. Despite the fact that percent based totally on Ni(OH)$_2$/CMG composites display excessive SCs, their electricity and electricity densities are strongly restricted through slim working voltage tiers (0–0.35 V). MnO$_2$ is anticipated to be a higher electrode fabric for pseudocapacitors due to its wider running voltage and power storage capacity through the mixture of surface redox reaction and electrostatic adsorption of ions. Zhu and coworkers synthesized a composition of needle-like MnO$_2$ supported through cross nanosheets (MnO$_2$/GO) in a water-isopropyl alcohol machine. It confirmed stability (84.1% at a 1,000 cycles) compared to pristine nano-MnO$_2$ (69.0% at 1,000 cycles). In every other study, supporters and coworkers deposited nanoscale MnO$_2$onto CMG sheets with the aid of a self-prescribing manner below microwave irradiation, exhibiting 2 mV s^{-1} and this charge is three times the pristine MnO$_2$ (103 F g^{-1}). It was additionally tested that the capacitance retention for natural MnO$_2$ is much less than 20%, while 74% of its capacitance nevertheless stays at 500mV s^{-1}. In addition, the capacitance of the MnO$_2$/CMG composite decreased by way of the handiest 4.6% at 15,000 scanning cycles, indicating excellent cycle stability. To further enhance the capacitive conduct of MnO$_2$/graphene, 3D macroporous CMG frameworks (3D-CMG) had been used to load MnO$_2$. 3D-CMGs had been produced with the use of polystyrene particles as sacrificial templates. A skinny layer of MnO$_2$ was then deposited onto the graphene framework to form a MnO$_2$/3D-CMG composite. A high SC of 389 F g^{-1} was obtained at 1 A g^{-1}, and 97.7% of the capacitance was retained when the present-day density turned into extended to 35 A g^{-1}. 3D-CMG provided both conductive channels for rate switch and macropores for fast ion diffusion. Graphene materials have shown the exceptional ability for programs in each EDLCs and PsCs. Those substances can be organized via a spread of techniques inclusive of CVD growth, self-meeting, electro-deposition, in situ increase, and even easy solution mixing. Graphene materials offer EC electrodes with sturdy scaffolds for loading big SSAs, conductive channels, and other additives.

Like this, they can substantially enhance the performance of ECs in terms of strength or power density, velocity functionality, and loop stability.

In recent years, a massive attempt has been put into fabricating graphene materials for programs in ECs, and significant progress has already been made. But, at the least, the subsequent difficulties or issues nonetheless remain. First, high-overall performance graphene electrodes require high-grade single-layer graphene to provide massive SSA, excessive conductivity, mechanical solid power, and pleasant electrochemical stability. Unfortunately, a cost-effective, clean, and environmentally friendly technique for the practice of massive-scale graphene has not but been evolved. Second, graphene substances with hierarchical porous systems are required to optimize the performance of ECs. As an example, the big pores of graphene hydrogels or N-doped graphene frameworks are appropriate for improving the gravimetric particular capacitances and rate capability of ECs. But, the massive pores bring about the tremendously low volumetric SC of the electrode and the extended undesirable weight of the tool due to filling the pores with excess electrolyte. To grow volumetric SCs, the amount and size of huge pores should be relatively decreased without compromising the performance of ECs. Using volumetric instead of gravimetric data is notably more fabulous and suitable for the estimation of tool performances because carbon-based electrodes best make up a small fraction of the burden of the entire tool [56]. It became used to evaluate the performance of ECs. Consequently, it is far difficult to consider one-of-a-kind graphene materials or analogs concerning unique organizations. Ruoff and coworkers have proposed a standardized method for the dependable dimension of an electrode cloth [57].

Fourth, the superb overall performance of graphene electrodes is regularly not maintained while all tool additives are protected [58]. As an example, a 2-μm-thick graphene film can also carry out excellently; but, average tool-primarily-based performances are probably now not very pleasant due to the low ratio of the electrode material to different tool additives. This problem isn't always effortlessly solved by increasing the majority loading of the electrode material because of re-stacking the graphene sheets, increasing the inner resistance, and proscribing the ion diffusion in a thick electrode. Therefore, scalability is one of the critical elements restricting the sensible software of graphene-primarily-based ECs. Huang et al. suggested a strategy to prevent re-stacking graphene sheets in the course of scale-up by transforming flat graphene sheets right into a crumpled paper ball shape. Graphene ball-primarily-based ECs exhibited improved performance after increasing the majority of charges of the electrode materials. Ultimately, other essential properties along with strength efficiency, self-discharge, and temperature working variety for graphene-primarily-based ECs have not often been studied. Transition to new realistic electricity garage structures with advanced performance.

10.29   HYBRID BATTERY/SUPERCAPACITOR ENERGY STORAGE DEVICE FOR THE ELECTRIC VEHICLES

The EV is developing quickly due to much fewer air pollutants, but it has to meet its energy needs and speedy charging. So this top-notch capacitor is mixed with a battery to enhance loop capacity and pace capability with a similar structure. The battery-SC combination is known as a hybrid energy storage system (HESS) EVs. A nicely-designed HESS with excessive power density, long-cycle existence, and excessive electricity density, and notably lighter than a stand-alone battery percent, can have an immediate impact on the general performance of EVs and, for that reason, help them penetrate the marketplace. Using HESS is maximum economically commonplace for regenerative braking energy healing in municipal shipping vehicles, inclusive of metropolis transit buses. We are suggesting one following layout in Figure 10.10 a and b.

This configuration offers a bidirectional DC/DC converter (BD-DC/DC-C) in a collection that can be positioned after the battery or SC; when placed after SC, this is known as constant DC link law; in this case, the dimensions of the SC are minimized [59]. In serial connection of BD-DC/DC-C, the first controls the battery, the second controls the SC; multi-enter 0-voltage switching BD-DC/DC-C.

This energy storage device (ESS) for electric-powered automobiles has attracted interest for its high energy storage and electricity density. But for the EV, using any of the storage systems became not practical, i.e., for the battery, the battery existence will regularly lower (because of unexpected modifications within the chemical conduct of the battery) due to expected changes for the duration of the battery discharge procedure, and for the terrific capacitor, in which the battery energy is not sufficient and less garage. It offers energy in brief periods while it has excessive potential and high electricity density. The practical solution for these negative aspects is to combine the battery with a tremendous capacitor to satisfy the power needs of EV electricity intake.

Normally, the SC is chargeable for the transient power call, and the battery is liable for the continuous power requirement. Distinctive kinds of batteries had been used for ESS, but the most promising is the LIB, that's maximum extensively utilized in transportable and stationary digital gadgets. New LIBs for EVs should be designed to satisfy high load calls for speed functionality or cyclability. The high-quality capacitor idea is mainly introduced to packages requiring high dynamic performance, voltage, and present-day transition conduct of the SC. Consequently, power storage gadget layout and manipulation are essential to make an electrical equivalent circuit for SC [60]. To manipulate the electricity stored in the notable capacitor financial institution, the voltage of the remarkable capacitor financial institution wishes to be managed. If not controlled, the SC voltage depends on the battery voltage, so there might be no opportunity to control power stored inside the SC financial institution. Exceptional topologies generally used to look at the conduct of hybrid strength garage gadget (HESS) are given, as an example (i) fundamental passive parallel; (ii) UC/battery configuration; (iii) battery/UC configuration; (iv) cascade configuration; (v) multi-converter configuration; and (vi) a couple of entering converter configuration. A few key design concerns ought to be taken into consideration within the development of battery/UC HESS topologies, e.g., (i) voltage method of two power resources; (ii) powerful use of UC saved power; and (iii) safety of battery from over current.

10.30 CONCLUSION

The arena is growing daily. Consequently, the development of strength garage materials is likewise growing very unexpectedly. Both environmental problems and marketplace calls have made renewable energy resources more popular than traditional electricity sources to meet the weight called for in real-time, reduce emissions, and protect herbal resources for the destiny. In actual time, renewable electricity sources are not a non-stop supply of electrical power. The quantity of strength generation from renewable sources relies explicitly upon the climate and weather, i.e., it could be cloudy or rainy for the solar and intermittent airflow for the wind. With the discontinuous nature of renewable strength assets, the researchers centered on electricity storage, which permits us to shop electrical energy by helping unexpected calls for modifications in balancing and preserving the grid. Gasoline cell, flywheel, batteries, top-notch capacitors, and so on. There are different garage technologies like to achieve the above properties, the electrode fabric of electrochemical energy storage devices performs a vital function. Usually, the SC is the answer for the excessive power density and temporary electricity demand, and the battery is chargeable for the extreme electricity density and non-stop power requirement. The brand new development of strength storage gadgets (ESDs) based totally on the fee garage mechanism of graphene oxide-based electrode substances has been reassembled.

REFERENCES

1. Dubal, D. P., Ayyad, O., Ruiz, V., & Gómez-Romero, P. (2015). Hybrid energy storage: The merging of battery and supercapacitor chemistries. *Chemical Society Reviews*, 44(7), 1777–1790. doi: 10.1039/c4cs00266k
2. Afif, A., Rahman, S. M., Tasfiah Azad, A., Zaini, J., Islan, M. A., & Azad, A. K. (2019). Advanced materials and technologies for hybrid supercapacitors for energy storage–A review. *Journal of Energy Storage*, 25, 100852. doi:10.1016/j.est.2019.100852
3. Zha, D., Xiong, P., & Wang, X. (2015). Strongly coupled manganese ferrite/carbon black/polyaniline hybrid for low-cost supercapacitors with high rate capability. *Electrochimica Acta*, 185, 218–228. doi:10.1016/j.electacta.2015.10.139
4. Zhou, Z., Gong, J., Guo, Y., & Mu, L. (2019). Multifunctional hybrid nanomaterials for energy storage. *Journal of Nanomaterials*, 2019, 3013594. doi:10.1155/2019/3013594
5. Peng, L., Zhu, Y., Li, H., & Yu, G. (2016). Chemically integrated inorganic-graphene two-dimensional hybrid materials for flexible energy storage devices. *Small*, 12(45), 6183–6199. doi:10.1002/smll.201602109
6. Gür, T. M. (2018). Review of electrical energy storage technologies, materials, and systems: Challenges and prospects for large-scale grid storage. *Energy & Environmental Science*. doi:10.1039/c8ee01419a
7. Sagadevan, S., Marlinda, A. R., Chowdhury, Z. Z., Wahab, Y. B. A., Hamizi, N. A., Shahid, M. M., ... & Johan, M. R. (2021). Fundamental electrochemical energy storage systems. In *Advances in Supercapacitor and Supercapattery* (pp. 27–43). Elsevier. doi:10.1016/B978-0-12-819897-1.00001-X
8. https://depts.washington.edu/matseed/batteries/MSE/classification.html
9. http://stealth316.3sg.org/2-dynabatt.htm
10. Ding, J., Hu, W., Paek, E., & Mitlin, D. (2018). Review of hybrid ion capacitors: From aqueous to lithium to sodium. *Chemical Reviews*, 118(14), 6457–6498. doi:10.1021/acs.chem rev.8b00116
11. Zhang, Y., Feng, H., Wu, X., Wang, L., Zhang, A., Xia, T., Zhang, L. (2009). Progress of electrochemical capacitor electrode materials: A review. *International Journal of Hydrogen Energy*, 34(11), 4889–4899. doi:10.1016/j.ijhydene.2009.04.005
12. Zuo, W., Li, R., Zhou, C., Li, Y., Xia, J., & Liu, J. (2017). Battery-supercapacitor hybrid devices: Recent progress and future prospects. *Advanced Science*, 4(7), 1600539. doi:10.1002/advs.201600539
13. Erdem, Emre; Najib, Sumaiyah (2019). Current progress achieved in novel materials for supercapacitor electrodes: Mini Review. *Nanoscale Advances*. doi:10.1039/C9NA00345B
14. Akinyele, D. O., & Rayudu, R. K. (2014). Review of energy storage technologies for sustainable power networks. *Sustainable Energy Technologies and Assessments*, 8, 74–91. doi:10.1016/j.seta.2014.07.004
15. Salinas-Torres, D., Sieben, J. M., Lozano-Castelló, D., Cazorla-Amorós, D., & Morallón, E. (2013). Asymmetric hybrid capacitors based on activated carbon and activated carbon fibre–PANI electrodes. *Electrochimica Acta*, 89, 326–333. doi:10.1016/j.electacta.2012.11.039
16. Sun, Y., Tang, J., Qin, F., Yuan, J., Zhang, K., Li, J., Qin, L.-C. (2017). Hybrid lithium-ion capacitors with asymmetric graphene electrodes. *Journal of Materials Chemistry A*, 5(26), 13601–13609. doi:10.1039/c7ta01113j
17. Wang, D.-G., Liang, Z., Gao, S., Qu, C., & Zou, R. (2020). Metal-organic framework-based materials for hybrid supercapacitor application. *Coordination Chemistry Reviews*, 404, 213093. doi:10.1016/j.ccr.2019.213093
18. Saraf, M., & Mobin, S. M. (2018). Metal-organic frameworks (MOFs) composited with nanomaterials for next-generation supercapacitive energy storage devices. *Journal: Handbook of Ecomaterials*, 811–831. doi:10.1007/978-3-319-48281-1_129-1

19. Campagnol, N., Romero-Vara, R., Deleu, W., Stappers, L., Binnemans, K., De Vos, D. E., & Fransaer, J. (2014). A hybrid supercapacitor based on porous carbon and the metal-organic framework MIL-100(Fe). *ChemElectroChem*, 1(7), 1182–1188. doi:10.1002/celc.201402022

20. An, C., Zhang, Y., Guo, H., & Wang, Y. (2019). Metal oxide-based supercapacitors: Progress and prospective. *Nanoscale Advances*. doi:10.1039/c9na00543a

21. Ho, M. Y., Khiew, P. S., Isa, D., Tan, T. K., Chiu, W. S., & Chia, C. H. (2014). A review of metal oxide composite electrode materials for electrochemical capacitors. *Nano*, 9(6), 1430002. doi: 10.1142/s1793292014300023

22. Cao, Z., & Wei, B. B. (2013). A perspective: Carbon nanotube macro-films for energy storage. *Energy & Environmental Science*, 2013, 6, 3183–3201. doi:10.1039/C3EE42261E

23. Miller, E. E., Hua, Y., & Tezel, F. H. (2018). Materials for energy storage: Review of electrode materials and methods of increasing capacitance for supercapacitors. *Journal of Energy Storage*, 20, 30–40. doi:10.1016/j.est.2018.08.009

24. Young, C., Kim, J., Kaneti, Y. V., & Yamauchi, Y. (2018). One-step synthetic strategy of hybrid materials from bimetallic metal–organic frameworks for supercapacitor applications. *ACS Applied Energy Materials*, 1(5), 2007–2015. doi:10.1021/acsaem.8b00103

25. Murugan, A. V., & Vijayamohanan, K. (2007). Applications of nanostructured hybrid materials for supercapacitors. In *Nanomaterials Chemistry: Recent Developments and New Directions*, pp. 219–248. doi:10.1002/9783527611362.ch7

26. Minakshi, M., Mitchell, D. R. G., Jones, R. T., Pramanik, N. C., Jean-Fulcrand, A., & Garnweitner, G. (2020). A hybrid electrochemical energy storage device using sustainable electrode materials. *ChemistrySelect*, 5(4), 1597–1606. doi:10.1002/slct.201904553

27. Zhong, X., Wu, Y., Zeng, S., & Yu, Y. (2018). Carbon and carbon hybrid materials as anodes for sodium-ion batteries. *Chemistry - An Asian Journal*, 13(10), 1248–1265. doi:10.1002/asia.201800132

28. Fleischmann, S., Tolosa, A., & Presser, V. (2018). Design of carbon/metal oxide hybrids for electrochemical energy storage. *Chemistry - A European Journal*, 24(47), 12143–12153. doi:10.1002/chem.201800772

29. Stevenson, A. J., Gromadskyi, D. G., Hu, D., Chae, J., Guan, L., Yu, L., & Chen, G. Z. (2015). Supercapatteries with hybrids of redox active polymers and nanostructured carbons. *Nanocarbons for Advanced Energy Storage*, 1, 179–210. doi:10.1002/9783527680054.ch6

30. Reddy, A. L. M., Gowda, S. R., Shaijumon, M. M., & Ajayan, P. M. (2012). Hybrid nanostructures for energy storage applications. Advanced Materials, 24(37), 5045–5064. doi:10.1002/adma.201104502

31. Arbizzani, C., Damen, L., Lazzari, M., Soavi, F., & Mastragostino, M. (2013). Lithium-ion batteries and supercapacitors for use in hybrid electric vehicles. In *Lithium Batteries: Advanced Technologies and Applications*, pp. 265–275. doi:10.1002/9781118615515.ch12

32. Veneri, O., Capasso, C., & Patalano, S. (2017). Experimental investigation into the effectiveness of a super-capacitor based hybrid energy storage system for urban commercial vehicles. *Applied Energy*. doi:10.1016/j.apenergy.2017.08.086

33. Guerrero, C. P. A., Ju, F., Li, J., Xiao, G., & Biller, S. (2015). Hybrid/electric vehicle battery manufacturing: The state-of-the-art. In *Contemporary Issues in Systems Science and Engineering*, pp. 795–815. doi:10.1002/9781119036821.ch24

34. Geetha, A., & Subramani, C. (2017). A comprehensive review on energy management strategies of hybrid energy storage system for electric vehicles. *International Journal of Energy Research*, 41(13), 1817–1834. doi:10.1002/er.3730

35. Lu, K. (2014). Energy storage and materials. In *Materials in Energy Conversion, Harvesting, and Storage*, pp. 323–386. doi:10.1002/9781118892374.ch11

36. Lu, K. (2014). Energy resources, greenhouse gases, and materials. In *Materials in Energy Conversion, Harvesting, and Storage*, pp. 1–10. doi:10.1002/9781118892374. ch1

37. Zito, R., & Ardebili, H. (2019). Fundamentals of energy. In *Energy Storage*, pp. 15–42. doi:10.1002/9781119083979.ch2

38. Sun, Y. Q., Wu, Q., & Shi, G. Q. (2011). Graphene based new energy materials. *Energy & Environmental Science*, 4, 1113–1132.

39. Stoller, M. D., Park, S., Zhu, Y., An, J., & Ruoff, R. S. (2008). Graphene based ultracapacitors. *Nano Letters*, 8, 3498–3502.

40. Chen, L., Hernandez, Y., Feng, X., & Müllen, K. (2012). From nanographene and graphene nanoribbons to graphene sheets: Chemical synthesis. *Angewandte Chemie International Edition*, 51, 7640–7654.

41. Bai, H., Li, C., & Shi, G. Q. (2011). Functional composite materials based on chemically converted graphene. *Advanced Materials*, 23, 1089–1115.

42. Stoller, M. D., & Ruoff, R. S. (2010). Best practice methods for determining an electrode material's performance for ultracapacitors. *Energy & Environmental Science*, 3, 1294–1301.

43. Zhu, Y., Murali, S., Stoller, M. D., Ganesh, K. J., Cai, W., Ferreira, P. J., Pirkle, A., Wallace, R. M., Cychosz, K. A., Thommes, M., et al. (2011). Carbon-based supercapacitors produced by activation of graphene. *Science*, 332, 1537–1541.

44. Stoller, M. D., Magnuson, C. W., Zhu, Y., Murali, S., Suk, J. W., Piner, R., & Ruoff, R. S. (2011). Interfacial capacitance of single layer graphene. *Energy & Environmental Science*, 4, 4685–4689.

45. Murali, S.; Potts, J. R.; Stoller, S.; Park, J.; Stoller, M. D.; Zhang, L. L.; Zhu, Y., & Ruoff, R. S. (2012). Preparation of activated graphene and effect of activation parameters on electrochemical capacitance. *Carbon*, 50, 3482–3485.

46. Zhang, L. L., Zhao, X., Stoller, M. D., Zhu, Y., Ji, H., Murali, S., Wu, Y., Perales, S., Clevenger, B., & Ruoff, R. S. (2012). Highly conductive and porous activated reduced graphene oxide films for high-power supercapacitors. *Nano Letters*, 12, 1806–1812.

47. Suriyakumar, S., Bhardwaj, P., Grace, A. N., & Stephan, A. M. (2021). Role of polymers in enhancing the performance of electrochemical supercapacitors: A review. *Batteries & Supercaps*. doi:10.1002/batt.202000272.

48. Xu, Y. X., Sheng, K. X., Li, C., & Shi, G. Q. (2010). Self-Assembled Graphene Hydrogel via a One-Step Hydrothermal Process. *ACS Nano*, 4, 4324–4330.

49. Zhang, L., & Shi, G. Q. (2011). Preparation of highly conductive graphene hydrogels for fabricating supercapacitors with high rate capability. *The Journal of Physical Chemistry C*, 115, 17206–17212.

50. Wang, D.-W., Li, F., Zhao, J., Ren, W., Chen, Z.-G., Tan, J., Wu, Z.-S., Gentle, I., Lu, G. Q., & Cheng, H.-M. (2009). Fabrication of graphene/polyaniline composite paper via in situ anodic electropolymerization for high-performance flexible electrode. *ACS Nano*, 3, 1745–1752.

51. Wu, Q., Xu, Y. X., Yao, Z. Y., Liu, A. R., & Shi, G. Q. (2010). Supercapacitors based on flexible graphene/polyaniline nanofiber composite films. *ACS Nano*, 4, 1963–1970.

52. Toupin, M., Brousse, T., & Belanger, D. (2004). Charge storage mechanism of MnO_2 electrode used in aqueous electrochemical capacitor. *Chemistry of Materials*, 16, 3184–3190.

53. Choi, B. G., Yang, M., Hong, W. H., Choi, J. W., & Huh, Y. S. (2012). 3D macroporous graphene frameworks for supercapacitors with high energy and power densities. *ACS Nano*, 6, 4020–4028.

54. Gogotsi, Y., & Simon, P. (2012). True performance metrics in electrochemical energy storage. *Science*, 334, 917–918.

55. Luo, J. Y., Jang, H. D., & Huang, J. X. (2013). Effect of sheet morphology on the scalability of graphene-based ultracapacitors. *ACS Nano*, 7, 1464–1471.

56. Aphale, A., Maisuria, K., Mahapatra, M. K., Santiago, A., Singh, P., & Patra, P. (2015). Hybrid electrodes by in-situ integration of graphene and carbon-nanotubes in polypyrrole for supercapacitors. *Scientific Reports*, 5(1). doi:10.1038/srep14445

57. Tie, D., Huang, S., Wang, J., Zhao, Y., Ma, J., & Zhang, J. (2018). Hybrid energy storage devices: Advanced electrode materials and matching principles. *Energy Storage Materials*. doi:10.1016/j.ensm.2018.12.018

58. Yang, D., & Ionescu, M. I. (2017). Metal oxide–carbon hybrid materials for application in supercapacitors. In *Metal Oxides in Supercapacitors*, pp. 193–218. doi:10.1016/b978-0-12-810464-4.00008-5

59. Kouchachvili, L., Yaïci, W., & Entchev, E. (2018). Hybrid battery/supercapacitor energy storage system for the electric vehicles. *Journal of Power Sources*, 374, 237–248. doi:10.1016/j.jpowsour.2017.11.040

60. Chen, J., Li, C., & Shi, G. (2013). Graphene materials for electrochemical capacitors. *The Journal of Physical Chemistry Letters*, 4(8), 1244–1253. doi:10.1021/jz400160k

11 The Science of High-Energy Graphene Oxide–Based Materials for Hybrid Energy Storage Applications

Moses Kigozi
Busitema University

Blessing N. Ezealigo
Universita Degli Studi di Cagliari via Marengo

Gabriel N. Kasozi, Emmanuel Tebandeke, and John Baptist Kirabira
Makerere University

CONTENTS

DOI: 10.1201/9781003215196-11

11.1 INTRODUCTION

11.2 BACKGROUND

The energy requirement and demand are on a high rise and are expected to double by 2050 (Tale et al., 2021). This is because the energy sector combines and unifies all the generations from industrialization to innovation. There is extended demand for energy in all sectors for effective, efficient and timely delivery, including even transportation like electric cars and trains. As the population increases, so does the energy demand, yet most depleted resources, are not sustainable like fossil fuels and are not environmentally friendly. A lot of work has been done towards energy generation and conversion from different sources like hydro, geothermal, wind, solar and nuclear, among others. The generation and conversion have created a gap due to lower cost and production time versing consumption time. The source of solar and geothermal are produced during the day at a lower cost, yet the demand (peak hours) is more at night. This creates a gap between energy conversion and energy storage for the appropriate time utilization. The generated and converted energy needs to be stored and used at an appropriate time.

This chapter will discuss the science of high energy storage using graphene oxide (GO)–based materials for application in hybrid energy storage (HES). Several attempts have been made to design systems for efficient energy storage in different devices, including batteries, supercapacitors (Scs), accumulators, fuel cells and other storage forms. Most of these devices are working at a lower efficiency due to several challenges that include materials, technology, availability, sustainability and stability, to mention but a few. The main challenge in energy storage innovation is the choice of materials to use and the cost so that the production is low and sustainable. Introducing new materials and technology is the best alternative for practical energy storage to solve peak hour demands. Improvement in materials by imparting different functions is the key to boosting the materials' properties and performance (Malik et al., 2021).

The application of different nanomaterials, specifically carbon-based nanomaterials, is the most promising innovation rapidly growing with the advancement of technology. The graphene/oxide materials have attracted massive attention from innovators and researchers because their properties and performance can be improved for an extensive potential applications. This can give high performance, low cost and eco-friendly energy conversion and storage (Tale et al., 2021).

11.3 ENERGY STORAGE

Sustainable energy storage systems such as fuels, batteries and Scs are well researched and developed towards innovative electrochemical applications. The Sc stands higher than batteries because of its rapid charge/discharge ability, higher power density and prolonged lifespan. Still, they are not commonly used because of their relatively lower energy density. The capacitive electrodes are used with battery-type electrodes to form HES for devices to incorporate high energy density and high power density. The HES devices combine both the properties of battery and Sc to provide excellent energy density and excellent power at once (Xuan et al., 2021).

According to Liu et al. 2013, there are mainly three different types of Scs categorised based on their energy storage mechanism, which include (Liu et al., 2013):

a. Electrochemical double-layer capacitor (EDLC), which stores energy through diffusion and accumulating electrostatic charge at the interface of electrolyte and electrode;

b. Pseudocapacitors, where energy storage is mainly by surface faradaic redox reactions on the surface of the electrode material; and

c. Battery-type electrodes where energy storage through diffusion-controlled faradaic electrochemical processes can enhance specific capacity and energy density. The battery-type Scs, also known as hybrid Sc (HSc), bridge the gap between Sc and batteries by combining the properties.

11.4 HYBRID ENERGY STORAGE (HES)

Hybrid Energy Storage (HES) has gained a lot of attention in energy storage due to its required multi-functionalities. There is also a need for fast charge, flexible, lightweight and high-performance energy storage due to our day-to-day demand for better energy systems. The hybrid supercapacitor (HSc) is the open centre system in HES because of its combined charge storage mechanism of surface charge storage and redox interaction that attribute to unique specific charge storage and high power capability (Liu et al., 2017). The general properties of the electrochemical behaviour of the HSc mainly depend on the electrode materials. This chapter discusses the different electrode materials synthesis used in HSc based on graphene/GO.

11.5 GRAPHENE-BASED MATERIALS

The carbon-based materials are in the advancing stages of research with the electro-chemical application. This is due to their unique properties like stability, abundance, eco-friendliness and ability to be improved. Graphene has gained a lot of attraction and attention because of its flexibility, unique structure, optical application and mechanical and electrical properties, among others. Graphene has a high specific surface area (SSA) of $2,630\,m^2g^{-1}$ (Allen et al., 2009), the exceptionally high carrier mobility of electrons of $200,000\,Vcm^2V^{-1}s^{-1}$ (Tale et al., 2021; Liu et al., 2021), and the excellent mechanical strength of 200 times more than steel and high flexibility (Pumera, 2010). Graphene has an external thermal conductivity ranging between 3,080 and $5,150\,Wm\,K^{-1}$ (Xu et al., 2013; Tale et al., 2021), optical transparency property of approximately 97% (Nair et al., 2008), with good chemical stability and electrical conductivity (Choi et al., 2012; Fang et al., 2021; Wu et al., 2021). Graphene, as a material at room temperature, is considered to have the best thermal conductivity of all other materials. The thickness of a single sheet layer of graphene is estimated to be 0.35 nm. The stability of graphene is due to its interatomic bonds present in the structure that helps to overcome the thermal strain with the only formation of crystal defects at high temperatures (Atif et al. 2016; Tale et al., 2020).

Graphene is the first 2D crystalline material synthesised in the laboratory. It has a zero-band gap semiconductor with higher conductivity than copper. It also stretches up to 120% of its length before recovering its original shape (Tale et al., 2020). To promote and improve the use of graphene materials, this chapter covers the approaches to producing highly performing graphene composites, including GO, metal oxides, non-metal composites and polymer composites. Graphene has a 2D hexagonal structure like a honeycomb containing carbon atoms in sp^2 hybrid orbitals. The groundbreaking 2D graphene material was discovered in 2004 (Wu et al., 2021), and other 2D materials like GO, MXenes and phosphorene followed. The properties of these materials give a promising start for practical application for the following, among others; photovoltaics, flexible and durable conductive electrodes, high-performance energy storage systems, micro/macro mechanics, memory devices, sensitive chemical sensors, twistoronics, photodetector and carbon-based electronic devices (Xu and Gao, 2014; Wang et al., 2013; Lozada-Hidalgo et al., 2018; Nedoliuk et al., 2019; Geim and Novoselov, 2007). Graphene is categorised into three carbon-based materials, as shown in Figure 11.1. The architectural design of graphene moves from 0D to 3D, which includes 0D (graphene quantum dots), 1D (graphene fibres), 2D (graphene films/membranes) and 3D (graphene aerogels, forms or fabrics). These graphene structures and their composite materials give high electrical transport efficiency, high mechanical strength, excellent thermal conductivity, magnetic and unique optical properties (Fang et al., 2021).

11.6 SYNTHESIS OF GRAPHENE MATERIALS

Graphene materials are synthesized by different methods, resulting in structural defects during the processing and growth stages, altering their properties from pure pristine. Various methods and innovations have been tried for high production and

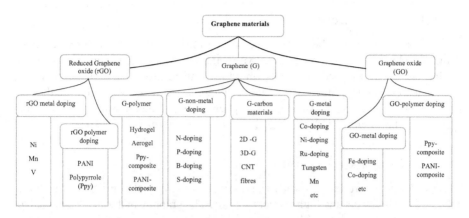

FIGURE 11.1 Group classification of graphene-based materials for energy storage application.

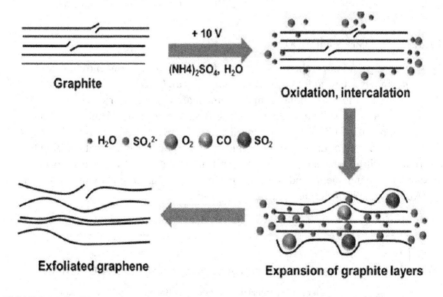

FIGURE 11.2 The conceptual diagram of exfoliation using ammonium sulphate as the electrolyte to insert SO_4^{2-} into the graphite layer (Liu et al., 2021).

low cost, giving minimal defects for different applications. The synthesis approach is either top to bottom or bottom to top approach. The stacked graphene (graphite) is dispersed into individual atomic layers of graphene sheets by breaking the van der Waals forces of attraction, as shown in Figure 11.2. The bottom-to-top approach where carbon molecules obtained from different sources are used to develop the structure of honeycomb graphene sheets. The most common methods include those highlighted in the following sections.

11.6.1 CHEMICAL EXFOLIATION

This method gives excellent graphene suspensions obtained from graphite using aqueous electrolytes like potassium hydroxide and sodium sulphate, as shown in Table 11.1. The produced materials can be used as electrode materials, conductive ink and transparent conducting oxides, to mention a few. However, the method is limited by the failure to produce large graphene sheets needed for device application (Jibrael and Mohammed, 2016; Tripathi et al., 2013; Parvez et al., 2007).

11.6.2 MECHANICAL EXFOLIATION

This is a top-to-bottom approach for preparation of high-quality graphene materials with minimal defects and high carrier mobility. The graphite material is subjected to shear stress, which peels off layers repeatedly by breaking down the van der Waals forces between the layers, producing a single individual layer. The method suffers a challenge of restacking of the separate layers, which causes the formation of different multi-layers and some carbon material formation hence reducing the purity of the materials (Shams et al. 2015; Lee et al., 2019; Choi et al. 2010; Yi and Shen, 2015).

11.6.3 HUMMER'S METHODS

This method is another chemical approach where strong concentrated acids and oxidising agents are used for exfoliation. Kigozi et al. (2020) reported using a ratio of nitric and sulphuric acids, potassium permanganate with sulphuric acid and $KMnO_7$, sodium nitrate and sulphuric acid as three different modified Hummers' methods for the synthesis of GO from graphite flakes. The materials were reported to have other performance for energy storage applications. This is a chemical reaction that is terminated with hydrogen peroxide. Several modifications have been reported with different oxidising agents and ratios to improve synthesis efficiency and sustainability (Alam et al., 2017; Krane, 2011).

TABLE 11.1

Different Electrolytes Used in Electrochemical Exfoliation of Graphene

Electrolyte	Electrode Material	Working Voltage (V)	Ref
0.1M $(NH_4)_2SO_4$	Graphite flakes	10.0	Parvez et al. (2014)
0.1M Na_2SO_4	Graphite flakes	10.0	Munuera et al. (2016)
0.1M $H_2C_2O_4$	Graphite rod	6.0–8.0	Liu et al. (2013)
0.1M H_3PO_4	Graphite rod	6.0–8.0	Liu et al. (2013)
0.5M $LiClO_4$	Graphite foil	10.0	Ambrosi and Pumera (2016)
0.5M H_2SO_4	Graphite foil	10.0	Ambrosi and Pumera (2016)
0.5M NaCl	Graphite foil	10.0	Munuera et al. (2017)
0.1M TAA cation	HOGP	−6.0 to 0	Cooper et al. (2014)

HOPG, highly oriented pyrolytic graphite; TAA, tetra-alkyl ammonium.

11.6.4 CHEMICAL VAPOUR DEPOSITION (CVD)

This is a bottom-to-top approach where pyrolysis of the precursor materials is done at a high temperature to form carbon which is then used to create the structured graphene materials. This process is carried out in an inert environment with a continuous flow of gases like nitrogen and argon (Tale et al., 2021). The method has the challenge of using high temperature and inert gas. Liang et al. (2021) reported the synthesis of graphene from CO_2 without graphitization (heating at high temperature in a controlled environment). They carried it at a lower temperature of 126°C. There are also other small-scale plants like MAPRONANO in Uganda, where GO is produced for the local and commercial market for different applications, including energy storage, water filtration and air/gas absorption. Also, GO is produced from agricultural waste biomass as a bottom-to-top approach.

11.7 CHALLENGES OF GRAPHENE AS A MATERIAL

i. The cost of graphene is very high, and the material is challenging to be produced in its pure form on a large scale.
ii. There is a challenge of obtaining a single monolayer that can easily restack together, and when more than ten layers stack together, the properties are similar to graphite.
iii. Graphene lacks off-state to be switched off entirely; hence cannot be used in transistors.
iv. Due to the hydrophobic nature of graphene, it cannot be used in water-oriented applications like water filters, moisture and humidity remote sensing.

11.8 GRAPHENE APPLICATION IN ENERGY STORAGE

Graphene is widely studied for different purposes, including energy applications. This is due to its excellent electrical, optical, conductivity and mechanical properties, as earlier explained. It can be used to improve different properties of different materials to improve product activities. The graphene materials and their composites are used in various energy storage applications as described in the following sections.

11.8.1 SUPERCAPACITOR (SC) APPLICATION

Different material forms of graphene used are discussed later in this chapter, which are applied for energy storage application for EDLCs Sc. Graphene is utilized because of its tremendously high SSA, as stated earlier in this chapter. This makes graphene an ideal material for application in electrostatic charge storage like EDLC. The EDLC devices store their charge on the interface between the electrode material surface and the electrolyte, influencing high SSA and high porosity. Compared to activated carbon (AC), graphene has higher SSA, which helps to exhibit higher specific capacitance of more than four times that of AC (Shams et al., 2015; Geim and Novoselov, 2010; Mattevi et al., 2012). When graphene hybrid materials are fabricated, they have higher performance than pure graphene due to the improvement of different properties. The addition of graphene in a material matrix creates advantages such as excellent electrical and thermal conductivity, high SSA, ultra-thickness and mechanical flexibility, among other electrochemical properties. The graphene hybrid structures can prevent particle agglomeration and volume change during the charge-discharge process in the electrochemical system.

The hybrid-structured graphene material with oxygen functional groups like GO, and reduced graphene oxide (rGO) command good interfacial bonding and electrical mobility within the materials. When the matrix includes a metal oxide, it prevents the graphene layers from restacking, which helps achieve a higher porous and continuous interconnection network within the structure. Thus, giving high capacitance and high power density in the Sc (Mattevi et al., 2012). Polymers like polyaniline (PANI) have a suitable electrical conductivity property and high pseudocapacitance in a nano-hybrid composite. The polymer hybrid structure of graphene has high capacitance performance due to the synergetic effect of the combination. The polymer hybrid structure also goes through the reduction-oxidation (redox) process for storage and release charges (Zhang and Samorì, 2017; Roy et al., 2015).

11.8.2 PHOTOVOLTAIC APPLICATION

Several materials with inorganic composites containing graphene are commonly used in fabrication of cells for free electrons and holes generated upon photon absorption. There is also a high concentration in research with polymer-graphene materials for application in photovoltaics with improved efficiency beyond 40% in semiconductors. The organic/polymer materials have gained considerable attention because of their lightweight, low cost, solution processability and high flexibility (Roy et al., 2015; Swanson, 2006).

11.8.3 DYE-SENSITIZED SOLAR CELL (DSSC) APPLICATION

Graphene hybrid structure materials are used in loading efficiency in the dye molecule by increasing the interface area, hence improving the conductivity of the electrons. When using graphene, there is caution on the ratio of TiO and graphene, which is crucial in achieving the system efficiency when the valence electron is excited from graphene to the TiO_2 conduction band to go through the graphene-TiO hybrid structure interface to enhance the separation of the holes and the electrons. Around 1% of the graphene is required for a perfect break where a higher concentration of graphene matrix may reduce transmittance. Graphene improves the light-scattering phenomenon in dye-sensitized solar cells due to its optical property hence higher efficiency than pure TiO_2 electrodes (Morais et al., 2015).

11.8.4 HYBRID SUPERCAPACITORS (HSc)/PSEUDOCAPACITORS

These are devices with which charge is not stored electrostatically but electrochemically as in conversion ionic batteries. They use hybrid-structured materials like conducting polymers, metal oxide or doped materials. These materials have very high capacitance compared to EDLC. The pseudocapacitive materials without graphene-based support have poor stability, short lifetime and they tend to be expensive in synthesis and hence need carbon back-born in the structure (Sugimoto et al., 2006). The HSc devices are made using similar hybrid electrodes or/two different hybrid or/and battery-like electrode materials. For example, they use self-assembled vanadium-graphene hydrogel electrodes (discussed later) to fabricate flexible high energy storage (HES) devices with two equal cells. The performance of the HES devices is improved by including GO materials. The HES devices act as batteries when storing charges. For the vanadium-graphene hydrogel, the storage mechanism depends on the redox reaction interaction of vanadium species. Also, the HES devices function as Scs at a high current density where the energy storage is in EDLC. The time scale of the transfer charge is attributed to the distribution of the energy between EDL and redox reactions.

11.9 GRAPHENE OXIDE (GO)/REDUCED GRAPHENE OXIDE (RGO)

GO is mainly produced by the chemical exfoliation of graphite and possesses a similar cyclohexane honeycomb network of carbon units in chair conformation like graphene. The GO hexagonal carbon network has got oxygen functional groups attached. Each 2D carbon layer of GO molecule has got an sp^2- and sp^3-bonded carbon atom. The topology of the 2D polymer of GO single molecule can easily be viewed and detected by optical microscope because of its sizeable lateral size with (10–100) µm compared to the conventional polymer. GO is superficial to form lamellar and nematic liquid crystals in water and polar organic solvents. This is due to the oxygen functional groups on the surface and their high aspect ratio (Xu and Gao, 2011; Kim et al., 2011). GO has high dispersibility and tuneable functionalization and is easily accessible compared to graphene. It can be applied either as a molecule, a particle or a soft polymer material. The ionic strength, pH and domain size are key determining parameters for colloidal behaviour in the dispersion of GO. This is used to help in the fabrication processes of GO-based materials for different applications. Both graphene and GO can be modified by atomic or molecular functionalization to improve their intrinsic properties and surface chemistry (Fang et al., 2021).

11.9.1 THE GEOMETRICAL STRUCTURE OF GRAPHENE OXIDE (GO)

The GO material is considered a 2D macromolecule that contains a graphene-line carbon panel arrangement with oxygen functionalized groups. The oxygen groups are covalently bonded to the carbon structural framework, which changes from sp^2- to sp^3-hybridized state. The oxygen functional groups are attached at different sites on the surface structure to mention but a few (Fang et al., 2021):

  i. Carbonyl, carboxyl and lactor groups at defects edges and holes
 ii. Hydroxyl and epoxide spices on the carbon planes
iii. Carboxyl and carbonyl groups formed via intense oxidation bond cleavage of C-C bond

Oxidation tunning parameters can control the functional groups of oxygen on GO sheets during preparation processes. Like the GO with nanopores in the sheet can be prepared by concentrated nitric acid solution. This gives a considerable number of nanopores created on the sheet (Wang et al., 2014).

The GO with defects and holes has intrinsic properties compared to pristine. The rich functional groups impart GO with amphiphilic behaviour and enhance reactivity, increasing solubility and liquid crystal behaviour. The functionalization of the GO is mainly tuned by hydrogen, oxygen and carbon as follows (Fang et al., 2021):

a. The absence of carbon atoms in the structure arrangement creates holes and defects.
b. Hydrogen atoms give tuneable layer distance, rich functional groups, amphiphilic behaviour and easy reactive doping.
c. The presence of oxygen imparts ton-grade production, high solubility in solvents, liquid crystal behaviours and easy reduction of graphene.

11.9.2 GRAPHENE OXIDE SYNTHESIS

GO is one of the most promising materials for carbon-based materials for energy applications. This is mainly produced by strong oxidants under acid condition methods, as shown in Figure 11.3. This is a chemical exfoliation method that consists primarily of three stages, namely;

i. Oxidation of graphite flakes
ii. Exfoliation of graphite oxide by mechanical process (sonication) to break the van der Waals forces to obtain individual sheets
iii. Separation by centrifugation to obtain pure GO

The chemical oxidation process of graphite creates oxygen functional groups on the surface, as mentioned earlier in this chapter. The thermal treatment process is included to promote the exfoliation of GO individual sheets to avoid restacking.

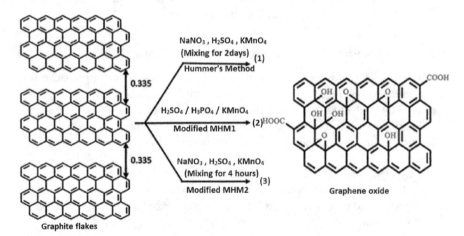

FIGURE 11.3 Graphene oxide synthesis by chemical oxidation method from graphite (Adetayo and Runsewe, 2019; Kigozi et al., 2020).

11.9.2.1   GO Hummers' Method

As highlighted earlier in the section for graphene, Hummer's method is the most commonly used approach. With Hummer's method, the GO synthesis exhibits a bright yellow colour and impacts the carbon to oxygen atomic ratio between 2.1 and 2.9 (Kigozi et al., 2020). The acidic oxidation (H_2SO_4-$KMnO_4$) helps to oxidise graphite after several hours completely. This method suffers the challenges of elimination of NO_2/N_2O_4 toxic gases and removal of Na^+ and NO_3^- ions. Benzoyl peroxide (BP) can be used as a strong oxidising agent that is effective within 10–15 minutes for GO preparation hence shorting the process (Shen et al., 2009). BP can be used as a single solvent with a direct reaction temperature of 110°C due to its low melting points.

In the conversion of graphite to GO, there is a challenge of separating GO and unoxidized graphite due to a decrease in oxidation with increased graphite flakes mass. The purity of produced GO is affected by oxidation and chemical impurities from the reactants and by-products (Chen et al., 2015).

11.9.2.2   GO Electrochemical Exfoliation Method

The electrochemical exfoliation oxidative method gives an alternative for GO synthesis. This method prevents the use of toxic chemicals, and it is a simple process. The process is environmentally friendly, low cost, less time consuming and gives controllable products. The system uses a cell with different electrolytes having exfoliation power and intercalation, as shown in Figure 11.2, which can provide a carbon to oxygen ratio of up to 7.6 (Abdelkader et al., 2014). The processing degree of oxidation to yield GO can easily be controlled by the change in the electrochemical parameters of the process. The properties of GO vary a lot with the produced size, which may be from 10 nm to several micrometres after the synthesis process. The large size of GO is highly considered for thermal, electrical and mechanical applications due to their aspect ratio. The molecular-size graphene sheets contribute a lot to forming functionalized biocompatible surfaces in drug delivery and biosensors. The main factors contributing to large-size GO sheets synthesis include oxidation conditions, starting graphite pH value treatment and centrifugation separation process (Wang et al., 2013; Pan and Aksay, 2011).

Increasing the oxidation efficiency can be done by increasing temperature or by excess oxidants that leads to the formation of epoxide and hydroxyl groups on the surface of the carbon network of GO. The formed oxygen functional groups on the GO sheets help improve the interlayer spacing and weaken the van der Waals forces between the GO sheets, hence easy cleaving. The strong excess ultra-sonication during the processing of GO sheets generates ultra-hot gas bubbles and sonochemical effects, which can cause the breaking of the C-C and C-O-C bonds, which causes cracks in the GO sheet. It is reported that the size of GO sheets decreases with an increase in sonication time; the rapid cooling, high temperature and high pressure created during sonication can influence the cleaving and exfoliation of GO sheets (Su et al., 2009).

As earlier discussed, the electrochemical exfoliation of graphene/GO is an effective method because it changes the electronic state of the material surface and its Fermi level (Fermi level is the thermodynamic work required to add one electron to the body). The electrochemical exfoliation method is simple, easy to control and effectively regulates the formation of oxygen functional group in the GO. The principle of electrochemical exfoliation is the insertion of ions into the layers of graphite materials, as shown in Figure 11.2. The insertion breaks the van der Waals forces holding/binding the layers together. The expansion weakens and damages the force between the layers, forming single or several layers of graphene/GO. The graphite can be exfoliated at either the cathode or anode with the voltage bias. The difference in the negative and positive cleaving voltage gives variation in the GO properties. The anode cleaving usually uses graphite material as the cleaving electrode with sulphate solution as an electrolyte. This is due to the lower redox potential of SO_4^{2-} as compared to other anions. The use of sulphate in exfoliation causes the liberation of SO_2 gas, which is also conducive to the GO layer's cleaving, as exhibited in Figure 11.2.

The system assembly is biased with a voltage to reduce the water at the cathode side, which creates a strong OH- nucleophile in the electrolyte, which was only at the grain boundary. The oxidation at the grain boundary and edges of graphite leads to depolarization and expansion of the GO layer. This promotes the intercalation of H_2O and SO_4^{2-} into the graphite layer. The action and interaction create a liberation of gases like SO_2, O_2, CO_2 and CO, among others, by reduction of the sulphate ions and self-oxidation of water. The intercalated gases can cause huge pressure between the layers of graphite that force the weakly bound layers to cleavage off by breaking the van der Waals binding forces (Munuera et al., 2017; Ambrosi and Pumera, 2016; Parvez et al., 2014).

There is considerable oxygen content in the anode stripping process for forming oxygen functional groups. Oxygen mainly comes from the electrolysis of water. If the electrolyte can effectively reduce the production of electrolytic water, it also causes an increase in the oxygen content in the process. The anode exfoliation can also be formed with halide-based electrolytes to produce GO with limited oxidation, resulting in low oxygen content. This is because halides have a lower redox potential, preventing the graphene lattice's oxidation. The pH of the electrolyte also controls the oxidation, where lower pH limits the degree of oxidation on the graphene. During exfoliation, some reducing agents can also be added to the electrolyte, effectively reducing the oxygen content in GO. Figure 11.4 illustrates the use of different agents to prevent oxygen functional groups from forming on the GO surfaces. Examples of sacrificial agents include sulphonated aromatic compounds: sodium benzene 1, 3-disulphnate and 1, 5-disulphanate aqueous solution (Munuera et al., 2016). The reducing agents have sodium borohydride as an antioxidant and ascorbic acid (Liu et al., 2021) and the protective agents include melamine (Chen et al., 2015). These agents are adsorbed on the surface of the graphite layer to prevent oxidation through pie-bonds.

When using cathode exfoliation, there is a limited form of oxygen functional groups compared to anode exfoliation. This is because no oxidation process occurs at the cathode, which can catalyse oxygen generation. The cathode exfoliation uses highly oriented pyrolytic graphite, graphite flakes or graphite rods as the cathode electrode and salts like alkyl ammonium salt, and molten salts as electrolytes (Abdelkader et al., 2014). The cathode exfoliation produces high-quality GO with embedded cations in between the sheets. The diameter of the cation is also a critical factor in the exfoliation of GO, where small cations are not as effective as large ones.

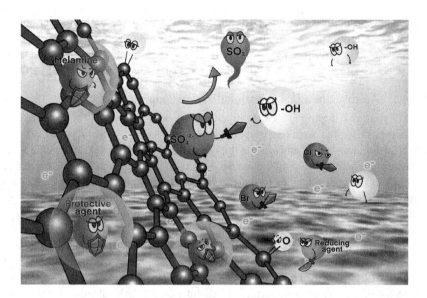

FIGURE 11.4 Schematic diagram showing reducing agent, sacrificial agent and protective agent protecting graphene sheet to reduce oxygen content (Liu et al. 2021).

TABLE 11.2
Reduced Graphene Oxide Methods

S. No.	Method	Conditions	Reference
1.	Micromechanical cleaving	It needs a highly oriented graphite	Ambrosi and Pumera (2016)
2.	Chemical oxidation and reduction exfoliation	Hydrazine metal hydride is needed	Lee and Seo (2014); Cao et al. (2017); Sheng et al. (2011); Alam et al. (2017)
3.	Electrochemical reduction	Biased potential and specific electrolyte	Nanakkal and Alexander (2017); Sharief et al. (2017)
4.	Thermal reduction	High temperature and time	Muszynski et al. (2008)
5.	Photocatalytic reduction	UV irradiation and catalyst	Ping Wang et al. (2013)

11.9.3 REDUCED GRAPHENE (rGO)

Advanced work has been covered using rGO to improve material from graphene and GO, and different techniques have been reported for rGO preparation, as listed in Table 11.2.

11.9.4 FUNCTIONALIZATION OF GRAPHENE AND GO

The functionalization of GO can be atomic or molecular and corresponds to polar surface properties. To achieve the desired designed material properties, the reactive target groups are usually utilized in the process of exfoliation—this help to reduce the contamination and cleaning. Different reactive groups are used to functionalize GO sheets, including atoms, small molecules and polymers. These species provide an efficient way to improve the performance of GO. The atoms and molecules cause the reconstruction of carbon lattice and surface chemistry modification of GO surfaces.

11.9.4.1 Atomic Chemical Functionalization/Doping

The heteroatoms doping of GO can efficiently restructure the sp^2-bonded carbon atoms, tailor the surfaces properties, induce electron density polarization of carbon-heteroatom bonds and change the chemical activity of the graphene/GO. The heteroatomic chemical doping can cause defects in the graphene plane, which help to provide additional activities to increase conductivity and chemical reaction sites. In atom doping, the atomic radicals can provide enough energy to overcome the kinetic barriers and thermodynamic to covalently bond with the basal plane of a network of carbon. The energy should not be a lot to break the C-C bond that can destroy the carbon lattice. For instance, the formation of the n-type electrical doping of material can be achieved by replacing carbon atoms with nitrogen atoms in the framework of graphene. There are three forms of N-doping mainly proposed in the framework of graphene: sp^2-hybridized graphitic N, sp^3-hybridized pyrrolic N and pyridinic N-doping. The pyridinic and pyrrolic N-doping creates active defect sites favour electrochemical and chemical processes. The graphitic N-doping induces the carbon atom replacement in the hexagonal ring structure, enhancing graphene's conductivity (Johns and Hersam, 2012).

11.9.4.2 Molecular Doping/Modification

The electronic structure of the graphene/GO is significant for developing carbon-based electronic devices. When GO directly interact with the electronic acceptors or electronic donors, there is induced modification of the electronic structure of GO. Doping GO chemically with gas molecules makes the material suitable for gas detection application due to the changes in the electrical conductivity and the behaviour of the adsorbed gases to act as donors like CO, NH_3 or acting as acceptors like NO_2, H_2O and iodine, which change the graphene local carrier concentration (Schedin et al., 2007). The availability of oxygen functional groups on the surface of the GO sheet makes it possible to assemble it on the other materials. The oxygen functional groups have two roles: dispersion of GO into the aqueous solution and modification of other bonding groups to enhance the chemical reactions like redox reactions. Several molecules are used to functionalize GO, including dopamine, quittor-thiophene molecules, amine-functionalized polyhedral oligomeric molecules and polymerized norepinephrine. For example, through a chemical reaction, GO sheets can be bonded with 3-aminopropyl-triethoxylsilane using the epoxy and carboxyl group with nucleophilic substitution and amidation, respectively (Fang et al., 2021).

11.9.5 IMPROVEMENT OF GO-BASED COMPOSITES FOR HYBRID ENERGY STORAGE

The GO-based composites are widely used in improved materials for high application. This is due to the lower performance of pure graphene than the theoretical value, including the difficulty of obtaining its pure form and limited synthesis. GO can be hybridized with other substances to form different functional groups complementary. The common GO hybrid materials include metal oxides, metal nanoparticles, polymers, non-metal composites, metal sulphides and carbon material composites. This section of the book chapter looks at the different GO/rGO material composites as described in the following sections.

11.9.5.1 G/GO-Metal Nanoparticles

When using metal catalysts, the shape and size of the particles play a vital role in the material's chemical and physical properties. Also, the large SSA and smaller particle volume are crucial in achieving the desired improvement because they control the interaction. To maximize the effective SSA of the nano-electrically active catalyst in terms of quality and electron transfer, the plane of a produced catalyst has to be uniformly immobilized on the excellent interaction and conductivity. GO is a suitable carrier for dispersing and fixing applied metal particles with a very high SSA (theoretical value of $2,630 \, m^2 g^{-1}$). Excellent conductivity can effectively distribute and support meta-nanoparticles in the structure. The electrode position method to prepare graphene-supported platinum nanoparticles can effectively prevent GO and Pt nanoparticles (Toh et al., 2018). Depositing metal nanoparticles on the GO with a large SSA facilitates the dispersion of the metal nanoparticle, improving catalytic activity and utilization rate. The typical single metal, bimetal and poly-metals used for doping of GO include Au, Ag, Pt, Cu, Pd, Pt/Co, Fe/Ni, Fe/Ni/Co, Fe/Cu/Pt and Bi, among others, as shown in Figure 11.5 (Toh et al., 2018; Wang et al., 2019; Wodarz et al., 2018).

There are different composites synthesized for the use in energy storage which includes:

   i. **Graphene-bismuth (G-Bi):** this composite is synthesized by Hummer's method. When GO is integrated with bismuth, the composite exhibits the specific capacitance three times. This means that Bi can enhance the capability of GO for electrode use in an electrochemical system at higher rates.
  ii. **GO-ruthenium oxide (GO-RuO):** using RuO highly improves the electrochemical performance of the material. RuO, in its hydrous structure, can render enhanced electrochemical performance. RuO consolidates reversible redox responses and conductivity of metallic materials, which occur at the terminal electrolyte and the elemental mass material. The material can be used in the HSc because of its complementary operating potential in the identical electrolyte. The layered nanostructure of the GO with RuO has better electrical conductivity and spaces for the electrochemical reaction, increasing the pseudocapacitance (Chen et al., 2015; Chodankar et al., 2019).
 iii. **GO-layered double hydroxide (GO-LDH):** the LDH materials are highly used in HScs because they provide multiple redox chemical reactions. For example, the GO-Fe LDH nanoflakes structure (Figure 11.5) undergo multiple redox reactions of Fe^{2+}/Fe^{3+} and Co^{2+}/Co^{3+} when using a scan rate of $5 \, mV \, s^{-1}$. The redox reaction is fast at a high and stable cycling process and

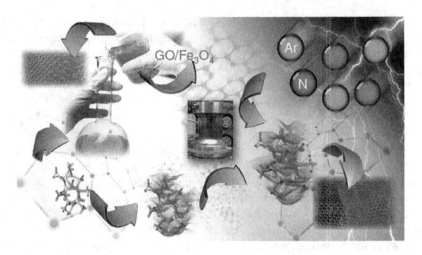

FIGURE 11.5 Schematic illustration of the synthetic procedures for NG/Fe$_3$O$_4$ (Zhang et al., 2016).

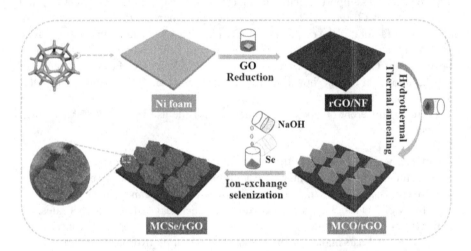

FIGURE 11.6 Schematic illustration of the synthesis procedure of GO-metal nanoparticle, with an example of hierarchical MCSe/rGO composites (Xuan et al., 2021).

more immediate with the metal ions interaction diffusion process than conventional batteries. When the GO-HDH is used in an HSc, it enhances fast charge with high retention and excellent cycle stability (Huang et al., 2020; Ji et al., 2019).

iv. **Mesoporous GO-V$_2$O$_5$:** when using a composite of materials, there is improvement in the cycle stability, and high-efficiency retention after 1,000s of cycles contributed by individual materials to form a combination. Different transition metals and their oxide are used in combination with GO (Figure 11.6). For example, using MnO$_2$/GO in the ratio of 3:1 is reported to have the highest specific capacitance (Miniach et al., 2017). The performance comes from the rust scaffold loading and conductivity of MnO_2. MnO_2 has a theoretical capacitance value of 1,386 F g^{-1}, and $Ni(OH)_2$ has 2,082 F g^{-1}. When these materials are put in a composite with GO, they convey high energy density and high power density. The performance is derived from the material because of 3D GO conductivity which leads to fast charge transportation and high electrolyte dissemination (Ede et al., 2017). The composite of $MnO_2/Ni(OH)_2/GO$ obtained by transition metal oxides and hydroxides is highly recommended for positive electrodes in the HSc or batteries.

v. **GO-metal sulphide:** The use of metal sulphides is a typical class of materials used to improve GO. The metal sulphides improve electron migration efficiency and higher electron capacity. This is mainly attributed to the emptier orbitals of S atoms in the metal sulphide. Several works have reported the use of different metal sulphides loaded onto the surface of GO nanosheets which include: MnS (Naveenkumar and Paruthimal Kalaignan, 2018), MoS_2 (Quan et al., 2018), Ni-Co-S (Wang et al., 2015), CdS (Tang et al., 2014), CdTeS (Liu et al., 2014), CoS (Pu et al., 2014) and $CuInS_2$ (Liu et al., 2013). The GO-metal sulphide can be prepared by use of the electro-deposition method. For example, CoS was deposited on 3D GO for the hybrid network that creates an electroactive area of the electrode material. The GO-metal sulphide composite can hold electrons, increase the mobility of electrons and improve the metal application's capacitance, efficiency and charge/discharge.

vi. **GO-metal oxide/hydroxide:** the metal oxides are widely applied in electrocatalysis, energy storage, energy conversion and optics. The nanometal oxides/hydroxides have strong aggregation characteristics and can effectively prevent disordered stacking and grouping of nanohybrid materials. The GO-metal oxide/hydroxides have a high specific surface with GO, which helps the material disperse to enhance activity. They also improve on maintaining redox reactions on the surface and enhance stable intercalation through electrochemical performance. Examples of metal oxides: (i) tungsten oxide (WO_3) improves the performance due to the change of oxidation state of W and WO_3 can prevent stacking and agglomeration of the GO sheet to create a more effective path for redox (Jo et al., 2011; Kim et al., 2015; Saha et al., 2014). (ii) Lanthanum oxide (LaO_3) is flexible and pliable, which quickly oxidizes in air and responds to water to form hydroxide with negligible internal resistance. From the electrochemical impedance spectroscopy, the metal oxide-GO composite exhibits capacitive behaviour of electron material at a lower frequency with high conductivity, high ion diffusion rate and good adsorption: the GO-La_2O_3 composite exhibits higher capacitance than the individual GO and La_2O_3. The amount of La_2O_3 loaded on the GO surface helps reduce restacking and defects of the GO surfaces, which lowers the internal resistance (Lin et al., 2015; Li et al., 2014).

11.9.5.2 GO-Non-Metal Doping

Due to different factors, the GO materials have low performance than the theoretical values. The restacking creates van der Waals' forces that tend to decrease the explicit surface territory and expand particle transport protection. When GO is doped with

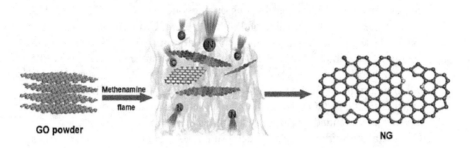

FIGURE 11.7 A schematic illustration of the NG synthesis method in HMTA flame (Liu et al., 2016).

nitrogen (N), as shown in Figure 11.7, it improves material wettability, increases electrical conductivity by achieving a lower thickness of the free charge transport and opens up the dynamic surface zone to electrolytic ion movement. The N-doping decreases the probability of restacking GO sheets, making a 3D permeable design that creates a pathway to charge transport and high porosity for productive particle adsorption/desorption (Xia et al., 2017; Xu et al., 2017; Kota et al., 2016). Non-metal doping is categorised as follows:

i. **Heteroatomic doping:** the GO network can also effectively improve the ionic conductivity of the material on the interface with the cathode/anode/electrolyte interaction. This capacitive performance leads to electron transfer reactions. Different studies have designed novel procedures to prepare 3D nitrogen-doped graphene aerogel nanomesh (GANM) materials from GO and NH_4OH, as shown in Table 11.3. For example, the $Fe(NO_3)_3.9H_2$-O-N-GANM material composite consists of a few GO layers and iron oxide that exist in the form of nanopores on the graphene sheets (Figure 11.5). This can stop the ion mobility path during the charge-discharge process. The N-GANM material can be used as an electrode that can improve the efficiency of electrochemical stability with a retention of up to 90% capacitance. This material contribution has better electrical conductivity and low resistance in electrodes than aerogel due to fast charge transfer and the material exhibits excellent performance as an HSc. The N-doping promotes the surface and electrical conductivity of GO hence promoting pseudocapacitance (Wei et al., 2016; Xu et al., 2015).

ii. **Carbon dots (CDs):** the CD has unique biocompatibility, tuneable and non-toxic properties. When CD is combined with GO, there is an improvement in cycle life and specific capacitance. The functional groups like carboxyl, hydroxyl and amide are on the surface of CD, which promotes the solubilization in aqueous electrolytes. The material is prepared by the hydrothermal synthesis method of GO and CD material. The CD materials with different functional groups on the surface prevent the restacking of GO nanosheets, which increase the SSA and pore volume that promote the mobility of the electrolytic ions. A higher ratio of CD to GO creates an

TABLE 11.3

Graphene/GO Composite Materials Performance for Hybrid Energy Storage

S. No.	Material/Composite	Synthesis Method	Electrolyte Used	Capacitance (F/g)	Retention	Ref.
A			Graphene/GO-Metal and Metal Oxide Composite			
1	rGO-V_2O_5	Hummers' method and salvo-thermal	1.0 M Na_2SO_4	466 at 1 A g^{-1}	83.3	Pandey et al. (2016)
2	Co-Fe LDH/Ppy	Modified Hummers' method	2 M KOH	728 at 1 A g^{-1}	83.51	Huang et al. (2012, 2020); Kwak et al. (2017)
3	Graphene V_2O_5 nano	Hydrothermal synthesis	1 M $LiClO_4$/PC	384 at 0.1 A g^{-1}	82.2	Liu et al. (2018)
4	Graphene/Fe_2O_3/MnS_2	Solvo-thermal method	3 M KOH	161 at 1 A g^{-1}	74.0	Meng et al. (2019)
5	Graphene/$FeCo_2O_4$ nano	Hummers' method	1 M Na_2SO_4	1,710 at 1 A g^{-1}	96.02	Chodankar et al. (2019)
6	Vanadium-graphene hydrogel	Hydrothermal synthesis	–	225 mAh at 5 mV s^{-1}	50	Malik et al. (2021)
7	N-graphene $CuCr_2O_4$	Sonication-assisted mechanical method	1 M H_2SO_4	530.6 at 0.5 A g^{-1}	83.7	Sarkar et al. (2020)
8	GO/ZnCo	Facile aqueous method	–	711.6 at 1 A g^{-1}	63.6	Yu et al. (2020)
9	$NaFe_2O_3$-GO	Hydrothermal synthesis	0.1 M $LiPF_4$	720 mAh at 50 mA	90	Kigozi et al. (2021)
10	Graphene WO_3 nanowire	Salvo-thermal method	0.1 M H_2SO_4	465 at 1 A g^{-1}	97.7	Nayak et al. 2017)
11	Graphene NiO	CVD	1 M $LiClO_4$	816 at 5 mV s^{-1}	100	Cao et al. (2011)
12	GO/Ni(OH)$_2$ nanoplates	–	–	1,335	100	Wang et al. (2010)
13	GO/Ni(OH)$_2$ nanoflakes	Facile chemical ppt method	6 M KOH	2,194	95.7	Yan et al. (2012)

(Continued)

TABLE 11.3 (Continued)

Graphene/GO Composite Materials Performance for Hybrid Energy Storage

S. No.	Material/Composite	Synthesis Method	Electrolyte Used	Capacitance (F/g)	Retention	Ref.
14	rGO/CNT/α-Ni(OH)$_2$	One-pot hydrothermal synthesis	2 M KOH	1,320 at 6 A g^{-1}	92.2	Yuan et al. (2011)
15	Graphene Bi$_2$O$_3$	Modified Hummers' method	6 M KOH	757 at 10 A g^{-1}	65	Wang et al. (2010)
16	GO/V$_2$O$_5$/CNT	—	1 M Na$_2$SO$_4$	1,848	66	Mtz-Enriquez et al. (2020)
17	Ni-Co LDH/rGO	Hydrothermal and salvo-thermal synthesis	3 M KOH	2,130 at 2 A g^{-1}	86.7	Le et al. (2019)
18	Graphene-WO$_3$	Salvo-thermal method	1 M H$_2$SO$_4$	465 at 1 A g^{-1}	97.7	Nayak et al. (2017)
19	Graphene-Ni(OH)$_2$	Salvo-thermal method	2 M KOH	1,632 at 1 A g^{-1}	95.2	Mi et al. (2012)
20	GO-MnCo$_2$O$_4$	Hummers' and hydrothermal methods	3 M KOH	503 at 1 A g^{-1}	73.4	Wang et al. (2019)
21	N/S-Co graphene hydrogel	Hummers' method	PVA/KOH	1,063 at 1 A g^{-1}	76	Zhang et al. (2018)
22	MnSe$_2$/CoSe/rGO	Hydrothermal approach	2 M KOH	1,138 C g^{-1} at 1 A g^{-1}	98.3	Xuan et al. (2021)
23	CoSe-rGO	Microwave-assisted	6 M KOH	761 at 1 A g^{-1}	90	Miao et al. (2021)
24	Ni/Al LDH-rGO	Electrochemical synthesis	6 M KOH	880 at 1 A g^{-1}	80	Musella et al. (2021)
B	**Graphene/GO Non-metal Composite**					
25	B-doped graphene aerogel	Hummers' method	—	308 at 1 A g^{-1}	92	Subramanian et al. (2005); Morais et al. (2015); Jow and Zheng (2011)

(Continued)

TABLE 11.3 (Continued)

Graphene/GO Composite Materials Performance for Hybrid Energy Storage

S. No.	Material/Composite	Synthesis Method	Electrolyte Used	Capacitance (F/g)	Retention	Ref.
26	N-doped porous graphene	Salvo-thermal and super-doping method	2 M KOH	390 at 5 mV s^{-1}	96	Dai et al. (2018)
27	N-doped 3D graphene	Hydrothermal and Hummers' method	6 M KOH	297	93.5	Zhang et al. (2017)
28	N/S-doped GO	Hummers' method	EMITFB	203.2 at 1 A g^{-1}	90	Chen et al. (2018)
29	N-doped graphene	Hydrothermal method	–	220 at 0.5 A g^{-1}	98.3	Dong et al. (2018)
30	N-doped rGO	Hummers' method	6 M KOH	244	92	Śliwak et al. (2017)
31	N-doped graphene hydrogel	Hummers' and hydrothermal method	6 M KOH	387.2 at 1 A g^{-1}	90	Liao et al. (2016)
32	Heteroatomic (N & S) doped graphene	Hummers' method	6 M KOH	227.9 at 10 A g^{-1}	83	Cheng et al. (2018)
33	3D N&B-doped graphene hydrogel	Hydrothermal synthesis	1 M H$_2$SO$_4$	239 at 1 mV s^{-1}	100	Niu et al. (2012)
C			Graphene/GO-Polymer Composites			
34	Graphene/PANI/CNT	In-situ deposition	1 M KCl	376.8 at 1 A g^{-1}	82	Cheng et al. (2013)
35	Aligning PANI/N-doped graphene	Modified Hummers method	1 M H$_2$SO$_4$	620 at 0.5 A g^{-1}	87.4	Ge et al. (2020)
35	Graphene indanthrone donor-π- acceptor	–	1 M H$_2$SO$_4$	535.5	87	Pan et al. (2020)
36	PANI-graphene paper	Polymerization of electro-vacuum filtration	1 M H$_2$SO$_4$	489 at 0.4 A g^{-1}	96	Yan et al. (2010)

(Continued)

TABLE 11.3 (Continued)
Graphene/GO Composite Materials Performance for Hybrid Energy Storage

S. No.	Material/Composite	Synthesis Method	Electrolyte Used	Capacitance (F/g)	Retention	Ref.
37	PANI-GNR-CNT	In-situ polymerization	2 M KOH	890	89	Mingkai Liu et al. (2013)
38	GH films	Hydrothermal method	H_2SO_4 PVA	372 mF cm^{-2}	91.6	Xu et al. (2013)
39	Film layer graphene	Filtration incorporated vacuum	$EMIMPF_6$	273 at 0.1 A g^{-1}	97	Yang et al. (2011)
40	Graphene-CNT-MnO$_2$	Hummers' and hydrothermal method	1.5 M Li$_2$SO$_4$	330.7	90	Xiong et al. (2016)
41	3D N-doped GA nanomesh	Hummers' method	1 M Na$_2$SO$_4$	290.03 at 1 A g^{-1}	98	Su et al. (2017)
42	rGO-PANI nano	Hydrothermal and salvo-thermal method	1 M H$_2$SO$_4$	1,084 at 1 A g^{-1}	86	Shabani-Nooshabadi and Zahedi (2017)
43	3D graphene metal organic framework	Hummers' and hydrothermal method	1 M Na$_2$SO$_4$	450 at 1 A g^{-1}	92	Miniach et al. (2017)
44	GO-poly-thiophene	Hummers' method	2 M KOH	296 at 0.3 A g^{-1}	91	Alabadi et al. (2016)
45	N/S co-doped 3D GH	Hummers' method	PVA/KOH	1,063 at 1 A g^{-1}	76	Zhang et al. (2018)
46	Graphene-PANI	Hummers' method	1 M H$_2$SO$_4$	912 at 1 A g^{-1}	86.5	Zheng et al. (2018)
47	Graphene-PANI layers and fibres	Interfacial polymerization	—	578 at 1 A g^{-1}	93	Li et al. (2019)
48	3D porous graphene	Hummers' method	6 M KOH	550 at 0.5 A g^{-1}	98	Xiong et al. (2019)
49	N-doped porous graphene	Solvothermal and super-doping method	2 M KOH	390 at 3.5 A g^{-1}	96	Dai et al. (2018)

(Continued)

TABLE 11.3 (*Continued*)
Graphene/GO Composite Materials Performance for Hybrid Energy Storage

S. No.	Material/Composite	Synthesis Method	Electrolyte Used	Capacitance (F/g)	Retention	Ref.
50	PANI-VG Ti multilayer	VG-DC arc plasma jet CVD	0.5M H_2SO_4	535.7 at 40 A g^{-1}	86	Shen et al. (2018)
51	AC with rGO	Chemical activation, pyrolyzed process, modified method	6M KOH	512 at 0.5 A g^{-1}	81	Morais et al. (2015); Huang et al. (2012); Wang et al. (2015)
52	Functionalized rGO aerogel	Hummers' method	1M Na_2SO_4	550 at 1 A g^{-1}	89.06	Bora et al. (2018)
53	Cu-PBA-GO	SILAR method	1M Na_2SO_4	611.6 at 0.5 A g^{-1}	86	Goda et al. (2021)
54	rGO/CoNi$_2$S$_4$/Ni-Co LDH	Hydrothermal process	2M KOH	1,310C g^{-1} at 1 A g^{-1}	77	Wang et al. (2010); Chang et al. (2021)
55	NiCo$_2$S$_2$ N/S-doped rGO	Modified Hummers' and hydrothermal	2M KOH	184.2 mAh g^{-1} at 1 A g^{-1}	64.9	Lai et al. (2012)

EMITFB, 1 ethyl-3-methyl imidazolium tetrafluoroborate; GA, graphene aerogel; GH, graphene hydrogel; GNR, graphene nanoribbon; PVA, poly(vinyl alcohol); VG, vertical graphene.

enriched amount of O and H and a limitation of carbon content (Xia et al., 2017; Hou et al., 2014; Kwak et al., 1996).

iii. **Doping with S, N, B or O:** using atoms to dope in GO improves electron acceptance and withdrawal. This is often applied to enhance the capabilities of GO-based HScs. Nitrogen is used in GO doping to amplify the number of charge carriers by increasing its P electrons to the π-system of GO. The pyrrolic/pyridic (N5), pyridinic (N6), and pyridine oxide (Nx) groups can be used to enhance the total energy density because they take part in the reversible redox reactions. Also, quaternary nitrogen (NQ) promotes the electrochemical performance of the material by improving its conductivity. The N-doped GO can be prepared by hydrothermal synthesis using hexamethylene tetramine because it's capable of producing nitrogen, as shown in Figure 11.8. The optimum temperature is around 180°C to ensure the N-reduced GO with the highest nitrogen content, leading to a very high capacitance value (Rao et al., 2014; Zhao et al. 2012; Śliwak et al., 2017). Boron-doped GO can be produced by Hummers' and hydrothermal methods (examples in Table 11.3). The material composite can improve retention by up to 96%. Although it is challenging to dope with boron on GO materials, the studies are still a complex energy storage mechanism. The boron particles in the GO cross-section are not arranged in their structure (Rao et al., 2014).

11.9.5.3 Go-Polymer Doping

The conductive polymer materials are used in energy storage to improve conductivity, capacitance and lower identical arrangement positions. The conducting polymers have redox stockpiling ability and give huge surface regions, perfect for application in HScs as electrode materials. The property of GO-polymer composite depends on the quality of GO as a filler in the polymer matrix, the interaction between the filler and matrix, the ratio of fill in the matrix, and the filler's dispersibility. The synthesis method of GO-polymer composite mainly includes solution mixing, melt blending, in-situ polymerization and electrochemical deposition, as shown in Figure 11.9 (Li and Xu 2015). The commonly used synergetic conductive polymer with GO sheets include PANI, polypyrrole (Ppy), poly 3,4-ethylene dioxythiophene and poly

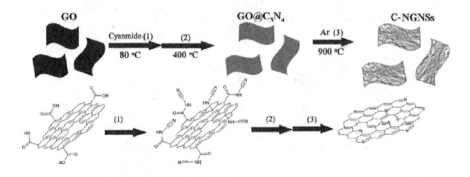

FIGURE 11.8 Schematic illustration for fabricating crumpled nitrogen-doped graphene (Wen et al., 2012).

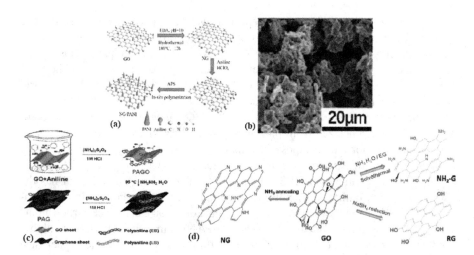

FIGURE 11.9 (a) Illustration for the synthetic mechanism of NG/PANI, (b) TEM morphology structure of NG/PANI nanocomposite (Ge et al., 2020), (c) the method to prepare polyaniline/GO nanocomposites and (d) structural images of NG, NH_2-G, RG and their synthetic root from GO (Lai et al., 2012).

diallyl dimethyl-ammonium chloride. These can improve the corrosion resistance, conductivity of materials and stability (Yang et al., 2015; Harfouche et al., 2017; Shabani-Nooshabadi and Zahedi, 2017; Wang et al., 2018). Using the electrodeposition method to produce the composite creates a unique shape and arrangement that increases the materials' SSA and creates a diffusion path for the electrolyte ions. For example, Ppy/GO composite was synthesized by electro-depositioning. It coated it on 304 stainless steel, which enhanced the decrease in corrosion resistance of the substance where current density needed to be at a minimum to increase open-circuit potential (Liu et al., 2021).

11.10 GRAPHENE OXIDE-POLYANILINE (GO-PANI) COMPOSITES

PANI is a promising material for energy storage applications because of its outstanding safety, minimal synthesis efforts, high pseudocapacitance, good conductivity and simple blend with other materials. PANI material requires protons to be well charged, and it needs a protic solvent to dissolve. PANI needs an acidic arrangement or a protic ionic solution as the electrolyte for better utilization for energy storage. To solve the challenge of cyclic stability of PANI, GO is needed in the matrix to support the framework. PANI has a broad application for pseudocapacitance (Li et al., 2019). PANI can be synthesized by the oxidative chemical polymerization of aniline and an oxidant. Alemayehu and Abiebie (2014) reported the synthesis of PANI, where 50 mL of 0.1 M aniline was mixed with 50 mL of 0.1 M H_2SO_4 and an appropriate 20 mL of 0.04 M $K_2Cr_2O_7$ aqueous solution used as oxidant by adding it dropwise with constant stirring under ice bath.

PANI-GO composites have been reported in different works of literature. The GO-PANI hybrid papers are produced by a simple, rapid polymerization mixture method where GO is the substrate. This synthesis method is unique because it can improve electrochemical application performance and biocompatibility, as shown in Figure 11.9. These kinds of properties make GO-PANI and graphene-PANI hybrid materials for batteries, flexible bio-electrodes, biosensors, electrochemical cells and bioengineering applications (Yan et al. 2010). The graphene-PANI nanofiber composites can be synthesized from pure graphene and PANI using ultrasonication, a unique approach for nanocomposites. This method gives a flexible graphene-PANI composite with different morphological structures (shown in Figure 11.10), giving high specific capacitance in HES, as shown in Table 11.3. PANI material has the advantage of environmental friendliness, high specific capacitance and simple operational costs. PANI contains atomic-size nitrogen atoms that can replace carbon atoms to improve the electrochemical properties. The PANI composite materials are highly conductive but suffer a challenge of lower stability and lower specific capacitance

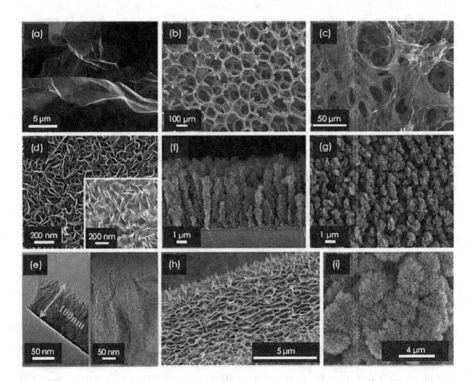

FIGURE 11.10 SEM morphology of (a) ordinary graphene (Munuera et al., 2017), (b) Co(OH)$_2$-loaded 3D multilayer graphene foam framework and (c) 3D graphene hydrogels (Pham et al., 2015). (d) SEM morphology and (e) HRTEM morphology of petal-like vertically oriented few-layered graphene (Peng et al., 2018). SEM morphology of (f) cross-section and (g) surface of forest-like vertically oriented graphene (Ma et al., 2015). SEM morphology of (h) maze-like (Zhang et al., 2013) and (i) cauliflower-like vertically oriented graphene (Seo et al., 2013).

due to arbitrarily connected PANI-nanoarrays on the GO surface stability. The challenge is overcome by using ethylenediamine as the nitrogen source and reductant in a dilute in-situ polymerization. Also, the composition ratio of the composite plays an essential role in electrochemical performance. It is reported that 35% of PANI composite exhibit significantly improved performance of up to 620 F g^{-1} of Sc performance, as highlighted in Table 11.3. This is due to the materials' prevention of various defect formations (Ge et al., 2020).

From literature, Shen et al. (2018) synthesized vertical graphene/PANI/Ti multi-layered electrode composite material by arc plasma jet CVD method that exhibited a specific capacitance of 535.7 F g^{-1} with retention of 86% (Shen et al., 2018). The graphene-PANI layers with PANI nanofiber were synthesized by interfacial polymerization and reported a specific capacitance of 578 F g^{-1} with a retention of 93% (Lai et al., 2012). The in-situ synthesis of a multi-growth site on PANI composite exhibited a specific capacitance of 86.4% (Zheng et al., 2018). Also, in-situ polymerization of GO-PANI was reported to record a specific capacitance of 890 F g^{-1} and its capacitive retention of 89% (Yu et al., 2014). This implies that the polymerization method of PANI composite favours electrochemical performance with the combination of the GO. This is due to the composite and nanosheets' permeability form that allows fast charge transfer and easy accessibility of the surface for absorption charges that enhance the excellent electrochemical performance, as shown in Figure 11.10.

When synthesising vertical graphene with PANI and TiO composite, the material forms a multilayer structure on the electrode that can enhance specific capacitance at even high current density. Luo et al. (2017) reported a specific capacitance of 535.7 F g^{-1} at a current density of 40 A g^{-1} and 461.2 F g^{-1} at a scan rate of 50 mV s^{-1}. This is attributed to the continuous distribution of subatomic levels between graphene sheets and PANI, forming an isolated structure of graphene-PANI that includes synergistically coordinating with electrolyte ions to promote self-release even at high current density (Kim et al., 2016; Kumar et al., 2018). PANI is a highly attractive electrode material, as shown in Table 11.3, but at times produces volume expansion and contraction during cycling stability and can easily collapse in the process of electrochemical charge/discharge.

11.11   GRAPHENE-POLYPYRROLE (GO-PPY)

Ppy is another excellent conducting polymer with excellent electrical conductivity, simplicity of structure arrangement and outstanding mechanical properties, among other properties. Ppy as a conducting polymer suffers similar dependability challenges to PANI. The Ppy material is commonly used as part of the electrode material ratio, usually 1%, to improve the conductivity of the materials for electrochemical activities. There is a lot of ongoing work to use Ppy as a doping material like PANI. Research shows that graphene/Ppy composite can enhance electrochemical performance, including its outstanding mechanical properties. The Ppy synthesis process and the natural availability without using oxidants is a great advantage compared to PANI. The few synthesized GO-Ppy composites morphologies exhibit proper structure arrangement that can absorb electrolytic ions and reduce particle transport path (Fang et al., 2021).

11.12   GRAPHENE-CARBON MATERIALS

Integrated V_2O_5/V_2O_5 and graphene with carbon nanotubes (CNTs) in electrode fabrication is the latest innovation in this sector. The material contains defects with carboxylic groups attributed to increased voltage. The carboxyl groups in the CNTs create a high SSA (Figure 11.10) and redox reaction sites which also cause delayed discharge. The composite containing MnO_2 with graphene sheets patched CNT exhibits excellent capabilities of high voltage, environmental safety and lower costs in the materials. The MnO_2 is incorporated in the composite to improve the electrochemical capacitance due to the conductivity network and robust scaffold from MnO_2 (Mtz-Enriquez et al., 2020). These material structures have different morphology, as shown in Figure 11.11. The material structure with high porosity allows the maximum interaction with electrolytes to enhance the pseudocapacitance leading to high performance. For example, nanocomposite of copper/graphene and CNT is one of the graphene-carbon materials with high performance. The synthesis of that composite is by electrodeposition which ensures constant accumulation, better repeatability, high dispersion rate and low energy requirement.

11.13   GRAPHENE HYDROGELS

The graphene hydrogel materials are used in hybrid Li-ion capacitors because of their porous nature in the structure. The devices from these kinds of materials render fast electron and ionic transportation in a non-aqueous electrode. These HScs have improved energy and power density compared to Li-ion batteries, making them

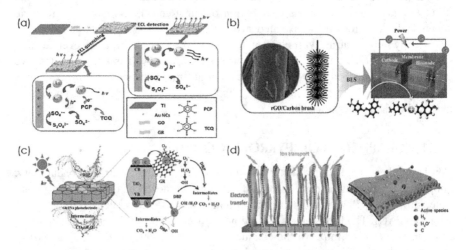

FIGURE 11.11   The electrochemiluminescence detection mechanism of pentachlorophenol (PCP) with Au NCs/GR in $S_2O_8^{2-}$ solution (Luo et al., 2014). (b) Schematic of the two-chamber bio-electro-catalysed system (BES) (Cui et al., 2018). (c) Schematic diagram of charge transfer and photocatalytic degradation mechanism of dibutyl phthalate (DBP) on graphene-loaded TiO_2 nanotube array (GR/TNA) photoelectrode. (d) Schematic illustration of the optimized ion transport and electron transfer pathways for hydrogen evolution in the $Fe_2P@rGO$ nanowall arrays (Liu et al. 2017, 2021).

potential materials for mobile devices (Wang et al., 2015). The graphene hydrogel materials are also used to produce flexible and ultrathin film-based HScs with an extended surface zone of up to $1,000\,m^2g^{-1}$. The hydrogels are highly conducting solid 3D graphene materials which can easily be used in fast charge devices and have a high-rate ability and cycle strength. They can be fabricated as a thin film with a thickness of $120\,\mu m$ on large surface area, which is reported to have excellent capability, specific capacitance, and cycle stability (Xu et al., 2013). A hydrothermal reduction mainly produces graphene hydrogel composite material for the uniform blending of GO with other materials (Kigozi et al., 2021). These HES devices should be charged using the pulse charging method. The devices continuously suspend charging to measure the open-circuit voltage at zero charging current (Khazaeli et al., 2020).

11.14   FUTURE OUTLOOK AND CONCLUSION

The graphene and GO-based composite materials exhibit very promising performances for energy storage, including hybrid applications. The main factors that prescribe the material of choice for HES application include material morphology, SSA, functionalization, electrical conductivity, electronic mobility, EDL and redox, high stability, low volume expansion and high mechanical strength, among others. These properties are determined by a combination of different materials as a way of improving performance. With all the excellent performance and progress of different materials, there are still some hindrances in the HES system technology and management. The cost of material production is still very high, and industrialization of the materials is still a big challenge compared to other energy storage devices. This makes the research focus redirect towards high performance with cost-effective alternative materials like graphene and GO. Therefore, there is a need to address and refocus on the following:

i. The carbon/oxygen content (C/O) ratio in GO materials makes a huge difference in performance and application. The GO with high oxygen in the ratio of C/O in the photocatalytic application is highly required for light response due to the oxygen functional groups on the structure. Low oxygen graphene with electron transportation efficiency and high current density is necessary for energy storage. Currently, there is little information on the regulation of oxygen and its functional groups on the surface structure of graphene and GO. There is a need for in-depth electrochemical and synthesis methods with a comprehensive mechanism for the regulation of oxygen content.

ii. In redesigning of the synthesis methods like the use of one-pot synthesis to reduce the pre-treatments, energy and time consumption.

iii. Expansion of graphene, GO sources are sustainable and cost-effective like GO from biomass as long as the material retains its physical and chemical properties.

iv. Remodifying materials with cost-effective substitutes gives mechanical, physical and chemical properties, which can be reflected in the energy storage system implementation for HES.

v. The future possibilities should also refocus on materials with the same properties as GO for cost-effective substitution with novel structure, porosity and functionalization that can be utilized for EDL and redox.
vi. There is a need to re-engineer the new high-performance materials to have adaptable commercialization and industrialization for large consumption.
vii. The engineering aspect and design plan of HES systems need to be refreshed to include the best current innovation to cut down on the expensive processes. This can be done by looking at minimal expenses, long life makeup, and low support for modified electrodes.
viii. There is a need to bridge the gap between researchers, innovators and industrialisation by creating collaborations. This can be done by assigning industrial challenges to innovators and researchers.

REFERENCES

Abdelkader, A. M., I. A. Kinloch, and Robert A. W. Dryfe. 2014. "High-Yield Electro-Oxidative Preparation of Graphene Oxide." *Chemical Communications* 50 (61): 8402–4. https://doi.org/10.1039/C4CC03260H.
Adetayo, Adeniji, and Damilola Runsewe. 2019. "Synthesis and Fabrication of Graphene and Graphene Oxide: A Review." *Open Journal of Composite Materials* 9 (2): 207–29. https://doi.org/10.4236/ojcm.2019.92012.
Alabadi, Akram, Shumaila Razzaque, Zehua Dong, Weixing Wang, and Bien Tan. 2016. "Graphene Oxide-Polythiophene Derivative Hybrid Nanosheet for Enhancing Performance of Supercapacitor." *Journal of Power Sources* 306 (February): 241–47. https://doi.org/10.1016/J.JPOWSOUR.2015.12.028.
Alam, SN, N Sharma, and L Kumar. 2017. "Synthesis of Graphene Oxide (GO) by Modified Hummers Method and Its Thermal Reduction to Obtain Reduced Graphene Oxide (RGO)." *Graphene.*
Alam, Syed Nasimul, Nidhi Sharma, Lailesh Kumar, Syed Nasimul Alam, Nidhi Sharma, and Lailesh Kumar. 2017. "Synthesis of Graphene Oxide (GO) by Modified Hummers Method and Its Thermal Reduction to Obtain Reduced Graphene Oxide (RGO)*." *Graphene* 6 (1): 1–18. https://doi.org/10.4236/GRAPHENE.2017.61001.
Alemayehu, Tassew, and Diribe Abiebie. 2014. "Preparation of Poly Aniline by Chemical Ox-Idative Method and Its Characterization." *OALib* 01 (06): 1–4. https://doi.org/10.4236/oalib.1100974.
Allen, Matthew J., Vincent C. Tung, and Richard B. Kaner. 2009. "Honeycomb Carbon: A Review of Graphene." *Chemical Reviews* 110 (1): 132–45. https://doi.org/10.1021/CR900070D.
Ambrosi, Adriano, and Martin Pumera. 2016. "Electrochemically Exfoliated Graphene and Graphene Oxide for Energy Storage and Electrochemistry Applications." *Chemistry-A European Journal* 22 (1): 153–59. https://doi.org/10.1002/CHEM.201503110.
Atif, Rasheed, Islam Shyha, and Fawad Inam. 2016. "Mechanical, Thermal, and Electrical Properties of Graphene-Epoxy Nanocomposites—A Review." *Polymers.* Accessed October 27, 2021. https://doi.org/10.3390/polym8080281.
Bora, Anindita, Kiranjyoti Mohan, Simanta Doley, and Swapan Kumar Dolui. 2018. "Flexible Asymmetric Supercapacitor Based on Functionalized Reduced Graphene Oxide Aerogels with Wide Working Potential Window." *ACS Applied Materials & Interfaces* 10 (9): 7996–8009. https://doi.org/10.1021/ACSAMI.7B18610.
Cao, Jianyun, Pei He, Mahdi A. Mohammed, Xin Zhao, Robert J. Young, Brian Derby, Ian A. Kinloch, and Robert A. W. Dryfe. 2017. "Two-Step Electrochemical Intercalation

and Oxidation of Graphite for the Mass Production of Graphene Oxide." *Journal of the American Chemical Society* 139 (48): 17446–56. https://doi.org/10.1021/JACS.7B08515.

Cao, Xiehong, Yumeng Shi, Wenhui Shi, Gang Lu, Xiao Huang, Qingyu Yan, Qichun Zhang, and Hua Zhang. 2011. "Preparation of Novel 3D Graphene Networks for Supercapacitor Applications." *Small* 7 (22): 3163–68. https://doi.org/10.1002/SMLL.201100990.

Chang, Jiuli, Shiqi Zang, Wenfang Liang, Dapeng Wu, Zhaoxun Lian, Fang Xu, Kai Jiang, and Zhiyong Gao. 2021. "Enhanced Faradic Activity by Construction of P-n Junction within Reduced Graphene Oxide@Cobalt Nickel Sulfide@Nickle Cobalt Layered Double Hydroxide Composite Electrode for Charge Storage in Hybrid Supercapacitor." *Journal of Colloid and Interface Science* 590: 114–24. https://doi.org/10.1016/j.jcis.2021.01.035.

Chen, Chia-Hsuan, Shiou-Wen Yang, Min-Chiang Chuang, Wei-Yen Woon, and Ching-Yuan Su. 2015. "Towards the Continuous Production of High Crystallinity Graphene via Electrochemical Exfoliation with Molecular in Situ Encapsulation." *Nanoscale* 7 (37): 15362–73. https://doi.org/10.1039/C5NR03669K.

Chen, Gao Feng, Zhao Qing Liu, Jia Ming Lin, Nan Li, and Yu Zhi Su. 2015. "Hierarchical Polypyrrole Based Composites for High Performance Asymmetric Supercapacitors." *Journal of Power Sources* 283: 484–93. https://doi.org/10.1016/j.jpowsour.2015.02.103.

Chen, Ji, Yingru Li, Liang Huang, Chun Li, and Gaoquan Shi. 2015. "High-Yield Preparation of Graphene Oxide from Small Graphite Flakes via an Improved Hummers Method with a Simple Purification Process." *Carbon* 81 (1): 826–34. https://doi.org/10.1016/J.CARBON.2014.10.033.

Chen, Yujuan, Zhaoen Liu, Li Sun, Zhiwei Lu, and Kelei Zhuo. 2018. "Nitrogen and Sulfur Co-Doped Porous Graphene Aerogel as an Efficient Electrode Material for High Performance Supercapacitor in Ionic Liquid Electrolyte." *Journal of Power Sources* 390 (June): 215–23. https://doi.org/10.1016/J.JPOWSOUR.2018.04.057.

Cheng, Lingli, Yiyang Hu, Dandan Qiao, Ying Zhu, Hao Wang, and Zheng Jiao. 2018. "One-Step Radiolytic Synthesis of Heteroatom (N and S) Co-Doped Graphene for Supercapacitors." *Electrochimica Acta* 259 (January): 587–97. https://doi.org/10.1016/J.ELECTACTA.2017.11.022.

Cheng, Qian, Jie Tang, Norio Shinya, and Lu Chang Qin. 2013. "Polyaniline Modified Graphene and Carbon Nanotube Composite Electrode for Asymmetric Supercapacitors of High Energy Density." *Journal of Power Sources* 241 (November): 423–28. https://doi.org/10.1016/J.JPOWSOUR.2013.04.105.

Chodankar, Nilesh R., Deepak P. Dubal, Su Hyeon Ji, and Do Heyoung Kim. 2019. "Highly Efficient and Stable Negative Electrode for Asymmetric Supercapacitors Based on Graphene/FeCo$_2$O$_4$ Nanocomposite Hybrid Material." *Electrochimica Acta* 295 (February): 195–203. https://doi.org/10.1016/J.ELECTACTA.2018.10.125.

Choi, Hyun-Jung, Sun-Min Jung, Jeong-Min Seo, Dong Wook Chang, Liming Dai, and Jong-Beom Baek. 2012. "Graphene for Energy Conversion and Storage in Fuel Cells and Supercapacitors." *Nano Energy*, 1 (4), 534–51.

Choi, Wonbong, Indranil Lahiri, Raghunandan Seelaboyina, and Yong Soo Kang. 2010. "Synthesis of Graphene and Its Applications: A Review." *Critical Reviews in Solid State and Materials Sciences* 35 (1): 52–71. https://doi.org/10.1080/10408430903505036.

Cooper, Adam J., Neil R. Wilson, Ian A. Kinloch, and Robert A. W. Dryfe. 2014. "Single Stage Electrochemical Exfoliation Method for the Production of Few-Layer Graphene via Intercalation of Tetraalkylammonium Cations." *Carbon* 66: 340–50. https://doi.org/10.1016/j.carbon.2013.09.009.

Cui, Dan, Li Ming Yang, Wen Zong Liu, Min Hua Cui, Wei Wei Cai, and Ai Jie Wang. 2018. "Facile Fabrication of Carbon Brush with Reduced Graphene Oxide (RGO) for Decreasing Resistance and Accelerating Pollutants Removal in Bio-Electrochemical Systems." *Journal of Hazardous Materials* 354 (July): 244–49. https://doi.org/10.1016/J.JHAZMAT.2018.05.001.

Dai, Shuge, Zhen Liu, Bote Zhao, Jianhuang Zeng, Hao Hu, Qiaobao Zhang, Dongchang Chen, Chong Qu, Dai Dang, and Meilin Liu. 2018. "A High-Performance Supercapacitor Electrode Based on N-doped Porous Graphene." *Journal of Power Sources* 387 (May): 43–48. https://doi.org/10.1016/J.JPOWSOUR.2018.03.055.

Dong, Jinyang, Gang Lu, Fan Wu, Chenxi Xu, Xiaohong Kang, and Zhiming Cheng. 2018. "Facile Synthesis of a Nitrogen-Doped Graphene Flower-like MnO_2 Nanocomposite and Its Application in Supercapacitors." *Applied Surface Science* 427 (January): 986–93. https://doi.org/10.1016/J.APSUSC.2017.07.291.

Ede, Sivasankara Rao, S. Anantharaj, K. T. Kumaran, Soumyaranjan Mishra, and Subrata Kundu. 2017. "One Step Synthesis of $Ni/Ni(OH)_2$ Nano Sheets (NSs) and Their Application in Asymmetric Supercapacitors." *RSC Advances* 7 (10): 5898–911. https://doi.org/10.1039/C6RA26584G.

Fang, Wen Zhang, Li Peng, Ying Jun Liu, Fang Wang, Zhen Xu, and Chao Gao. 2021. "A Review on Graphene Oxide Two-Dimensional Macromolecules: From Single Molecules to Macro-Assembly." *Chinese Journal of Polymer Science (English Edition)* 39 (3): 267–308. https://doi.org/10.1007/s10118-021-2515-1.

Ge, Manman, Huilian Hao, Qiu Lv, Jianghong Wu, and Wenyao Li. 2020. "Hierarchical Nanocomposite That Coupled Nitrogen-Doped Graphene with Aligned PANI Cores Arrays for High-Performance Supercapacitor." *Electrochimica Acta* 330 (January): 135236. https://doi.org/10.1016/J.ELECTACTA.2019.135236.

Geim, AK, and KS Novoselov. 2010. "The Rise of Graphene." *Nanoscience and Technology: World Scientific.* https://doi.org/10.1142/9789814287005_0002.

Geim, AK., and KS. Novoselov. 2007. "The Rise of Graphene." *Nature Materials* 6: 183–191.

Goda, Emad S., Sang Eun Hong, and Kuk Ro Yoon. 2021. "Facile Synthesis of Cu-PBA Nanocubes/Graphene Oxide Composite as Binder-Free Electrodes for Supercapacitor." *Journal of Alloys and Compounds* 859: 157868. https://doi.org/10.1016/j.jallcom.2020.157868.

Harfouche, Nesrine, Natalia Gospodinova, Belkacem Nessark, and François Xavier Perrin. 2017. "Electrodeposition of Composite Films of Reduced Graphene Oxide/Polyaniline in Neutral Aqueous Solution on Inert and Oxidizable Metal." *Journal of Electroanalytical Chemistry* 786 (February): 135–44. https://doi.org/10.1016/J.JELECHEM.2017.01.030.

Hou, Jianhua, Chuanbao Cao, Xilan Ma, Faryal Idrees, Bin Xu, Xin Hao, and Wei Lin. 2014. "From Rice Bran to High Energy Density Supercapacitors : A New Route to Control Porous Structure of 3D Carbon." *Scientific Reports* 4 (7260): 1–6. https://doi.org/10.1038/srep07260.

Huang, Chengxiang, Chenghao Ni, Lingze Yang, Tiantian Zhou, Chen Hao, Xiaohong Wang, Cunwang Ge, and Linli Zhu. 2020. "High-Performance Supercapacitor Based on Graphene Oxide through in-Situ Polymerization and Co-Precipitation Method." *Journal of Alloys and Compounds* 829 (July): 154536. https://doi.org/10.1016/J.JALLCOM.2020.154536.

Huang, Xiaodan, Kun Qian, Jie Yang, Jun Zhang, Li Li, Chengzhong Yu, and Dongyuan Zhao. 2012. "Functional Nanoporous Graphene Foams with Controlled Pore Sizes." *Advanced Materials* 24 (32): 4419–23. https://doi.org/10.1002/ADMA.201201680.

Ji, Zhenyuan, Na Li, Yifei Zhang, Minghua Xie, Xiaoping Shen, Lizhi Chen, Keqiang Xu, and Guoxing Zhu. 2019. "Nitrogen-Doped Carbon Dots Decorated Ultrathin Nickel Hydroxide Nanosheets for High-Performance Hybrid Supercapacitor." *Journal of Colloid and Interface Science* 542 (April): 392–99. https://doi.org/10.1016/J.JCIS.2019.02.037.

Jibrael, Raneen Imad, and Mustafa KA Mohammed. 2016. "Structural and the Optical Properties of Graphene Prepared by Electrochemical Exfoliation Technique." *Al-Nahrain Journal of Science* 19 (4): 71–77. https://www.anjs.edu.iq/index.php/anjs/article/view/181.

Jo, Changshin, Ilkyu Hwang, Jinwoo Lee, Chul Wee Lee, and Songhun Yoon. 2011. "Investigation of Pseudocapacitive Charge-Storage Behavior in Highly Conductive Ordered Mesoporous Tungsten Oxide Electrodes." *Journal of Physical Chemistry C* 115 (23): 11880–86. https://doi.org/10.1021/JP2036982.

Johns, James E., and Mark C. Hersam. 2012. "Atomic Covalent Functionalization of Graphene." *Accounts of Chemical Research* 46 (1): 77–86. https://doi.org/10.1021/AR300143E.

Jow, T. R., and J. P. Zheng. 2011. "Amorphous Thin Film Ruthenium Oxide as an Electrode Material for Electrochemical Capacitors." *MRS Online Proceedings Library* 393 (1): 433–38. https://doi.org/10.1557/PROC-393-433.

Khazaeli, Ali, Gabrielle Godbille-Cardona, and Dominik P. J. Barz. 2020. "A Novel Flexible Hybrid Battery–Supercapacitor Based on a Self-Assembled Vanadium-Graphene Hydrogel." *Advanced Functional Materials* 30 (21): 1910738. https://doi.org/10.1002/ADFM.201910738.

Kigozi, Moses, Blessing N Ezealigo, Azikiwe Peter Onwualu, and Nelson Y Dzade. 2021. "Hydrothermal Synthesis of Metal Oxide Composite Cathode Materials for High Energy Application." In *Chemically Deposited Nanocrystalline Metal Oxide Thin Films*, edited by Ezema, Fabian I., Chandrakant D. Lokhande, and Rajan Jose, 1st ed., 489–508. Springer, Cham. https://doi.org/10.1007/978-3-030-68462-4_19.

Kigozi, Moses, Richard K Koech, Orisekeh Kingsley, Itohan Ojeaga, Emmanuel Tebandeke, Gabriel Kasozi, and Peter A Onwualu. 2020. "Synthesis and Characterization of Graphene Oxide from Locally Mined Graphite Flakes and Its Supercapacitor Applications." *Results in Materials* 7: 100113. https://doi.org/10.1016/j.rinma.2020.100113.

Kim, Da Mi, Si Jin Kim, Young Woo Lee, Da Hee Kwak, Han Chul Park, Min Cheol Kim, Bo Mi Hwang, et al. 2015. "Two-Dimensional Nanocomposites Based on Tungsten Oxide Nanoplates and Graphene Nanosheets for High-Performance Lithium-Ion Batteries." *Electrochimica Acta* 163 (May): 132–39. https://doi.org/10.1016/J.ELECTACTA.2015.02.121.

Kim, Hyun-Kyung, Ali Reza Kamali, Kwang Chul Roh, Kwang-Bum Kim, and Derek John Fray. 2016. "Dual Coexisting Interconnected Graphene Nanostructures for High Performance Supercapacitor Applications." *Energy & Environmental Science* 9 (7): 2249–56. https://doi.org/10.1039/C6EE00815A.

Kim, Ji Eun, Tae Hee Han, Sun Hwa Lee, Ju Young Kim, Chi Won Ahn, Je Moon Yun, and Sang Ouk Kim. 2011. "Graphene Oxide Liquid Crystals." *Angewandte Chemie International Edition* 50 (13): 3043–47. https://doi.org/10.1002/ANIE.201004692.

Kota, Manikantan, Xu Yu, Sun Hwa Yeon, Hae Won Cheong, and Ho Seok Park. 2016. "Ice-Templated Three Dimensional Nitrogen-Doped Graphene for Enhanced Supercapacitor Performance." *Journal of Power Sources* 303 (January): 372–78. https://doi.org/10.1016/J.JPOWSOUR.2015.11.006.

Krane, N. 2011. "Preparation of Graphene." In *Selected Topics in Physics: Physics of Nanoscale Carbon*, pp. 872–76.

Kumar, Sachin, Ghuzanfar Saeed, Nam Hoon Kim, and Joong Hee Lee. 2018. "Hierarchical Nanohoneycomb-like $CoMoO_4$–MnO_2 Core-Shell and Fe_2O_3 Nanosheet Arrays on 3D Graphene Foam with Excellent Supercapacitive Performance." *Journal of Materials Chemistry A* 6 (16): 7182–93. https://doi.org/10.1039/C8TA00889B.

Kwak, B.S., J.B. Bates, and F.X. Hart. 1996. *Thin-Film Solid Ionic Devices and Materials*. Electrochemical Society.

Kwak, Myung-Jun, Ananthakumar Ramadoss, Ki-Yong Yoon, Juhyung Park, Pradheep Thiyagarajan, and Ji-Hyun Jang. 2017. "Single-Step Synthesis of N-Doped Three-Dimensional Graphitic Foams for High-Performance Supercapacitors." *ACS Sustainable Chemistry and Engineering* 5 (8): 6950–57. https://doi.org/10.1021/ACSSUSCHEMENG.7B01132.

Lai, Linfei, Liang Wang, Huanping Yang, Nanda Gopal Sahoo, Qian Xin Tam, Jilei Liu, Chee Kok Poh, San Hua Lim, Zexiang Shen, and Jianyi Lin. 2012. "Tuning Graphene Surface Chemistry to Prepare Graphene/Polypyrrole Supercapacitors with Improved Performance." *Nano Energy* 1 (5): 723–31. https://doi.org/10.1016/J.NANOEN.2012.05.012.

Le, Kai, Zhou Wang, Fenglong Wang, Qi Wang, Qian Shao, Vignesh Murugadoss, Shide Wu, et al. 2019. "Sandwich-like NiCo Layered Double Hydroxide/Reduced Graphene Oxide Nanocomposite Cathodes for High Energy Density Asymmetric Supercapacitors." *Dalton Transactions* 48 (16): 5193–5202. https://doi.org/10.1039/C9DT00615J.

Lee, Dongwook, and Jiwon Seo. 2014. "Three-Dimensionally Networked Graphene Hydroxide with Giant Pores and Its Application in Supercapacitors." *Scientific Reports* 4 (December). https://doi.org/10.1038/SREP07419.

Lee, Xin Jiat, Billie Yan Zhang Hiew, Kar Chiew Lai, Lai Yee Lee, Suyin Gan, Suchithra Thangalazhy-Gopakumar, and Sean Rigby. 2019. "Review on Graphene and Its Derivatives: Synthesis Methods and Potential Industrial Implementation." *Journal of the Taiwan Institute of Chemical Engineers*, 98, 163–80.

Li, Jianpeng, Dingshu Xiao, Yaqi Ren, Huiru Liu, Zhenxuan Chen, and Jiaming Xiao. 2019. "Bridging of Adjacent Graphene/Polyaniline Layers with Polyaniline Nanofibers for Supercapacitor Electrode Materials." *Electrochimica Acta* 300 (March): 193–201. https://doi.org/10.1016/J.ELECTACTA.2019.01.089.

Li, Qi, Michael Horn, Yinong Wang, Jennifer MacLeod, Nunzio Motta, and Jinzhang Liu. 2019. "A Review of Supercapacitors Based on Graphene and Redox-Active Organic Materials." *Materials*, 12 (5): 703. https://doi.org/10.3390/MA12050703.

Li, Yu, and Hui Xu. 2015. "Development of a Novel Graphene/Polyaniline Electrodeposited Coating for on-Line in-Tube Solid Phase Microextraction of Aldehydes in Human Exhaled Breath Condensate." *Journal of Chromatography A* 1395 (May): 23–31. https://doi.org/10.1016/J.CHROMA.2015.03.058.

Li, Zhiwei, Xiao Hou, Laigui Yu, Zhijun Zhang, and Pingyu Zhang. 2014. "Preparation of Lanthanum Trifluoride Nanoparticles Surface-Capped by Tributyl Phosphate and Evaluation of Their Tribological Properties as Lubricant Additive in Liquid Paraffin." *Applied Surface Science* 292 (February): 971–77. https://doi.org/10.1016/J.APSUSC.2013.12.089.

Liang, Chu, Yun Chen, Min Wu, Kai Wang, Wenkui Zhang, Jian Chen, Yang Xia, Jun Zhang, Shiyou Zheng, and Hongge Pan. 2021. "Green Synthesis of Graphite from CO_2 without Graphitization Process of Amorphous Carbon." *Nature Communications* 12 (119): 1–9. https://doi.org/10.1038/s41467-020-20380-0.

Liao, Yuqing, Yulan Huang, Dong Shu, Yayun Zhong, Junnan Hao, Chun He, Jie Zhong, and Xiaona Song. 2016. "Three-Dimensional Nitrogen-Doped Graphene Hydrogels Prepared via Hydrothermal Synthesis as High-Performance Supercapacitor Materials." *Electrochimica Acta* 194 (March): 136–42. https://doi.org/10.1016/J.ELECTACTA.2016.02.067.

Lin, Dingchang, Wei Liu, Yayuan Liu, Hye Ryoung Lee, Po-Chun Hsu, Kai Liu, and Yi Cui. 2015. "High Ionic Conductivity of Composite Solid Polymer Electrolyte via In Situ Synthesis of Monodispersed SiO_2 Nanospheres in Poly(Ethylene Oxide)." *Nano Letters* 16 (1): 459–65. https://doi.org/10.1021/ACS.NANOLETT.5B04117.

Liu, Guangzhen, Zhensheng Xiong, Liming Yang, Hui Shi, Difan Fang, Mei Wang, Penghui Shao, and Xubiao Luo. 2021. "Electrochemical Approach toward Reduced Graphene Oxide-Based Electrodes for Environmental Applications: A Review." *Science of the Total Environment* 778: 146301. https://doi.org/10.1016/j.scitotenv.2021.146301.

Liu, Jilei, Chee Kok Poh, Da Zhan, Linfei Lai, San Hua Lim, Liang Wang, Xiaoxu Liu, et al. 2013. "Improved Synthesis of Graphene Flakes from the Multiple Electrochemical

Exfoliation of Graphite Rod." *Nano Energy* 2 (3): 377–86. https://doi.org/10.1016/J.
NANOEN.2012.11.003.

Liu, Meijun, Liming Yang, Tian Liu, Yanhong Tang, Shenglian Luo, Chengbin Liu, and
Yunxiong Zeng. 2017. "Fe$_2$P/Reduced Graphene Oxide/Fe$_2$P Sandwich-Structured
Nanowall Arrays: A High-Performance Non-Noble-Metal Electrocatalyst for
Hydrogen Evolution." *Journal of Materials Chemistry A* 5 (18): 8608–15. https://doi.
org/10.1039/C7TA01791J.

Liu, Mingkai, Yue-E. Miao, Chao Zhang, Weng Weei Tjiu, Zhibin Yang, Huisheng Peng, and
Tianxi Liu. 2013. "Hierarchical Composites of Polyaniline–Graphene Nanoribbons–
Carbon Nanotubes as Electrode Materials in All-Solid-State Supercapacitors."
Nanoscale 5 (16): 7312–20. https://doi.org/10.1039/C3NR01442H.

Liu, Wei, Min-Sang Song, Biao Kong, and Yi Cui. 2017. "Flexible and Stretchable Energy
Storage: Recent Advances and Future Perspectives." *Advanced Materials* 29 (1):
1603436. https://doi.org/10.1002/ADMA.201603436.

Liu, Wei-wen, Siang-piao Chai, Abdul Rahman, and U. Hashim. 2014. "Synthesis and
Characterization of Graphene and Carbon Nanotubes : A Review on the Past and
Recent Developments." *Journal of Industrial and Engineering Chemistry* 20 (4): 1171–
85. https://doi.org/10.1016/j.jiec.2013.08.028.

Liu, Xi, Pei Lin, Xiaoqin Yan, Zhuo Kang, Yanguang Zhao, Yang Lei, Chuanbao Li, Hongwu
Du, and Yue Zhang. 2013. "Enzyme-Coated Single ZnO Nanowire FET Biosensor for
Detection of Uric Acid." *Sensors and Actuators B: Chemical* 176 (January): 22–27.
https://doi.org/10.1016/J.SNB.2012.08.043.

Liu, Yan Zhen, Yong Feng Li, Fang Yuan Su, Li Jing Xie, Qing Qiang Kong, Xiao Ming Li,
Jian Guo Gao, and Cheng Meng Chen. 2016. "Easy One-Step Synthesis of N-Doped
Graphene for Supercapacitors." *Energy Storage Materials* 2 (January): 69–75. https://
doi.org/10.1016/J.ENSM.2015.09.006.

Liu, Zhangming, Haiyan Zhang, Qiao Yang, and Yaowu Chen. 2018. "Graphene/V$_2$O$_5$ Hybrid
Electrode for an Asymmetric Supercapacitor with High Energy Density in an Organic
Electrolyte." *Electrochimica Acta* 287 (October): 149–57. https://doi.org/10.1016/J.
ELECTACTA.2018.04.212.

Lozada-Hidalgo, Marcelo, Sheng Zhang, Sheng Hu, Vasyl G. Kravets, Francisco J. Rodriguez,
Alexey Berdyugin, Alexander Grigorenko, and Andre K. Geim. 2018. "Giant Photoeffect
in Proton Transport through Graphene Membranes." *Nature Nanotechnology* 13 (4):
300–303. https://doi.org/10.1038/s41565-017-0051-5.

Luo, Shenglian, Hua Xiao, Shanli Yang, Chengbin Liu, Jiesheng Liang, and Yanhong
Tang. 2014. "Ultrasensitive Detection of Pentachlorophenol Based on Enhanced
Electrochemiluminescence of Au Nanoclusters/Graphene Hybrids." *Sensors and
Actuators B: Chemical* 194 (April): 325–31. https://doi.org/10.1016/J.SNB.2013.12.108.

Luo, Yangxi, Qin'e Zhang, Wenjing Hong, Zongyuan Xiao, and Hua Bai. 2017. "A High-
Performance Electrochemical Supercapacitor Based on a Polyaniline/Reduced
Graphene Oxide Electrode and a Copper(II) Ion Active Electrolyte." *Physical Chemistry
Chemical Physics* 20 (1): 131–36. https://doi.org/10.1039/C7CP07156F.

Ma, Yifei, Mei Wang, Namhun Kim, Jonghwan Suhr, and Heeyeop Chae. 2015. "A Flexible
Supercapacitor Based on Vertically Oriented 'Graphene Forest' Electrodes." *Journal of
Materials Chemistry A* 3 (43): 21875–81. https://doi.org/10.1039/C5TA05687J.

Malik, Md Tanvir Uddin, Aditya Sarker, S. M. Sultan Mahmud Rahat, and Sanzeeda
Baig Shuchi. 2021. "Performance Enhancement of Graphene/GO/RGO Based
Supercapacitors: A Comparative Review." *Materials Today Communications* 28 (July):
102685. https://doi.org/10.1016/j.mtcomm.2021.102685.

Mattevi, Cecilia, Florian Colléaux, HoKwon Kim, Yen-Hung Lin, Kyung T. Park,
Manish Chhowalla, and Thomas D. Anthopoulos. 2012. "Solution-Processable

Organic Dielectrics for Graphene Electronics." *Nanotechnology* 23 (34). https://doi.org/10.1088/0957-4484/23/34/344017.

Meng, Jiangyan, Yunying Wang, Xiaolin Xie, and Hongying Quan. 2019. "High-Performance Asymmetric Supercapacitor Based on Graphene-Supported Iron Oxide and Manganese Sulfide." *Ionics* 25 (10): 4925–33. https://doi.org/10.1007/S11581-019-03061-X.

Mi, Juan, Xiao-Rong Rong Wang, Rui-Jun Jun Fan, Wen-Hui Hui Qu, and Wen-Cui Cui Li. 2012. "Coconut-Shell-Based Porous Carbons with a Tunable Micro/Mesopore Ratio for High-Performance Supercapacitors." *Energy & Fuels* 26 (8): 5321–29. https://doi.org/10.1021/ef3009234.

Miao, Chenxu, Xianzhi Yin, Genglei Xia, Kai Zhu, Ke Ye, Qian Wang, Jun Yan, Dianxue Cao, and Guiling Wang. 2021. "Facile Microwave-Assisted Synthesis of Cobalt Diselenide/Reduced Graphene Oxide Composite for High-Performance Supercapacitors." *Applied Surface Science* 543 (October 2020). https://doi.org/10.1016/j.apsusc.2020.148811.

Miniach, Ewa, Agata Śliwak, Adam Moyseowicz, Laura Fernández-Garcia, Zoraida González, Marcos Granda, Rosa Menendez, and Grażyna Gryglewicz. 2017. "MnO$_2$/Thermally Reduced Graphene Oxide Composites for High-Voltage Asymmetric Supercapacitors." *Electrochimica Acta* 240 (June): 53–62. https://doi.org/10.1016/J.ELECTACTA.2017.04.056.

Morais, Andreia, João Paulo C. Alves, Francisco Anderson S. Lima, Monica Lira-Cantu, and Ana Flavia Nogueira. 2015. "Enhanced Photovoltaic Performance of Inverted Hybrid Bulk-Heterojunction Solar Cells Using TiO$_2$/Reduced Graphene Oxide Films as Electron Transport Layers." *Journal of Photonics for Energy* 5. https://doi.org/10.1117/1.JPE.5.057408.

Mtz-Enriquez, A. I., C. Gomez-Solis, A. I. Oliva, A. Zakhidov, P. M. Martinez, C. R. Garcia, A. Herrera-Ramirez, and J. Oliva. 2020. "Enhancing the Voltage and Discharge Times of Graphene Supercapacitors Depositing a CNT/V$_2$O$_5$ Layer on Their Electrodes." *Materials Chemistry and Physics* 244 (April): 122698. https://doi.org/10.1016/J.MATCHEMPHYS.2020.122698.

Munuera, J. M., J. I. Paredes, M. Enterría, A. Pagán, S. Villar-Rodil, M. F. R. Pereira, J. I. Martins, et al. 2017. "Electrochemical Exfoliation of Graphite in Aqueous Sodium Halide Electrolytes toward Low Oxygen Content Graphene for Energy and Environmental Applications." *ACS Applied Materials and Interfaces* 9 (28): 24085–99. https://doi.org/10.1021/ACSAMI.7B04802.

Munuera, J. M., J. I. Paredes, S. Villar-Rodil, M. Ayán-Varela, A. Martínez-Alonso, and J. M. D. Tascón. 2016. "Electrolytic Exfoliation of Graphite in Water with Multifunctional Electrolytes: En Route towards High Quality, Oxide-Free Graphene Flakes." *Nanoscale* 8 (5): 2982–98. https://doi.org/10.1039/C5NR06882G.

Musella, Elisa, Isacco Gualandi, Giacomo Ferrari, Davide Mastroianni, Erika Scavetta, Marco Giorgetti, Andrea Migliori, et al. 2021. "Electrosynthesis of Ni/Al Layered Double Hydroxide and Reduced Graphene Oxide Composites for the Development of Hybrid Capacitors." *Electrochimica Acta* 365: 137294. https://doi.org/10.1016/j.electacta.2020.137294.

Muszynski, Ryan, Brian Seger, and Prashant V. Kamat. 2008. "Decorating Graphene Sheets with Gold Nanoparticles." *Journal of Physical Chemistry C* 112 (14): 5263–66. https://doi.org/10.1021/JP800977B.

Nair, R. R., P. Blake, A. N. Grigorenko, K. S. Novoselov, T. J. Booth, T. Stauber, N. M. R. Peres, and A. K. Geim. 2008. "Fine Structure Constant Defines Visual Transparency of Graphene." *Science* 320 (5881): 1308. https://doi.org/10.1126/SCIENCE.1156965.

Nanakkal, A. R., and L. K. Alexander. 2017. "Photocatalytic Activity of Graphene/ZnO Nanocomposite Fabricated by Two-Step Electrochemical Route." *Journal of Chemical Sciences* 129 (1): 95–102. https://doi.org/10.1007/S12039-016-1206-X.

Naveenkumar, P., and G. Paruthimal Kalaignan. 2018. "Electrodeposited MnS on Graphene Wrapped Ni-Foam for Enhanced Supercapacitor Applications." *Electrochimica Acta* 289 (November): 437–47. https://doi.org/10.1016/J.ELECTACTA.2018.09.100.

Nayak, Arpan Kumar, Ashok Kumar Das, and Debabrata Pradhan. 2017. "High Performance Solid-State Asymmetric Supercapacitor Using Green Synthesized Graphene–WO$_3$ Nanowires Nanocomposite." *ACS Sustainable Chemistry and Engineering* 5 (11): 10128–38. https://doi.org/10.1021/ACSSUSCHEMENG.7B02135.

Nedoliuk, Ievgeniia O., Sheng Hu, Andre K. Geim, and Alexey B. Kuzmenko. 2019. "Colossal Infrared and Terahertz Magneto-Optical Activity in a Two-Dimensional Dirac Material." *Nature Nanotechnology* 14 (8): 756–61. https://doi.org/10.1038/s41565-019-0489-8.

Niu, Zhiqiang, Jianjun Du, Xuebo Cao, Yinghui Sun, Weiya Zhou, Huey Hoon Hng, Jan Ma, Xiaodong Chen, and Sishen Xie. 2012. "Electrophoretic Build-Up of Alternately Multilayered Films and Micropatterns Based on Graphene Sheets and Nanoparticles and Their Applications in Flexible Supercapacitors." *Small* 8 (20): 3201–8. https://doi.org/10.1002/SMLL.201200924.

Pan, Bingyige, Li Bai, Cheng-Min Hu, Xinping Wang, Wei-Shi Li, and Fu-Gang Zhao. 2020. "Graphene-Indanthrone Donor–π–Acceptor Heterojunctions for High-Performance Flexible Supercapacitors." *Advanced Energy Materials* 10 (18): 2000181. https://doi.org/10.1002/AENM.202000181.

Pan, Shuyang, and Ilhan A. Aksay. 2011. "Factors Controlling the Size of Graphene Oxide Sheets Produced via the Graphite Oxide Route." *ACS Nano* 5 (5): 4073–83. https://doi.org/10.1021/NN200666R.

Pandey, Gaind P., Tao Liu, Emery Brown, Yiqun Yang, Yonghui Li, Xiuzhi Susan Sun, Yueping Fang, and Jun Li. 2016. "Mesoporous Hybrids of Reduced Graphene Oxide and Vanadium Pentoxide for Enhanced Performance in Lithium-Ion Batteries and Electrochemical Capacitors." *ACS Applied Materials and Interfaces* 8 (14): 9200–9210. https://doi.org/10.1021/ACSAMI.6B02372.

Parvez, Khaled, Sheng Yang, Xinliang Feng, and Klaus Müllen. 2007. "Exfoliation of Graphene via Wet Chemical Routes." *Metals Synthetic.*

Parvez, Khaled, Zhong-Shuai Wu, Rongjin Li, Xianjie Liu, Robert Graf, Xinliang Feng, and Klaus Müllen. 2014. "Exfoliation of Graphite into Graphene in Aqueous Solutions of Inorganic Salts." *Journal of the American Chemical Society* 136 (16): 6083–91. https://doi.org/10.1021/JA5017156.

Peng, Lin, Yeru Liang, Hanwu Dong, Hang Hu, Xiao Zhao, Yijing Cai, Yong Xiao, Yingliang Liu, and Mingtao Zheng. 2018. "Super-Hierarchical Porous Carbons Derived from Mixed Biomass Wastes by a Stepwise Removal Strategy for High-Performance Supercapacitors." *Journal of Power Sources* 377 (December 2017): 151–60. https://doi.org/10.1016/j.jpowsour.2017.12.012.

Pham, Viet Hung, Tesfaye Gebre, and James H. Dickerson. 2015. "Facile Electrodeposition of Reduced Graphene Oxide Hydrogels for High-Performance Supercapacitors." *Nanoscale* 7 (14): 5947–50. https://doi.org/10.1039/C4NR07508K.

Pu, Zonghua, Qian Liu, Abdullah M. Asiri, Abdullah Y. Obaid, and Xuping Sun. 2014. "One-Step Electrodeposition Fabrication of Graphene Film-Confined WS2 Nanoparticles with Enhanced Electrochemical Catalytic Activity for Hydrogen Evolution." *Electrochimica Acta* 134 (July): 8–12. https://doi.org/10.1016/J.ELECTACTA.2014.04.092.

Pumera, Martin. 2010. "Graphene-Based Nanomaterials and Their Electrochemistry." *Chemical Society Reviews* 39 (11): 4146–57. https://doi.org/10.1039/C002690P.

Quan, Quan, Shunji Xie, Bo Weng, Ye Wang, and Yi-Jun Xu. 2018. "Revealing the Double-Edged Sword Role of Graphene on Boosted Charge Transfer versus Active Site Control in TiO$_2$ Nanotube Arrays@RGO/MoS$_2$ Heterostructure." *Small* 14 (21): 1704531. https://doi.org/10.1002/SMLL.201704531.

Rao, Chintamani Nagesa Ramachandra, Kailash Gopalakrishnan, and A. Govindaraj 2014. "Synthesis, Properties and Applications of Graphene Doped with Boron, Nitrogen and Other Elements." *Nano Today* 9 (3): 324–43. https://doi.org/10.1016/J. NANTOD.2014.04.010.

Roy, Sunanda, Xiuzhi Tang, Tanya Das, Liying Zhang, Yongmei Li, Sun Ting, Xiao Hu, and C. Y. Yue. 2015. "Enhanced Molecular Level Dispersion and Interface Bonding at Low Loading of Modified Graphene Oxide to Fabricate Super Nylon 12 Composites." *ACS Applied Materials and Interfaces* 7 (5): 3142–51. https://doi.org/10.1021/AM5074408.

Saha, Dipendu, Yunchao Li, Zhonghe Bi, Jihua Chen, Jong K Keum, Dale K Hensley, Hippolyte A Grappe, et al. 2014. "Studies on Supercapacitor Electrode Material from Activated Lignin- Derived Mesoporous Carbon." *Langmuir* 30: 900–910.

Sarkar, Suprabhat, R. Akshaya, and Sutapa Ghosh. 2020. "Nitrogen-Doped Graphene/ $CuCr_2O_4$ Nanocomposites for Supercapacitors Application: Effect of Nitrogen Doping on Coulombic Efficiency." *Electrochimica Acta* 332 (February): 135368. https://doi.org/ 10.1016/J.ELECTACTA.2019.135368.

Schedin, Fredrik, Andrei Konstantinovich Geim, Sergei Vladimirovich Morozov, Ew W. Hill, Peter Blake, Mi I. Katsnelson, and Kostya Sergeevich Novoselov. 2007. "Detection of Individual Gas Molecules Adsorbed on Graphene." *Nature Materials* 6 (9): 652–55. https://doi.org/10.1038/nmat1967.

Seo, Dong Han, Zhao Jun Han, Shailesh Kumar, and Kostya (Ken) Ostrikov. 2013. "Structure-Controlled, Vertical Graphene-Based, Binder-Free Electrodes from Plasma-Reformed Butter Enhance Supercapacitor Performance." *Advanced Energy Materials* 3 (10): 1316–23. https://doi.org/10.1002/AENM.201300431.

Shabani-Nooshabadi, Mehdi, and Fatemeh Zahedi. 2017. "Electrochemical Reduced Graphene Oxide-Polyaniline as Effective Nanocomposite Film for High-Performance Supercapacitor Applications." *Electrochimica Acta* 245 (August): 575–86. https://doi. org/10.1016/J.ELECTACTA.2017.05.152.

Shams, S. Saqib, Ruoyu Zhang, and Jin Zhu. 2015. "Graphene Synthesis: A Review." *Materials Science Poland* 33 (3): 566–78. https://doi.org/10.1515/msp-2015-0079.

Sharief, Saad Asadullah, Rahmat Agung Susantyoko, Mayada Alhashem, and Saif Almheiri. 2017. "Synthesis of Few-Layer Graphene-like Sheets from Carbon-Based Powders via Electrochemical Exfoliation, Using Carbon Black as an Example." *Journal of Materials Science* 52 (18): 11004–13. https://doi.org/10.1007/S10853-017-1275-3.

Shen, Haode, Hongji Li, Mingji Li, Cuiping Li, Lirong Qian, Lin Su, and Baohe Yang. 2018. "High-Performance Aqueous Symmetric Supercapacitor Based on Polyaniline/Vertical Graphene/Ti Multilayer Electrodes." *Electrochimica Acta* 283 (September): 410–18. https://doi.org/10.1016/J.ELECTACTA.2018.06.182.

Shen, Jianfeng, Yizhe Hu, Min Shi, Xin Lu, Chen Qin, Chen Li, and Mingxin Ye. 2009. "Fast and Facile Preparation of Graphene Oxide and Reduced Graphene Oxide Nanoplatelets." *Chemistry of Materials* 21 (15): 3514–20. https://doi.org/10.1021/CM901247T.

Sheng, Kai Xuan, Yu Xi Xu, Chun Li, and Gao Quan Shi. 2011. "High-Performance Self-Assembled Graphene Hydrogels Prepared by Chemical Reduction of Graphene Oxide." *New Carbon Materials* 26 (1): 9–15. https://doi.org/10.1016/S1872-5805(11)60062-0.

Śliwak, Agata, Bartosz Grzyb, Noel Díez, and Grażyna Gryglewicz. 2017. "Nitrogen-Doped Reduced Graphene Oxide as Electrode Material for High Rate Supercapacitors." *Applied Surface Science* 399 (March): 265–71. https://doi.org/10.1016/J.APSUSC.2016.12.060.

Su, Ching-Yuan, Yanping Xu, Wenjing Zhang, Jianwen Zhao, Xiaohong Tang, Chuen-Horng Tsai, and Lain-Jong Li. 2009. "Electrical and Spectroscopic Characterizations of Ultra-Large Reduced Graphene Oxide Monolayers." *Chemistry of Materials* 21 (23): 5674–80. https://doi.org/10.1021/CM902182Y.

Su, Xiao Li, Lin Fu, Ming Yu Cheng, Jing He Yang, Xin Xin Guan, and Xiu Cheng Zheng. 2017. "3D Nitrogen-Doped Graphene Aerogel Nanomesh: Facile Synthesis and Electrochemical Properties as the Electrode Materials for Supercapacitors." *Applied Surface Science* 426 (December): 924–32. https://doi.org/10.1016/J. APSUSC.2017.07.251.

Subramanian, V., Hongwei Zhu, Robert Vajtai, P. M. Ajayan, and Bingqing Wei. 2005. "Hydrothermal Synthesis and Pseudocapacitance Properties of MnO_2 Nanostructures." *Journal of Physical Chemistry B* 109 (43): 20207–14. https://doi.org/10.1021/JP0543330.

Sugimoto, Wataru, Katsunori Yokoshima, Yasushi Murakami, and Yoshio Takasu. 2006. "Charge Storage Mechanism of Nanostructured Anhydrous and Hydrous Ruthenium-Based Oxides." *Acta Electrochimica.* https://www.sciencedirect. com/science/article/pii/S0013468606004178.

Swanson, Richard M. 2006. "A Vision for Crystalline Silicon Photovoltaics." *Progress in Photovoltaics: Research and Applications* 14 (5): 443–53. https://doi.org/10.1002/pip.709.

Tale, Bhagyashri, K. R. Nemade, and P. V. Tekade. 2020. "Graphene-Based Nano-Composites for Efficient Energy Conversion and Storage in Solar Cells and Supercapacitors: A Review." *Polymer-Plastics Technology and Materials* 60 (7): 784–97. https://doi. org/10.1080/25740881.2020.1851378.

Tang, Yanhong, Xu Hu, and Chengbin Liu. 2014. "Perfect Inhibition of CdS Photocorrosion by Graphene Sheltering Engineering on TiO_2 Nanotube Array for Highly Stable Photocatalytic Activity." *Physical Chemistry Chemical Physics* 16 (46): 25321–29. https://doi.org/10.1039/C4CP04057K.

Toh, S. Y., K. S. Loh, S. K. Kamarudin, and W. R. W. Daud. 2018. "Facile Preparation of Ultra-Low Pt Loading Graphene-Immobilized Electrode for Methanol Oxidation Reaction." *International Journal of Hydrogen Energy* 43 (33): 16005–14. https://doi. org/10.1016/J.IJHYDENE.2018.07.016.

Tripathi, Prashant, Ch Patel, Ravi Prakash, M. A. Shaz, and O. N. Srivastava. 2013. "Synthesis of High-Quality Graphene through Electrochemical Exfoliation of Graphite in Alkaline Electrolyte." *Arxiv.Org.* https://arxiv.org/abs/1310.7371.

Wang, Hailiang, Hernan Sanchez Casalongue, Yongye Liang, and Hongjie Dai. 2010. "$Ni(OH)_2$ Nanoplates Grown on Graphene as Advanced Electrochemical Pseudocapacitor Materials." *Journal of the American Chemical Society* 132 (21): 7472–77. https://doi. org/10.1021/JA102267J.

Wang, Hongzhi, Chen Shen, Jin Liu, Weiguo Zhang, and Suwei Yao. 2019. "Three-Dimensional $MnCo_2O_4$/Graphene Composites for Supercapacitor with Promising Electrochemical Properties." *Journal of Alloys and Compounds* 792 (July): 122–29. https://doi.org/ 10.1016/J.JALLCOM.2019.03.405.

Wang, Huan Wen, Zhong Ai Hu, Yan Qin Chang, Yan Li Chen, Zi Qiang Lei, Zi Yu Zhang, and Yu Ying Yang. 2010. "Facile Solvothermal Synthesis of a Graphene Nanosheet– Bismuth Oxide Composite and Its Electrochemical Characteristics." *Electrochimica Acta* 55 (28): 8974–80. https://doi.org/10.1016/J.ELECTACTA.2010.08.048.

Wang, Huanwen, Cao Guan, Xuefeng Wang, and Hong Jin Fan. 2015. "A High Energy and Power Li-Ion Capacitor Based on a TiO_2 Nanobelt Array Anode and a Graphene Hydrogel Cathode." *Small* 11 (12): 1470–77. https://doi.org/10.1002/SMLL.201402620.

Wang, Ming-Hao, Bo-Wen Ji, Xiao-Wei Gu, Hong-Chang Tian, Xiao-Yang Kang, Bin Yang, Xiao-Lin Wang, Xiang Chen, Cheng-Yu Li, and Jing-Quan Liu. 2018. "Direct Electrodeposition of Graphene Enhanced Conductive Polymer on Microelectrode for Biosensing Application." *Biosensors.* https://www.sciencedirect. com/science/article/pii/S0956566317304785.

Wang, Na, Bairui Tao, Fengjuan Miao, and Yu Zang. 2019. "Electrodeposited Pd/Graphene/ ZnO/Nickel Foam Electrode for the Hydrogen Evolution Reaction." *RSC Advances* 9 (58): 33814–22. https://doi.org/10.1039/C9RA05335B.

Wang, Pengfei, Yuxing Xu, Hui Liu, Yunfa Chen, Jun Yang, and Qiangqiang Tan. 2015. "Carbon/Carbon Nanotube-Supported RuO2 Nanoparticles with a Hollow Interior as

Excellent Electrode Materials for Supercapacitors." *Nano Energy* 15 (July): 116–24. https://doi.org/10.1016/J.NANOEN.2015.04.006.

Wang, Ping, Zhong-Gang Liu, Xing Chen, Fan-Li Meng, Jin-Huai Liu, and Xing-Jiu Huang. 2013. "UV Irradiation Synthesis of an Au–Graphene Nanocomposite with Enhanced Electrochemical Sensing Properties." *Journal of Materials Chemistry A* 1 (32): 9189–95. https://doi.org/10.1039/C3TA11155E.

Wang, Xiluan, Liying Jiao, Kaixuan Sheng, Chun Li, Liming Dai, and Gaoquan Shi. 2013. "Solution-Processable Graphene Nanomeshes with Controlled Pore Structures." *Scientific Reports* 3 (1): 1–5. https://doi.org/10.1038/srep01996.

Wang, Xiwen, Zhian Zhang, Yaohui Qu, Yanqing Lai, Jie Li, Z Zhang, Y Qu, Y Lai, and J Li. 2014. "Nitrogen-Doped Graphene/Sulfur Composite as Cathode Material for High Capacity Lithium-Sulfur Batteries." *Journal of Power Sources* 14. https://doi.org/10.1016/j.jpowsour.2014.01.093.This.

Wang, Yang, Lidong Wang, Bing Wei, Qinghua Miao, Yinan Yuan, Ziyue Yang, and Weidong Fei. 2015. "Electrodeposited Nickel Cobalt Sulfide Nanosheet Arrays on 3D-Graphene/Ni Foam for High-Performance Supercapacitors." *RSC Advances* 5 (121): 100106–13. https://doi.org/10.1039/C5RA20898J.

Wei, Xianjun, Suige Wan, and Shuyan Gao. 2016. "Self-Assembly-Template Engineering Nitrogen-Doped Carbon Aerogels for High-Rate Supercapacitors." *Nano Energy* 28 (October): 206–15. https://doi.org/10.1016/J.NANOEN.2016.08.023.

Wen, Zhenhai, Xinchen Wang, Shun Mao, Zheng Bo, Haejune Kim, Shumao Cui, Ganhua Lu, Xinliang Feng, and Junhong Chen. 2012. "Crumpled Nitrogen-Doped Graphene Nanosheets with Ultrahigh Pore Volume for High-Performance Supercapacitor." *Advanced Materials* 24 (41): 5610–16. https://doi.org/10.1002/ADMA.201201920.

Wodarz, Siggi, Shogo Hashimoto, Mana Kambe, Giovanni Zangari, and Takayuki Homma. 2018. "Fabrication of Electrodeposited FeCuPt Nanodot Arrays Toward L10 Ordering." *IEEE Transactions on Magnetics* 54 (2). https://doi.org/10.1109/TMAG.2017.2746741.

Wu, Jiayang, Linnan Jia, Yuning Zhang, Yang Qu, Baohua Jia, and David J. Moss. 2021. "Graphene Oxide for Integrated Photonics and Flat Optics." *Advanced Materials* 33 (3): 1–29. https://doi.org/10.1002/adma.202006415.

Xia, Kaisheng, Zhiyuan Huang, Lin Zheng, Bo Han, Qiang Gao, Chenggang Zhou, Hongquan Wang, and Jinping Wu. 2017. "Facile and Controllable Synthesis of N/P Co-Doped Graphene for High-Performance Supercapacitors." *Journal of Power Sources* 365 (October): 380–88. https://doi.org/10.1016/J.JPOWSOUR.2017.09.008.

Xia, Wei, Chong Qu, Zibin Liang, Bote Zhao, Shuge Dai, Bin Qiu, Yang Jiao, et al. 2017. "High-Performance Energy Storage and Conversion Materials Derived from a Single Metal-Organic Framework/Graphene Aerogel Composite." *Nano Letters* 17 (5): 2788–95. https://doi.org/10.1021/ACS.NANOLETT.6B05004.

Xiong, Chuanyin, Bingbing Li, Xin Lin, Heguang Liu, Yongjian Xu, Junjie Mao, Chao Duan, Tiehu Li, and Yonghao Ni. 2019. "The Recent Progress on Three-Dimensional Porous Graphene-Based Hybrid Structure for Supercapacitor." *Composites Part B: Engineering* 165 (May): 10–46. https://doi.org/10.1016/J.COMPOSITESB.2018.11.085.

Xiong, Chuanyin, Tiehu Li, Alei Dang, Tingkai Zhao, Hao Li, and Huiqin Lv. 2016. "Two-Step Approach of Fabrication of Three-Dimensional MnO_2-Graphene-Carbon Nanotube Hybrid as a Binder-Free Supercapacitor Electrode." *Journal of Power Sources* 306 (February): 602–10. https://doi.org/10.1016/J.JPOWSOUR.2015.12.056.

Xu, Hui, Qing Yang, Fan Fen Li, Linsheng Tang, Shanmin Gao, Bowei Jiang, C. Sekhar, et al. 2015. "Will Advanced Lithium- Alloy Anodes Have a Chance in Lithium-Ion Batteries?" *AIP* 2 (032146): 200–247. https://doi.org/10.1016/j.jpowsour.2013.03.160.

Xu, Yue, Ying Tao, Huan Li, Chen Zhang, Donghai Liu, Changsheng Qi, Jiayan Luo, Feiyu Kang, and Quan Hong Yang. 2017. "Dual Electronic-Ionic Conductivity of Pseudo-Capacitive Filler Enables High Volumetric Capacitance from Dense

Graphene Micro-Particles." *Nano Energy* 36 (June): 349–55. https://doi.org/10.1016/J.NANOEN.2017.04.054.

Xu, Yuxi, Zhaoyang Lin, Xiaoqing Huang, Yuan Liu, Yu Huang, and Xiangfeng Duan. 2013. "Flexible Solid-State Supercapacitors Based on Three-Dimensional Graphene Hydrogel Films." *ACS Nano* 7 (5): 4042–49. https://doi.org/10.1021/NN4000836.

Xu, Zhen, and Chao Gao. 2011. "Aqueous Liquid Crystals of Graphene Oxide." *ACS Nano* 5 (4): 2908–15. https://doi.org/10.1021/NN200069W.

Xu, Zhen, and Chao Gao. 2014. "Graphene in Macroscopic Order: Liquid Crystals and Wet-Spun Fibers." *Accounts of Chemical Research* 47 (4): 1267–76. https://doi.org/10.1021/AR4002813.

Xuan, Haicheng, Guohong Zhang, Xiaokun Han, Rui Wang, Xiaohong Liang, Yuping Li, and Peide Han. 2021. "Construction of MnSe$_2$/CoSe$_2$/Reduced Graphene Oxide Composites with Enhanced Electrochemical Performance as the Battery-like Electrode for Hybrid Supercapacitors." *Journal of Alloys and Compounds* 863: 158751. https://doi.org/10.1016/j.jallcom.2021.158751.

Yan, Jun, Wei Sun, Tong Wei, Qiang Zhang, Zhuangjun Fan, and Fei Wei. 2012. "Fabrication and Electrochemical Performances of Hierarchical Porous Ni(OH)2 Nanoflakes Anchored on Graphene Sheets." *Journal of Materials Chemistry* 22 (23): 11494–502. https://doi.org/10.1039/C2JM30221G.

Yan, Xingbin, Jiangtao Chen, Jie Yang, Qunji Xue, and Philippe Miele. 2010. "Fabrication of Free-Standing, Electrochemically Active, and Biocompatible Graphene Oxide–Polyaniline and Graphene–Polyaniline Hybrid Papers." *ACS Applied Materials and Interfaces* 2 (9): 2521–29. https://doi.org/10.1021/AM100293R.

Yang, Liming, Yanhong Tang, Dafeng Yan, Tian Liu, Chengbin Liu, and Shenglian Luo. 2015. "Polyaniline-Reduced Graphene Oxide Hybrid Nanosheets with Nearly Vertical Orientation Anchoring Palladium Nanoparticles for Highly Active and Stable Electrocatalysis." *ACS Applied Materials and Interfaces* 8 (1): 169–76. https://doi.org/10.1021/ACSAMI.5B08022.

Yang, Xiaowei, Junwu Zhu, Ling Qiu, and Dan Li. 2011. "Bioinspired Effective Prevention of Restacking in Multilayered Graphene Films: Towards the Next Generation of High-Performance Supercapacitors." *Advanced Materials* 23 (25): 2833–38. https://doi.org/10.1002/ADMA.201100261.

Yi, Min, and Zhigang Shen. 2015. "A Review on Mechanical Exfoliation for the Scalable Production of Graphene." *Journal of Materials Chemistry A* 3: 11700–15. https://doi.org/10.1039/x0xx00000x.

Yu, Dingshan, Kunli Goh, Hong Wang, Li Wei, Wenchao Jiang, Qiang Zhang, Liming Dai, and Yuan Chen. 2014. "Scalable Synthesis of Hierarchically Structured Carbon Nanotube–Graphene Fibres for Capacitive Energy Storage." *Nature Nanotechnology* 9 (7): 555–62. https://doi.org/10.1038/nnano.2014.93.

Yu, Jianhua, Zhenxing Cui, Xu Li, Di Chen, Jiawen Ji, Qian Zhang, Jing Sui, Liyan Yu, and Lifeng Dong. 2020. "Facile Fabrication of ZIF-Derived Graphene-Based 2D Zn/Co Oxide Hybrid for High-Performance Supercapacitors." *Journal of Energy Storage* 27 (February): 101165. https://doi.org/10.1016/J.EST.2019.101165.

Yuan, Li Xia, Zhao Hui Wang, Wu Xing Zhang, Xian Luo Hu, Ji Tao Chen, Yun Hui Huang, and John B. Goodenough. 2011. "Development and Challenges of LiFePO$_4$ Cathode Material for Lithium-Ion Batteries." *Energy and Environmental Science* 4 (2): 269–84. https://doi.org/10.1039/c0ee00029a.

Zhang, Haiyang, Bin Dai, Xugen Wang, Wei Li, You Han, Junjie Gu, and Jinli Zhang. 2013. "Non-Mercury Catalytic Acetylene Hydrochlorination over Bimetallic Au–Co(III)/SAC Catalysts for Vinyl Chloride Monomer Production." *Green Chemistry* 15 (3): 829–36. https://doi.org/10.1039/C3GC36840H.

Zhang, Weijie, Zhongtao Chen, Xinli Guo, Kai Jin, Yi Xuan Wang, Long Li, Yao Zhang, Zengmei Wang, Litao Sun, and Tong Zhang. 2018. "N/S Co-Doped Three-Dimensional

Graphene Hydrogel for High Performance Supercapacitor." *Electrochimica Acta* 278 (July): 51–60. https://doi.org/10.1016/J.ELECTACTA.2018.05.018.

Zhang, Weili, Chuan Xu, Chaoqun Ma, Guoxian Li, Yuzuo Wang, Kaiyu Zhang, Feng Li, et al. 2017. "Nitrogen-Superdoped 3D Graphene Networks for High-Performance Supercapacitors." *Advanced Materials* 29 (36): 1701677. https://doi.org/10.1002/ADMA.201701677.

Zhang, Xiaoyan, and Paolo Samorì. 2017. "Graphene/Polymer Nanocomposites for Supercapacitors." *ChemNanoMat* 3 (6): 362–72. https://doi.org/10.1002/cnma.201700055.

Zhang, Xue-Yu, Shi-Han Sun, Xiao-Juan Sun, Yan-Rong Zhao, Li Chen, Yue Yang, Wei Lü, and Da-Bing Li. 2016. "Plasma-Induced, Nitrogen-Doped Graphene-Based Aerogels for High-Performance Supercapacitors." *Light: Science & Applications* 5 (10): e16130. https://doi.org/10.1038/lsa.2016.130.

Zhao, Yang, Chuangang Hu, Yue Hu, Huhu Cheng, Gaoquan Shi, and Liangti Qu. 2012. "A Versatile, Ultralight, Nitrogen-Doped Graphene Framework." *Angewandte Chemie International Edition* 51 (45): 11371–75. https://doi.org/10.1002/ANIE.201206554.

Zheng, Xuewen, Huitao Yu, Ruiguang Xing, Xin Ge, He Sun, Ruihong Li, and Qiwei Zhang. 2018. "Multi-Growth Site Graphene/Polyaniline Composites with Highly Enhanced Specific Capacitance and Rate Capability for Supercapacitor Application." *Electrochimica Acta* 260 (January): 504–13. https://doi.org/10.1016/J.ELECTACTA.2017.12.100.

12 Recent Advances in Composites of Mixed Transition Metal Oxides and Graphene Oxide– Based Anode Materials for Lithium-Ion Batteries

Onyekachi Nwakanma
University of Nigeria

Fabian I. Ezema
University of Nigeria
iThemba LABS-National Research Foundation
University of South Africa (UNISA)

CONTENTS

DOI: 10.1201/9781003215196-12

12.1 INTRODUCTION

12.1.1 General Outlook and Need for Alternative Sources

The abundance of renewable energy sources (i.e., solar, geothermal, wind, bio-mass, hydropower) provides viable alternatives to non-renewable energy resources (e.g., crude oil, coal, diesel), a potential solution to environmental pollution adverse effects from greenhouse gases (GHCs) and growing energy demands. The global growth in energy demand is closely associated with the increase in the human population, technological advancements, and urbanization, among other factors. Statistics from the World Population Prospects show that the world's population is growing, albeit at a slowing rate (Figure 12.1); this implies an accompanying growth in energy demands to meet the growing populace's requirements. The energy consumption globally and accompanying growth account for 82% of total GHG emissions, with CO_2 from fuel combustion responsible for the most substantial fraction, up to 92% in most developing countries, though with varying percentages according to the economic structure of the country [1]. With a global energy-related CO_2 emissions projection at 0.6% per year from 2018 to 2050 [2] and associated population growth, these spell unprecedented pollution across the globe and massive dangers to lives.

The dangers posed by these pollutions also affect other areas of the environment, and not only the atmosphere [2,4,5]. Faced with the dangers of pollution posed by GHG and other nuisances, the decarbonization of the energy sector, especially in the developing world, is one of the most critical challenges facing the global energy system. Furthermore, in an age where the energy scenery is constantly changing, the stakeholders in the energy sectors pay attention to many different change signals, facilitating a shared understanding of successful energy transitions and viable

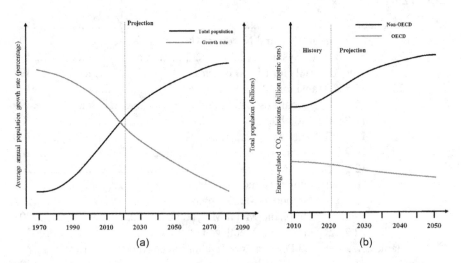

FIGURE 12.1   (a) Population size and projected annual growth rate for the world between 1950 and 2020, after [3], and (b) CO_2 global energy-related emissions estimated by international energy outlook [1].

strategies. Among the strategies many world councils and stakeholders adopted to combat pollution is electrification to reduce GHG from the energy sector.

An alternative to the use of fossil fuels could be nuclear energy. The World Nuclear Association (2019) report shows that nuclear energy supplies about 10.5% of global demand and is the second-largest low-carbon energy source after hydro, aiming to 25% of global electricity in 2050 delivered by nuclear energy [6]. These amounts of energy from nuclear energy sources signify a tremendous amount of energy, which goes a long way to supplement energy needs. However, although carbon-free, atomic sources are not the safest form of energy. Many safety concerns are raised by nuclear power, including harmful radiation, especially from the waste products that are very hazardous to health, the potential to be used as a source of nuclear weapons, and the possibility of a meltdown of the plants. Therefore, the search for a clean, eco-friendly, and safe form of energy made solar energy the most viable alternative to address the world's energy crisis.

Consequently, the rapid development of the modern economy and demands for renewable energy resources led to the development of robust, environmentally benign, cost-effective, and renewable energy alternatives with considerable research enthusiasm as next-generation energy resources to tackle these urgent environmental issues [7]. Moreover, these renewable energy sources also leave no damaging fingerprints on the ecosystem since they occur mostly naturally, are environmentally benign, very robust, cost-effective, and thus with resources to tackle urgent environmental issues [8]. Moreover, there is a flourishing abundance of emerging and competitive cost-efficient energy technologies, notably renewable energy, accelerating the transition toward carbon-free, decentralized, and digitalized energy systems.

Along with the need to trap these energies from renewable sources and address the ever-increasing energy demands and associated crisis, many developed energy storage appliances provide a means to store these energies. In addition, energies generated from other sources (e.g., nuclear) can also be stored for future use with the right tools.

Examples of the storage devices include myriads of energy storage appliances such as conventional capacitors, lithium-ion (Li-ion) batteries, and supercapacitors developed and widely employed in high-technology industries [9,10]. Among these energy storage devices, batteries and supercapacitors are the most sought-after candidate for power and energy storage owing to their unique electrochemical performances (Table 12.1). These devices and technologies possess various advantages, though with some associated disadvantages.

However, a significant limitation of the capacitors arises from their unsatisfactory energy density, restricting their practical applications [12]. Characteristically, their energy density substantially depends on the specific capacitance value and the potential operating range of the devices. Comparatively, batteries offer a much higher energy density, well-suited for long-term energy storage. Therefore, they are better suited for higher energy density applications where devices may need to run on a single charge for extended periods. In addition, the batteries also provide a near-constant voltage output or gradual voltage loss until they are completely spent, in contrast to capacitors' voltage output, which declines linearly with their charge [13,14].

TABLE 12.1

Comparison between supercapacitor (i.e., electrical double-layer capacitors (EDLCs)) and battery (i.e., Li-Ion batteries) [11]

Features	Supercapacitors	Battery
Storage mechanism for energy	Physical	Chemical
Storage determinants for the charge	Microstructure and electrolyte	Active mass and thermodynamics
Power density	5–10 kW kg^{-1}	0.5–1 kW kg^{-1}
Energy density	1–10 Wh kg^{-1}	20–100 Wh kg^{-1}
Power limitation	Electrolyte conductivity	Depends on the kinetics of the reaction and mass transport
Charge time	1–30 seconds	1–3 hours
Discharge time	1–30 seconds	1–5 hours
Cycle life	>500,000 (input and output, ease in charge)	<2,000 (irreversible redox reaction and phase change)
Cycle life limitation	Owing to the occurrence of side reactions	Physically stable, chemically reversible
Configurations	Bipolar	Bipolar
Internal Potential	Owing to the high area platform and electrolyte	Owing to the active material and electrolyte

12.1.2 THE BATTERY SYSTEM

Since the energy from renewable sources may not be available at all times (e.g., no sunlight at night), batteries may store excess power in the battery system. The storage allows for pulling from the batteries later instead of the typical power grid, ensuring the power will be available when needed. Additionally, the energy storage using batteries allows for potential delinking from low-consumption periods, and the systems overcome obstacles caused by the intermittent production of this energy [14]. For instance, solar battery systems enable building a far more resilient home and are less dependent on the grid. Also, installing solar batteries reduces domestic carbon footprint and moves closer to self-sufficiency, ensuring that any energy produced is not wasted [15].

Among the several battery systems, Li-ion batteries have grown significantly in recent years, owing to their distinct advantages and improvements over other forms of battery technology. It is the most widely used energy storage system for portable electronic devices (e.g., laptops, mobile phones, microelectronic devices, and electric vehicles). Its excellent features include low maintenance, high energy density, no memory effect, and negligible self-discharge [16,17]. Recent studies show that Li-ion battery technology advances quickly, aiming to address the possible limitations of the battery system (improved safety, longer lifetime, smaller size, lighter weight, and lower cost) and improve the overall technology [18,19]. These studies aim to explore these properties, improve the existing alluring properties, and even hybridize the new systems for better optimizations (Figure 12.2). A key aspect of enhancing the Li-ion battery operation involves the improved performance of the electrode materials which

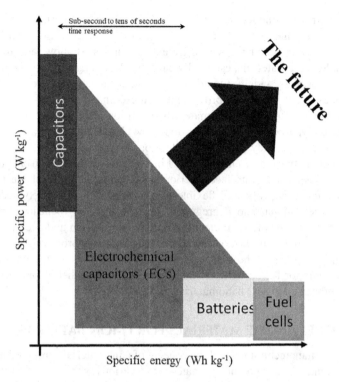

FIGURE 12.2 Ragone plot comparing the specific power and a specific energy for various energy storage systems [20].

may be limited in terms of energy and power density due to the low theoretical capacity and low Li-ion transport rate [19].

The numerous metal oxide materials used as potential anodes for Li-ion batteries have excellent theoretical capacities, high power density, and broad usefulness. They also have relative natural abundance, lucrativeness, and are environmentally benign [11]. Furthermore, mixed transition metal oxides employed as electrode materials exhibit some unique properties, such as changeable valence allowing for intercalation of ions and electrons into the lattice of the metallic compounds and their inherent high stability owing to the coexistence of two distinct metal species in a single crystal structure [21,22]. Furthermore, the cooperative effects between the unique metal oxides in mixed transition metal oxide (MTMO) can enlarge the potential working window and develop more electroactive sites than single-component metal oxide [11,21,23].

However, MTMO may show poor conductivity, aggregation through the continuous charge–discharge processes, and inferior cycling stability [24]. Further challenges suffered by the metal oxides include severe volume changes during the alloying–dealloying processes, pulverization, agglomeration of primitive particles, and poor-electronic conductivity, which hinders the reaction with lithium during electrochemical reactions [19].

One approach to addressing these challenges involves developing the metal oxide materials into nanostructures [25], which may allow for coating or combining the buffering matrix or conductive materials with metal oxide materials as another way to alleviate challenges [26,27]. It also allows for designing various nanostructures of metal oxide nanomaterials with different dimensions, hollow, and hierarchical structures. Several materials, especially carbon-related, e.g., carbon nanotubes and graphene nanosheets, widely studied as the buffering and conductive agent for metal oxide anodes, allow for the improvement of these anodes, including MTMO graphene oxide–based (GOB) anode materials for Li-ion batteries [28,29].

Consequently, a potential hybridization of MTMO with carbonaceous material, essentially graphene, may provide and develop a synergistic effect to increase the electrical conductivity and enhance the kinetics of ions and electron diffusion at the electrode/electrolyte interface interior of the electrodes. Graphene provides a highly conductive network and flexible supporting layer, effectively reducing the volume-change and particle aggregation. Furthermore, its inclusion can provide more electroactive sites for sufficient faradaic redox reaction and accommodate the volume expansion during the continuous charge–discharge process [30].

This chapter presents the recent advances in composites of MTMOs and GOB anode materials for Li-ion batteries and the numerous synthetic approaches developed to synthesize MTMO nanomaterial.

12.2   NOVEL ANODE MATERIALS FOR LI-ION BATTERIES

The advent of nanotechnology and the concept of nanomaterials provided made various revolutionary developments on materials of various types at the nanoscale level [31], providing a novel method to address the challenges involved in developing the materials for anode applications [25]. Additionally, functionalization can enhance the properties and characteristics of nanoparticles through surface modification, enabling them to play significant roles in their applications [32].

For the anode, which is the negative electrode of a primary cell and is consistently associated with the oxidation or the release of electrons into the external circuit [33], the surface and other properties influence the performances of the anode and, consequently, the entire battery. In rechargeable cells, the anodes function as the negative pole during discharge and the positive pole during charging [33,34]. Some requirements for the choice of anode materials include the following properties good conductivity, high coulombic output, relatively high stability, and efficient, reducing agent. In addition, the choice of materials, especially for large-scale and industrial applications, consider the cost and ease of fabrication.

This novel nanotechnology techniques are very instrumental in the development of MTMOs as anodes in Li-ion batteries, which in comparison to traditional metal oxides, offer the following advantages [19]:

a. Two types of metal elements in MTMOs have different expansion coefficients, which results in a synergistic effect.
b. MTMOs can alloy with more Li ions than mixed and typical metal oxides, owing to the complex chemical compositions and higher reversible capacities. In addition, both elements in MTMOs are electrochemically active metals to Li metal, resulting in better electrochemical performance.

c. The MTMOs exhibit higher electrical conductivity significantly than simple metal oxides due to the comparatively low activation energy for transferring electrons between cations.

d. MTMOs are often more environmentally benign than traditional metal oxides.

MTMOs refer to ternary metal oxides with two different metal cations, including stannates, ferrites, cobaltates, and nickelates.

12.2.1 MIXED TRANSITION METAL OXIDE (MTMO)

Recent studies explore the metal and their composites as anode materials for Li-ion batteries, considering the advantages. They generally possess the chemical composition, XM_2O_4 (M = Mo, Co, V, W, Fe, Mn), and the studies include their synthesis methods, Li-storage mechanisms, and good structural design. Moreover, these metal oxides tend to possess a higher diffusion coefficient for the electrode material leading to rapid migration of electrolyte ions to the surface of the bulk electrode hence ameliorating the electrochemical activities [35].

One of the advantages of MTMOs may be due to the lower charge/discharge voltages leading to high energy density, and they are mostly environmentally friendly. Generally, the electrochemical reactions of XM_2O_4 during the conversion process with lithium can be represented as [19]:

$$XM_2O_4 + 8Li^+ + 8e^- \rightarrow X + 2M + 4Li_2O$$
$$X + Li_2O \rightarrow XO + 2Li^+ + 2e^- \tag{12.1}$$
$$M + Li_2O \rightarrow MO + 2Li^+ + 2e^-$$

12.2.1.1 Cobaltites

The cobalt oxide materials show tremendous research interest as anode materials for LIBs mainly due to their high reversible capacity in chemical reactions. It, however, exhibits relatively high toxicity and cost, prompting the efforts of several researchers for other eco-friendly and low-cost alternative elements for doping or total replacement. The standard choice of dopants (e.g., Zn) present bivalent ions during doping to the parent cobalt oxide material, primarily occupying the tetrahedral sites in the cubic spinel structure, while the trivalent Co-ions occupy the octahedral sites [36]. For instance, doping with Zn provides additional capacity due to the alloying process between Zn and Li. The challenges posed by the resultant combined volume of $ZnCo_2O_4$ may be addressed by synthesizing structures with hollow structures or yolk-shell arrangements [36]. The equation describes the electrochemical reaction for $ZnCo_2O_4$ formation [37]:

$$ZnCo_2O_4 + 8Li^+ + 8e^- \leftrightarrow Zn + 2Co + 4Li_2O$$
$$Zn + Li^+ + e^- \leftrightarrow LiZn$$
$$Zn + Li_2O \leftrightarrow ZnO + 2Li^+ + 2e^- \tag{12.2}$$
$$2Co + 2Li_2O \leftrightarrow 2CoO + 4Li^+ + 4e^-$$
$$2CoO + 2/3Li_2O \leftrightarrow 2/3Co_3O_4 + 4/3Li^+ + 4/3e^-$$

The set of equations shows that the complex reaction mechanism of $ZnCo_2O_4$ involves multiple steps due to the multielement formation. Some studies from researchers also attempt to design nanostructures with different dimensions with similar electrochemical mechanisms to $ZnCo_2O_4$ using other cobaltates, such as $NiCo_2O_4$ [38], $MnCo_2O_4$ [39], and $FeCo_2O_4$ [40]. These nanostructures present high surface-to-volume ratios and excellent electronic transport properties needed to enhance LIB capacity and cycling performance. They also have a high surface area, low density, and high loading capacity.

The positive effects on the electrochemical reactions account for the conversion reaction or alloying/de-alloying reaction of the dopants with Li_2O, increasing the capacity compared to only Co_3O_4. Further studies on nanostructured cobaltates and hierarchical structures with highly conductive substrates such as carbon cloth, Ni foam, or graphene nanosheets aim to address the challenges associated with conductivity and volume exchange, limiting the performance.

12.2.1.2 Molybdates

Like the cobaltites, recent studies on metal molybdates ($XMoO_4$, X = Ni, Co, Zn) aim to explore their several oxidation states ranging from +3 to +6 for Mo [41], which potentially enhances the reversible capacity in Li-ion battery applications where they can serve both as cathode and anode materials [42]. The electrochemical reaction of $XMoO_4$ with lithium may be described by:

$$XMoO_4 + 8Li \rightarrow X + Mo + 4Li_2O$$
$$X + Li_2O \leftrightarrow XO + 2Li \qquad (12.3)$$
$$Mo + 3Li_2O \leftrightarrow MoO_3 + 6Li$$

Compared with just Mo oxide, the molybdates show great potential as alternative anodes for LIBs due to the addition of metal X, which is electrochemically active and provides extra capacity for reversible capacities by the reaction between metal X and Mo with Li_2O. A good design during nanoparticle synthesis provides high surface area, short ion diffusion lengths, and effective electron transport pathways, resulting in improved cycling and rate performance.

Further studies on molybdate nanoparticles with graphene reportedly exhibit enhanced electrochemical performance, with good rate capability and impressive cycling stability. In addition, synergetic chemical coupling effects between the graphene conductive network and nanoparticles significantly impact the performance [41].

12.2.1.3 Vanadates

Vanadium elements have a high natural abundance and are lucrative, presenting some fascinating electrochemical results (e.g., wide-operating potential window due to multiple oxidation states) [43]. These metal vanadates' electrochemical behaviors depend on the synthetic methods and the developed morphologies, making them stand out from the other MTMOs.

For example, the 3D porous nanoroses with an approximate average size of 474 nm reportedly possessed high specific capacitance with good charge storage capability and superior rate capability [44]. In addition, the large specific surface area of 3D mesoporous and well-defined pore structure allows for intimate contact between the electroactive sites and electrolytes, beneficial for pronounced redox reactions.

12.2.1.4 Ferrites

Ferrites (XFe_2O_4), as alternative anode materials to iron oxides, have been studied for their application in LIBs. For example, the theoretical capacity of Fe_3O_4 is 926 mAh g^{-1}, assuming the completely reversible formation of four Li_2O per formula unit, in comparison with $ZnFe_2O$, which can achieve 1,000.5 mAh g^{-1} [45].

Various reported examples of 3D nanostructures of XFe_2O_4, e.g., spheres, cubes, and ordered macroporous, yolk-shell structures, provide buffering spaces for the electroactive core material and improve the surface area [19]. Their reported properties suggested enhancing the surface areas, shortening lithium transport distances, and creating a volume-change buffering system. A typical challenge bedeviling the XFe_2O_4 material is the conductivity, commonly addressed using carbon materials, e.g., graphene or other conductive materials, such as conductive polymers, due to their high conductivity and stability. However, introducing carbon nanomaterials into the XFe_2O_4 structures reportedly rapidly improved the electrochemical performance due to the carbons' highly conductive and buffering matrix [46].

12.2.1.5 Manganates

Manganates (XMn_2O_4), among the MTMOs, offer some advantages, including low toxicity, low cost, and reasonably low operating voltages (0.5 V for charging and 1.2 V for discharging) compared with Co and Fe [47]. Also, their lower charge/discharge voltages as anode materials can effectively increase the energy density of LIBs. In addition, one-dimensional manganates nanomaterials, such as nanowires and nanotubes, reportedly can supply fast Li^+ and electron transport pathways, a large contact area with electrolyte, and a buffer zone for mechanical stress during electrochemical processes. They also show superior electrochemical lithium-storage performance with a high specific capacity, good rate behavior, and excellent cyclability. Additionally, they can provide a tremendous buffering matrix that alleviates the pulverization problem, making the electrode stable and enhancing the cycling performance.

However, MTMOs may still suffer from low electrical conductivity, poor rate performance, inferior cycling stability, and non-effective utilization of the active material during the electrochemical reaction compared to the hybrid electrode materials (combination of MTMO with carbonaceous material, e.g., graphene).

12.2.2 GRAPHENE OXIDE–BASED MATERIALS

Graphene inclusion, among other carbonaceous materials, offers one option to improve the existing advantages of MTMO materials. Consequently, the hybridization of carbonaceous materials, in this case, graphene, with various MTMOs, offers a functional and effective strategy to enhance the electrochemical performances of anode materials. Compared with other carbonaceous materials, e.g., activated carbon nanotubes, graphene possesses a sheet of sp^2-bonded carbon atoms with one atom thick in a honeycomb crystal structure [11].

Although activated carbon reportedly has a large specific area, it has limited practical application because of the poor-electrical conductivity. Similarly, carbon nanotubes, which reportedly possess a larger specific surface area and better electrical conductivity than activated carbon, have a high contact resistance with the current collector.

The interest from researchers in graphene may be due to its intriguing physical and chemical properties, e.g., significant theoretical surface area, superior electrical conductivity, outstanding electrochemical stability, high flexibility, and excellent structural tenacity. Furthermore, graphene can exist in one, two or three dimensions, and the hybridization of graphene sheets with nanoscale MTMO potentially enlarges the accessible specific surface of the active materials. At the same time, the synergistic effect and contact between them promote the diffusion process of electrons and ions. Furthermore, the various synthetic approaches integrated with 2D graphene sheets lead to a facile and controllable nanocomposite formation with distinct nanostructures, aiming to utilize supercapacitors. Also, the free-standing 2D graphene nanosheets consist of 2D edge plane sites that accelerate the ions adsorption process and enhance the specific energy and specific power of graphene-MTMO nanocomposite.

12.2.2.1 Graphene-Metal Cobaltites

Studies on the integration of graphene with metal cobaltite as nanocomposites draw tremendous research enthusiasm owing to the improved physical and electrochemical properties compared to lone metal cobaltites. Some studies show that reduced graphene oxide (rGO) minimized the ions migration pathway and eased the diffusion process of electrolyte ions to the redox centers [48,49]. For instance, rGO hinders the agglomeration of the cobaltite particles, which assures effective reversible redox reactions at the electrode/electrolyte interface. It also improves the cycling stability due to the peculiar nanoarchitecture, improving the electrical conductivity and serving as a buffer matrix to adapt the volume expansion or variation during the continuous charge–discharge process.

12.2.2.2 Graphene-Metal Molybdates

Like graphene/metal cobaltite, graphene/metal molybdate also has considerable attention as a potential charge storage system due to its advanced electrochemical properties, good mechanical stability, and electronic conductivity [50]. For example, flower-like nanoflake $CoMoO_4$-rGO nanocomposites have a larger specific surface area and average pore diameter than the pristine $CoMoO_4$ nanoparticles, suggesting surface area enhancement through integration with graphene sheets [51]. The superior electrochemical properties may be due to the assumed reasons: (i) The flower-like nanoarchitecture encouraged the ions (OH^-) and electrons transportation which accelerated the reaction rate. (ii) The graphene networks offered adequate electroactive spaces, which enabled effective utilization of the graphene/molybdate nanocomposite electrode material. (iii) Graphene served as a cushion to absorb or reduce the structural stress or volume variation during the continuous charge–discharge process and thus prevented the structural deformation.

Comparatively, the direct growth of $NiMoO_4$ nanoparticles and quantum dots on the reduced nanohole graphene oxide (rNHGO) allowed intimate contact between the active material and the current collector, resulting in effective ions diffusion. Moreover, the larger specific surface area of NiMoO4@rNHGO could offer more electroactive spaces for the reversible redox reaction, which prolonged the cycle life and reinforced the structural tenacity [11]. The synergistic effect between quantum dots, $NiMoO_4$ nanoparticles, and rNHGO curtailed the electrolyte ions' diffusion route to the bulk of the electrode during the charge–discharge process.

12.2.2.3 Graphene-Metal Vanadates

Compared to the traditional faradaic bimetallic vanadates, incorporating bimetallic vanadates with graphene demonstrates better electrochemical activities of the synergistic effect between the graphene nanosheet and metal vanadates. Hence, many researchers are devoted to evaluating and determining the feasibility of graphene-metal vanadate nanocomposites as electrode material [52]. Using SEM analysis, the morphologies of integrated bismuth vanadate with rGO ($BiVO_4$/rGO) indicated that the $BiVO_4$ nanoparticles decorated thin rGO nanosheets. The unique electrochemical properties of $BiVO_4$/rGO nanocomposite could be due to (i) increased electrical conductivity due to the incorporation of rGO and thus accelerating the ionic transportation and (ii) optimized synthesis parameters for $BiVO_4$/rGO nanocomposite, which effectively prevented the agglomeration of nanomaterials [53].

12.3 CHARACTERIZATIONS AND PERFORMANCE EVALUATION OF ANODE MATERIALS

The synthesis methods used to prepare the nanoparticles, calcination temperatures, and their properties (e.g., structural, morphological, topographical) may significantly affect their composition. Various methods used to synthesize nanoparticles of MTMOs and to dope with other materials (e.g., graphene) are classified broadly into vacuum and non-vacuum techniques [54]. Although the vacuum techniques potentially yield nanoparticles that may have purer qualities with lesser contaminations, the allures of non-vacuum techniques depend on their cost-to-efficiency ratios. Also, some non-vacuum techniques appear to be easily scalable for industrial applications for preparing nanomaterials; their purity is improved by working in controlled environments, e.g., in closed chambers with an argon environment [55]. Comparing some deposition techniques (Table 12.2) shows their advantages and disadvantages.

The characteristics and electrochemical performances of nanomaterials for electrode applications depend on the properties of the synthesized materials, e.g., structure, morphology, electrical conductivity, and some post-synthesis processes, such as thermal annealing.

Morphology has a high impact on the specific surface area of the electrode material, i.e., grain size, specific surface area, and pore size distribution. For example, the electrode materials' porosity and specific surface area are interrelated: the pore sizes, pore volume generated, and the pore size distribution for a given overall specific surface area of the active material. The availability of the electroactive sites and accessibility to electrolyte ions are the most crucial factors determining an electrode material's capacitive performance.

TABLE 12.2

Comparison of Some Synthesis Techniques with Their Advantages and Disadvantages of MTMO Preparation [11]

Synthesis Technique	Advantages	Disadvantages
Metal-Organic Framework (MOF)	• Nanocomposite with 1D, 2D, and 3D can be synthesized • High porosity and large specific surface area of the nanocomposite	• Challenging to produce electrodes with enhanced electrochemical performance
Hydro/ Solvothermal	• Possible for large-scale synthesis • High-purity material can be produced • Facile method • Monodispersed nanoparticles with manageable morphology can be acquired	• Involves relatively high temperature and pressure
Sol-gel	• Low reaction temperature • High-purity materials can be produced • Possible for large-scale synthesis	• Not feasible to fabricate film with porous structures
Microwave-assisted	• Possible for large-scale synthesis • Speeding up the rate of reaction and hence reducing the duration of reactions	• Challenging morphological and structural control
Electrochemical chemical precipitation	• Producing composite with consistent morphology • Possible for large-scale synthesis • Simple route with mild reaction temperatures	• Challenging large-scale synthesis • Difficult to control the structural and morphological behaviors of the nanoparticles

Calcination is an essential process employed to improve the crystallinity of the nanomaterial and induce the formation of MTMO nanocomposite. However, calcination of MTMO-based materials at high temperatures may lead to the agglomeration of the active composite, which adversely affects its structural and electrochemical properties. Consequently, several studies aim to optimize the temperatures for the best possible results. The electrical conductivity and the associated resistance may also affect an electrode material's performance. For instance, the material's resistance to electron conduction dominates material with poor conductivity, especially with a high charge–discharge rate. It confines the charge–discharge process to a finite volume near the current collector and may hinder the practical applications of the oxide-based material as a functional electrode.

REFERENCES

[1] S. Nalley, A. Larose, IEO2021 Highlights. U.S. Energy Information Administration, (2021). https://www.eia.gov/outlooks/ieo/pdf/IEO2021_ReleasePresentation.pdf.

[2] Ebel, R.E., Croissant, M.P., Masih, J.R., Calder, K.E., Thomas, R.G., International energy outlook: US Department of Energy. *Wash. Q.*, 19 (1996) 70–99. https://doi.org/10.1080/01636609609550217.

[3] Parant, A., World population prospects. *Futuribles (Paris, France: 1981)*, (141) (1990) 49–78. http://www.ncbi.nlm.nih.gov/pubmed/12283219.

[4] J.H. Kim, BS. Shim, H.S. Kim, Y.J. Lee, S.K. Min, D. Jang, Z. Abas, J. Kim, Review of nanocellulose for sustainable future materials, *Int. J. Precis. Eng. Manuf. - Green Technol.* 2 (2015) 197–213. https://doi.org/10.1007/s40684-015-0024-9.

[5] Y. Su, Y. Liang, L. Chai, Z. Han, S. Ma, J. Lyu, Z. Li, L. Yang, Water degradation by China's fossil fuels production: A life cycle assessment based on an input–output model, *Sustainability.* 11 (2019) 4130. https://doi.org/10.3390/su11154130.

[6] World Nuclear Association, At Work 2019 Edition, (2019), pp. 1–32. https://www.world-nuclear.org/getattachment/Our-Association/Publications/Annual-Reports-and-Brochures/At-Work-Annual-Report-2019/at-work-2019-may-edition.pdf.aspx.

[7] P. Breeze, The environmental impact of energy storage technologies, in: *Power System Energy Storage Technologies*, Elsevier, 2018: pp. 79–84. https://doi.org/10.1016/B978-0-12-812902-9.00009-2.

[8] E.T. Sayed, T. Wilberforce, K. Elsaid, M.K.H. Rabaia, M.A. Abdelkareem, K.J. Chae, A.G. Olabi, A critical review on environmental impacts of renewable energy systems and mitigation strategies: Wind, hydro, biomass and geothermal, *Sci. Total Environ.* 766 (2021) 144505. https://doi.org/10.1016/J.SCITOTENV.2020.144505.

[9] K. Chen, S. Song, F. Liu, D. Xue, Structural design of graphene for use in electrochemical energy storage devices, *Chem. Soc. Rev.* 44 (2015) 6230–6257. https://doi.org/10.1039/C5CS00147A.

[10] M. Chen, Y. Zhang, G. Xing, S.L. Chou, Y. Tang, Electrochemical energy storage devices working in extreme conditions, *Energy Environ. Sci.* 14 (2021) 3323–3351. https://doi.org/10.1039/D1EE00271F.

[11] WH Low, P.S. Khiew, S.S. Lim, C.W. Siong, E.R. Ezeigwe, Recent development of mixed transition metal oxide and graphene/mixed transition metal oxide based hybrid nanostructures for advanced supercapacitors, *J. Alloys Compd.* 775 (2019) 1324–1356. https://doi.org/10.1016/j.jallcom.2018.10.102.

[12] I.N. Jiya, N. Gurusinghe, R. Gouws, Electrical circuit modelling of double layer capacitors for power electronics and energy storage applications: A review, *Electronics* 7, (2018) 268. https://doi.org/10.3390/ELECTRONICS7110268.

[13] B. Lokeshgupta, S. Sivasubramani, Multi-objective home energy management with battery energy storage systems, *Sustain. Cities Soc.* 47 (2019) 101458. https://doi.org/10.1016/J.SCS.2019.101458.

[14] Y. Yang, S. Bremner, C. Menictas, M. Kay, Battery energy storage system size determination in renewable energy systems: A review, *Renew. Sustain. Energy Rev.* 91 (2018) 109–125. https://doi.org/10.1016/J.RSER.2018.03.047.

[15] Environmental Energies, Benefits of Battery Storage - Environmental Energies, (2020). https://www.environmentalenergies.co.uk/news/benefits-of-battery-storage (accessed January 15, 2022).

[16] X. Liang, J. Yun, Y. Wang, H. Xiang, Y. Sun, Y. Feng, Y. Yu, A new high-capacity and safe energy storage system: Lithium-ion sulfur batteries, *Nanoscale.* 11 (2019) 19140–19157. https://doi.org/10.1039/C9NR05670J.

[17] M. Killer, M. Farrokhseresht, N.G. Paterakis, Implementation of large-scale Li-ion battery energy storage systems within the EMEA region, *Appl. Energy.* 260 (2020) 114166. https://doi.org/10.1016/J.APENERGY.2019.114166.

[18] S. Anuphappharadorn, S. Sukchai, C. Sirisamphanwong, N. Ketjoy, Comparison the economic analysis of the battery between lithium-ion and lead-acid in PV stand-alone application, *Energy Procedia.* 56 (2014) 352–358. https://doi.org/10.1016/j.egypro.2014.07.167.

[19] Y. Zhao, X. Li, B. Yan, D. Xiong, D. Li, S. Lawes, X. Sun, Recent developments and understanding of novel mixed transition-metal oxides as anodes in lithium ion batteries, *Adv. Energy Mater.* 6 (2016) 1502175. https://doi.org/10.1002/aenm.201502175.

[20] D.R. Rolison, O.F. Nazar, Electrochemical energy storage to power the 21st century, *MRS Bull.* 36 (2011) 486–493. https://doi.org/10.1557/mrs.2011.136.

[21] S. Terny, M.A. Frechero, Understanding how the mixed alkaline-earth effect tunes transition metal oxides-tellurite glasses properties, *Phys. B Condens. Matter.* 583 (2020) 412054. https://doi.org/10.1016/J.PHYSB.2020.412054.

[22] J. Li, Z. Liu, Q. Zhang, Y. Cheng, B. Zhao, S. Dai, H.H. Wu, K. Zhang, D. Ding, Y. Wu, M. Liu, M.S. Wang, Anion and cation substitution in transition-metal oxides nanosheets for high-performance hybrid supercapacitors, *Nano Energy.* 57 (2019) 22–33. https://doi.org/10.1016/J.NANOEN.2018.12.011.

[23] A.S. Yasin, A.Y. Mohamed, I.M.A. Mohamed, D.Y. Cho, C.H. Park, C.S. Kim, Theoretical insight into the structure-property relationship of mixed transition metal oxides nanofibers doped in activated carbon and 3D graphene for capacitive deionization, *Chem. Eng. J.* 371 (2019) 166–181. https://doi.org/10.1016/J.CEJ.2019.04.043.

[24] Y. Xu, J. Wei, L. Tan, J. Yu, Y. Chen, A Facile approach to NiCoO$_2$ intimately standing on nitrogen doped graphene sheets by one-step hydrothermal synthesis for supercapacitors, *J. Mater. Chem. A.* 3 (2015) 7121–7131. https://doi.org/10.1039/C5TA00298B.

[25] P.G. Bruce, B. Scrosati, J.M. Tarascon, Nanomaterials for rechargeable lithium batteries, *Angew. Chemie - Int. Ed.* 47 (2008) 2930–2946. https://doi.org/10.1002/ANIE.200702505.

[26] T. Yang, H. Zhang, Y. Luo, L. Mei, D. Guo, Q. Li, T. Wang, Enhanced electrochemical performance of CoMoO$_4$ nanorods/reduced graphene oxide as anode material for lithium-ion batteries, *Electrochim. Acta.* 158 (2015) 327–332. https://doi.org/10.1016/J.ELECTACTA.2015.01.154.

[27] X. Cao, Z. Yin, H. Zhang, Three-dimensional graphene materials: Preparation, structures and application in supercapacitors, *Energy Environ. Sci.* 7 (2014) 1850–1865. https://doi.org/10.1039/C4EE00050A.

[28] N. Mahmood, C. Zhang, H. Yin, Y. Hou, Graphene-based nanocomposites for energy storage and conversion in lithium batteries, supercapacitors and fuel cells, *J. Mater. Chem. A.* 2 (2013) 15–32. https://doi.org/10.1039/C3TA13033A.

[29] S. Han, D. Wu, S. Li, F. Zhang, X. Feng, Graphene: A two-dimensional platform for lithium storage, *Small.* 9 (2013) 1173–1187. https://doi.org/10.1002/smll.201203155.

[30] S. Al-Rubaye, R. Rajagopalan, S.X. Dou, Z. Cheng, Facile synthesis of a reduced graphene oxide wrapped porous NiCo$_2$O$_4$ composite with superior performance as an electrode material for supercapacitors, *J. Mater. Chem. A.* 5 (2017) 18989–18997. https://doi.org/10.1039/C7TA03251J.

[31] I. Khan, K. Saeed, I. Khan, Nanoparticles: Properties, applications and toxicities, *Arab. J. Chem.* 12 (2019) 908–931. https://doi.org/10.1016/j.arabjc.2017.05.011.

[32] R. Thiruppathi, S. Mishra, M. Ganapathy, P. Padmanabhan, B. Gulyás, nanoparticle functionalization and its potentials for molecular imaging, *Adv. Sci.* 4 (2017) 1600279. https://doi.org/10.1002/advs.201600279.

[33] J. Ryu, S. Park, Nanoscale anodes for rechargeable batteries: Fundamentals and design principles, *Nanobatteries Nanogener.* (2021) 91–157. https://doi.org/10.1016/B978-0-12-821548-7.00007-5.

[34] G. Yasin, M. Arif, MA Mushtaq, M. Shakeel, N. Muhammad, M. Tabish, A. Kumar, T.A. Nguyen, Nanostructured anode materials in rechargeable batteries, *Nanobatteries Nanogener.* (2021) 187–219. https://doi.org/10.1016/B978-0-12-821548-7.00009-9.

[35] J. Bhagwan, S. Rani, V. Sivasankaran, K.L. Yadav, Y. Sharma, Improved energy storage, magnetic and electrical properties of aligned, mesoporous and high aspect ratio nanofibers of spinel-NiMn$_2$O$_4$, *Appl. Surf. Sci.* 426 (2017) 913–923. https://doi.org/10.1016/J.APSUSC.2017.07.253.

[36] J. Bai, X. Li, G. Liu, Y. Qian, S. Xiong, J. Bai, X.G. Li, G.Z. Liu, Y.T. Qian, S.L. Xiong, Unusual formation of ZnCo$_2$O$_4$ 3D hierarchical twin microspheres as a high-rate and

ultralong-life lithium-ion battery anode material, *Adv. Funct. Mater.* 24 (2014) 3012–3020. https://doi.org/10.1002/ADFM.201303442.

[37] L. Hu, B. Qu, C. Li, Y. Chen, L. Mei, D. Lei, L. Chen, Q. Li, T. Wang, Facile synthesis of uniform mesoporous $ZnCo_2O_4$ microspheres as a high-performance anode material for Li-ion batteries, *J. Mater. Chem. A.* 1 (2013) 5596–5602. https://doi.org/10.1039/C3TA00085K.

[38] G. Gao, H. B. Wu, S. Ding, X.W. Lou, Preparation of carbon-coated $NiCo_2O_4@SnO_2$ hetero-nanostructures and their reversible lithium storage properties, *Small.* 11 (2015) 432–436. https://doi.org/10.1002/SMLL.201400152.

[39] AK Mondal, D. Su, S. Chen, A. Ung, H.S. Kim, G. Wang, Mesoporous $MnCo_2O_4$ with a flake-like structure as advanced electrode materials for lithium-ion batteries and supercapacitors, *Chem. – A Eur. J.* 21 (2015) 1526–1532. https://doi.org/10.1002/CHEM.201405698.

[40] S.G. Mohamed, C.-J. Chen, C.K. Chen, S.-F. Hu, R.-S. Liu, High-performance lithium-ion battery and symmetric supercapacitors based on $FeCo_2O_4$ nanoflakes electrodes, *ACS Appl. Mater. Interfaces.* 6 (2014) 22701–22708. https://doi.org/10.1021/am5068244.

[41] T. Yang, H. Zhang, Y. luo, L. Mei, D. Guo, Q. Li, T. Wang, Enhanced electrochemical performance of $CoMoO_4$ nanorods/reduced graphene oxide as anode material for lithium-ion batteries, *Electrochimica Acta* 158 (2015) 327–332.

[42] J. Haetge, I. Djerdj, T. Brezesinski, Nanocrystalline $NiMoO_4$ with an ordered mesoporous morphology as potential material for rechargeable thin film lithium batteries, *Chem. Commun.* 48 (2012) 6726–6728. https://doi.org/10.1039/C2CC31570J.

[43] Y. Yan, B. Li, W. Guo, H. Pang, H. Xue, Vanadium based materials as electrode materials for high performance supercapacitors, *J. Power Sources.* 329 (2016) 148–169. https://doi.org/10.1016/J.JPOWSOUR.2016.08.039.

[44] J. Zhang, B. Yuan, S. Cui, N. Zhang, J. Wei, X. Wang, D. Zhang, R. Zhang, Q. Huo, Facile synthesis of 3D porous $Co_3V_2O_8$ nanoroses and 2D $NiCo_2V_2O_8$ nanoplates for high performance supercapacitors and their electrocatalytic oxygen evolution reaction properties, *Dalt. Trans.* 46 (2017) 3295–3302. https://doi.org/10.1039/C7DT00435D.

[45] D. Bresser, E. Paillard, R. Kloepsch, S. Krueger, M. Fiedler, R. Schmitz, D. Baither, M. Winter, S. Passerini, Carbon Coated $ZnFe_2O_4$ Nanoparticles for Advanced Lithium-Ion Anodes, *Adv. Energy Mater.* 3 (2013) 513–523. https://doi.org/https://doi.org/10.1002/aenm.201200735.

[46] B. Wang, S. Li, B. Li, J. Liu, M. Yu, Facile and large-scale fabrication of hierarchical $ZnFe_2O_4$/graphene hybrid films as advanced binder-free anodes for lithium-ion batteries, *New J. Chem.* 39 (2015) 1725–1733. https://doi.org/10.1039/C4NJ01802H.

[47] W. Kang, Y. Tang, W. Li, X. Yang, H. Xue, Q. Yang, C.S. Lee, High interfacial storage capability of porous $NiMn_2O_4$/C hierarchical tremella-like nanostructures as the lithium ion battery anode, *Nanoscale.* 7 (2014) 225–231. https://doi.org/10.1039/C4NR04031G.

[48] C. Zhang, C. Lei, C. Cen, S. Tang, M. Deng, Y. Li, Y. Du, Interface polarization matters: Enhancing supercapacitor performance of spinel $NiCo_2O_4$ nanowires by reduced graphene oxide coating, *Electrochim. Acta.* 260 (2018) 814–822. https://doi.org/10.1016/J.ELECTACTA.2017.12.044.

[49] S. Al-Rubaye, R. Rajagopalan, S.X. Dou, Z. Cheng, Facile synthesis of a reduced graphene oxide wrapped porous $NiCo_2O_4$ composite with superior performance as an electrode material for supercapacitors, *J. Mater. Chem. A.* 5 (2017) 18989–18997. https://doi.org/10.1039/C7TA03251J.

[50] L. Jinlong, Y. Meng, L. Tongxiang, Enhanced performance of $NiMoO_4$ nanoparticles and quantum dots and reduced nanohole graphene oxide hybrid for supercapacitor applications, *Appl. Surf. Sci.* 419 (2017) 624–630. https://doi.org/10.1016/J.APSUSC.2017.05.115.

[51] L. Jinlong, Y. Meng, K. Suzuki, H. Miura, Synthesis of $CoMoO_4$@RGO nanocomposites as high-performance supercapacitor electrodes, *Microporous Mesoporous Mater.* 242 (2017) 264–270. https://doi.org/10.1016/J.MICROMESO.2017.01.034.

[52] R. Kumar, P.K. Gupta, P. Rai, A. Sharma, Free-standing $Ni_3(VO_4)_2$ nanosheet arrays on aminated r-GO sheets for supercapacitor applications, *New J. Chem.* 42 (2018) 1243–1249. https://doi.org/10.1039/C7NJ03862C.

[53] S. Dutta, S. Pal, S. De, Hydrothermally synthesized $BiVO_4$-reduced graphene oxide nanocomposite as a high performance supercapacitor electrode with excellent cycle stability, *New J. Chem.* 42 (2018) 10161–10166. https://doi.org/10.1039/C8NJ00859K.

[54] K. Seshan, *Handbook of Thin-film Deposition Processes and Techniques: Principles, Methods, Equipment and Applications*, Noyes Publications/William Andrew Pub., 2002. https://books.google.com.mx/books?id=72BEzQEACAAJ.

[55] Y. Qi, Y. Huang, D. Jia, S.J. Bao, Z.P. Guo, Preparation and characterization of novel spinel $Li_4Ti_5O_{12-x}Br_x$ anode materials, *Electrochim. Acta.* 54 (2009) 4772–4776. https://doi.org/10.1016/j.electacta.2009.04.010.

13 Flexible Supercapacitors Based on Graphene Oxide

Swati N. Pusawale

Rajarambapu Institute of Technology,
affiliated to Shivaji University

CONTENTS

13.1 INTRODUCTION

Energy is always a major concern for every nation as it is directly related with its development. With the growing demand for energy sources, different approaches of converting the energy into useful form as well as of storing of the energy in efficient way so it can be utilized whenever requires is the major concern [1]. Extensive research is going on developing new electrode materials for energy storage devices such as batteries, supercapacitors, and fuel cells to improve their performance. Among all this, supercapacitors are promising energy storage devices due to their higher power density, cyclic stability, and less toxicity compared with battery. There are two types of charge storage mechanisms in supercapacitors: one is electric double-layer capacitance (EDLC), and the other is pseudo-capacitance arising from faradic oxidation and reduction reaction. The charge storage mechanism in carbon-based material is mainly of EDLC type, whereas in metal oxides and polymers it is of pseudocapacitive type. Both the type of materials has some advantages and disadvantages as EDLC-based materials have high power density and excellent cyclic stability, but their energy density is low; however, the pseudo-capacitors have high energy density compared with EDLC type, but low conductivity, rate capability, and cyclic stability limit their performance [2–4]. Researchers overcome this problem to some extent by combing both types of materials. Recently flexible supercapacitors are gaining a lot of attention as compared to normal supercapacitors, and flexible supercapacitors are useful considering in practical applications, convenience,

DOI: 10.1201/9781003215196-13

and safety requirements [5–7]. The flexible supercapacitors provide advantages of high energy and power density in addition to shape adaptability making them highly useful in wearable electronics, smart phones, portable displays, etc. [8]. Extensive research has been going on the preparation and fabrication of flexible supercapacitor devices. The electrochemical performance, design, and mechanical properties are important aspects in flexible supercapacitors considering its use in smart devices [9]. The performance of flexible supercapacitor depends on current collector, electrolyte and device structure [10].

Among all the materials studied for flexible supercapacitors, graphene is one of the popular electrode materials. Graphene is a popular allotrope of carbon in two dimensions with defect-free single layer of sp^2-bonded carbon atoms in honeycomb lattice structure [11]. It has properties such as good flexibility, higher surface area, high electrical conductivity, and chemical stability, which makes it a promising material for flexible supercapacitors [12]. The intrinsic electrochemical double-layer capacitance of single-layer graphene was observed as 21 mF cm^{-2} [13]. If entire surface area of the graphene is utilized, it can achieve a higher capacitance of 550 F g^{-1} which is highest in carbon materials [14].

The properties of graphene are strongly dependent on the size of sheets, the number of layers, and the presence of defects [15]. However, the production method for graphene is also a concern considering its mass production as mostly chemical vapor deposition or mechanical exfoliation are used for this purpose, but it results into the high cost of production with less yield. The other form of 2D carbon is reduced graphene oxide (rGO) whose production is comparatively cheaper, and whose yield is high and is also attracting attention as electrode material for supercapacitor. rGO is commonly produced by the reduction of graphene oxide (GO) with chemical, thermal or electrochemical methods [16–18].

In this book chapter, we describe the current progress in graphene-based flexible supercapacitors with much focus on the metal oxide and polymer-based graphene electrode, synthesis strategies, electrochemical performance, and flexibility properties. In addition, we also give the summary of graphene-based flexible supercapacitors. This will bring new insights into for flexible supercapacitors.

13.2　GRAPHENE

As stated earlier, graphene is an excellent material for flexible supercapacitor; however, graphene sheets suffer the disadvantage of π–π interaction that results in bulk and compact structure, making difficult electrolyte ions diffusion and charge transfer, resulting in watered down EDLC of graphene [19,20]. Another problem with graphene is its complicated synthesis procedure that limits its mass production. In view of this D. He and his group used two-step top-down strategy of electrochemical exfoliation to prepare paper-like electrodes. A micro solid-state supercapacitor device was assembled using this electrode showed volumetric capacitance of 3.6 F cm^{-3} with good flexibility [21]. A cable-type supercapacitor using 3D porous graphene on carbon thread was synthesized by Park and Choi by microwave-assisted method. The microwave irradiation reduced graphene to rGO. The two wire electrodes are coated with a polyvinyl alcohol-Na$_2$SO$_4$ gel electrolyte to form the cable-like structure.

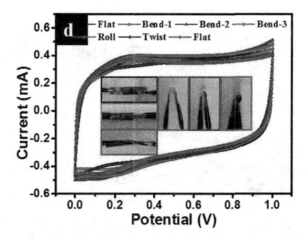

FIGURE 13.1 CV curves of supercapacitor of 3D graphene under different states. Inset show a photograph of the device under different bending conditions. (Reprinted with permission from Ref. [24], Copyright 2017, Elsevier.)

The supercapacitor showed a maximum energy density of 56.5 mWh cm^{-1} with capacitance retention of 96.5% under bent conditions [22].

Raha et al. prepared forest-like 3D carbon structure of rGO for supercapacitor. The rGO surface was modified using UV ozone treatment to improve surface energy by incorporating pore-like defects. About 150% increase in specific capacitance of rGO was observed after introducing the defects. The flexible supercapacitor on scotch tape using PVA–KOH gel electrolyte showed 96% capacitance retention after 1,000 bending cycles [23]. Ramadoss et al. used chemical vapor deposition for deposition 3D graphene on graphite paper. A symmetric supercapacitor was fabricated using PVA–H$_2$SO$_4$ gel electrolyte showed excellent flexibility with no change in CV curve after rolling, bending, and twisting conditions as shown in Figure 13.1 [24].

The pore size of electrode material is an important aspect for the electrochemical performance as it can impact the ion electrolyte interaction. Considering this Sari et al. demonstrated a way to control the pore size of electrochemically exfoliated graphene by freeze drying. The variation in pore diameter from 5 to 0.39 μm was observed with an increase in graphene concentration. A high capacitance of 45.40 F g^{-1} was observed for electrode having 10 mg mL^{-1} graphene concentration. The supercapacitor showed high specific capacitance of 45.4 F g^{-1} in aqueous electrolytes compared to 23.89 F g^{-1} in gel electrolytes [25]. Micro supercapacitor using CVD-grown 3D graphene using polymer binder was studied for flexible supercapacitors. The electrode was tailored by atmospheric pressure plasma functionalization to tune the surface wettability. The water contact angle decreases with an increase in plasma treatment time. The electrode functionalized with 30 second plasma treatment showed good electrochemical performance compared with pristine and 60 second plasma treated graphene structure [26].

It is possible to improve the mechanical and chemical properties of graphene using chemical doping on an sp^2-hybridized carbon sheet. Asymmetric supercapacitor

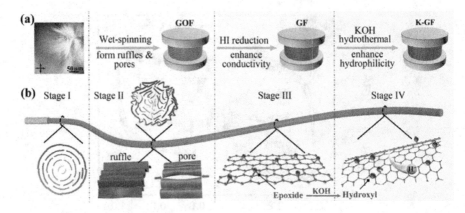

FIGURE 13.2 Schematic illustrations of (a) synthesis of KOH-treated graphene fiber using GO liquid crystal (8 g L^{-1}) and (b) the microstructure and surface functional changes of the fiber during synthesis. (Reprinted with permission from Ref. [29], Copyright 2019, American Chemical Society.)

device was fabricated with H$_2$SO$_4$/PVA solid gel electrolyte and boron-doped reduced graphene as positive and reduced GO as negative electrodes. Boric acid, boron powder, and boron trioxide were used as three distinct precursors with various weight ratios for the deposition of B-doped rGO. As per FESEM studies the rod-like morphology of the boron changed to nanoballs when precursor of boron is varied from boric acid to boron trioxide. The boron trioxide found as useful precursor for B doping in rGO as it shows the specific capacitance of 266 F g^{-1} which was highest compared with other precursors [27]. The preparation cost of the graphene is highly dependent on the raw materials. Considering this, Singh et al. used cheap materials such as charcoal for the preparation graphene quantum dots of average particle size of 5 nm. The specific capacitance of 257 F g^{-1} was observed at 3 A g^{-1}. A flexible symmetric supercapacitor fabricated using these quantum dots showed a specific capacitance 55 F g^{-1} at a current density of 3 A g^{-1}. Even at a higher current density, the electrode showed good capacitance retention [28]. Graphene fibers suffer the disadvantage of hydrophobicity for making it hydrophilic; mostly, non-conducting additives need to be added. Guan et al. applied a different strategy for the synthesis of hydrophilic graphene nanofibers, and they improved the hydrophilicity using hydrothermal treatment in KOH. The schematic of this synthesis is shown in Figure 13.2a. Figure 13.2b shows the microstructure and functional changes during its production. The synthesized graphene fibers showed a porous ruffle structure, good hydrophilicity, and conductivity. The yarn supercapacitor showed a capacitance of 145.6 mF cm^{-2} with an energy density of 3.23 µW h cm^{-2} and a power density of 0.017 mW cm^{-2} [29].

13.3 GRAPHENE WITH POLYMER

Conducting polymer (CP)–based electrodes have high potential in supercapacitors due to their good conductivity, flexibility, simple synthesis methods, and relatively low cost of production [30]. Despite CPs having good properties it cannot alone be considered as an active material for supercapacitors because of their poor cyclic stability. To overcome this problem, it is advised to combine CPs with other materials such as carbon and metal oxides to improve their electrochemical performance. In one such attempt, carbon nanoparticles were added on reduced GO nanosheets using hydrothermal method by Liu et al. PANI was then electrodeposited on the composite material to form free-standing rGO-carbon nanoparticle-PANI paper-like film. An increase in capacitance of composite electrode was observed compared with pure electrode with a specific capacitance of 738.7 F g^{-1} at 1 A g^{-1} owning to the better spacing between graphene nanosheets facilitating electrolyte access in the inner surface of the nanocomposite [31]. Prasit Pattananuwat and Duangdao Aht-ong used cyclic voltammetry technique for the co-deposition of graphene-polypyrrole composite with use of sodium dodecylbenzenesulfonate (SDBS) and poly(styrene sulfonate) (PSS) surfactants. The 3D graphene porous structure with polypyrrole is used to avoid the restriction of ion transportation. The addition of PSS surfactant resulted in less restacking, and agglomeration resulted in low value of ESR. The particle size distribution of graphene hydrogel with and without surfactant is shown in Figure 13.3 which shows formation of fragmented graphene sheets for pure sample having particle size of 120 μm which then reduced to 33 μm for SDBS-graphene and bimodal

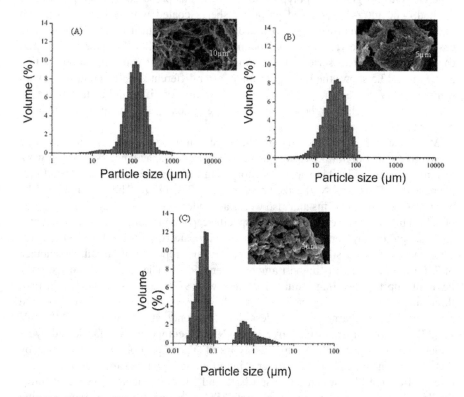

FIGURE 13.3 Particle size distribution curves of (a) graphene hydrogel, graphene hydrogel synthesized using (b) SDBS and (c) PSS surfactant. Inset shows the SEM of each sample. (Reproduced reprinted with permission [32], Copyright 2017, Elsevier.)

size distribution in the range from nanometer to submicrometers in PSS-graphene. The specific capacitance of 640.8 F g^{-1} was observed with good cyclic stability [32].

Flexible free-standing supercapacitor electrode of graphene paper and polypyrrole composite was prepared via green strategy by Wang et al. Polypyrrole was electrodeposited on the surface of graphene paper, and the best electrode performance was observed with eight deposition cycles of polypyrrole. An asymmetric device was fabricated using graphene paper-polypyrrole and polypyrrole as positive and negative electrodes respectively. The device showed rectangular CV curves with an aerial capacitance of 128.8 mF cm^{-2}. The electrochemical performance of the device is shown in Figure 13.4. The device operated up to a potential of 1.4 V with good flexibility with no much change in capacitance after sharp bending to 90° and 180° [33].

A facile electrochemical approach was used by Yu et al. for a metallic fabric supercapacitor comprising a composite of graphene and polyaniline. The graphene microstructure in this work withstands 8.2 mg cm^{-2} of maximum PANI loading with good electrochemical performance. The maximum specific capacitance of 1,506.6 mF cm^{-2} was observed with 92% capacity retention after 5,000 cycles. The electrode showed good mechanical stability even after bending, the capacitance retained to 95.8% of its original value after 1,000 bending cycles [34]. A flexible solid-state supercapacitor of graphene-polypyrrole hydrogel was prepared by a simple heating method. The ultra-thin stacked graphene sheets coated with polypyrrole-formed highly porous 3D microstructure that provides an active surface for adsorption and desorption of electrolyte ions with good mechanical flexibility. The fabricated supercapacitor was used in series combination to power a light-emitting diode (LED, 3V). Figure 13.5 shows the LED glow under after different bending conditions, with symmetric CV curves and promising cyclic stability for different bending states for 12,000 cycles which is highest in the case of graphene-polypyrrole-based supercapacitor [35].

Many researchers modified the graphene using nitrogen doping as it modifies the conductivity, hydrophilicity as well as induces pseudo-capacitance as nitrogen work as a redox center. Flexible supercapacitor containing hybrid films of N-doped graphene and PEDOT was synthesized by Teng et al. The PEDOT helps to improve the conductivity of hybrid films and also works as binder to form free-standing electrode of hybrid films. The electrode showed a specific capacitance of 206 F g^{-1} [36]. Kumar et al. adopted spray deposition technique to fabricate rGO-PEDOT:PSS composite electrode for flexible solid-state supercapacitor and reported an aerial capacitance of 246 mF cm^{-2} which is highest among other ASSCs. The electrode showed good flexibility up to a bending radius of 0.5 mm with 95% capacity retention. Further, three supercapacitors connected in series showed rectangular CV's and also good rate capability to charge 4.8 V for less than 5 seconds at a scan rate of 2,000 mV s^{-1} [37]. Flexible supercapacitor of rGO-PANI nanocomposite was fabricated by an inkjet method by diao and group. Two types of supercapacitors were fabricated by this method one is sandwich and other is interdigitated type. In sandwich-type structure gold-coated PET was used as substrate and current collector however it limits the flexibility of the device. In interdigitated SC, only one substrate is required, and so it provides better flexibility. Although both type of SCs showed better capacitive performance, the capacitive performance and flexibility of interdigitated SC

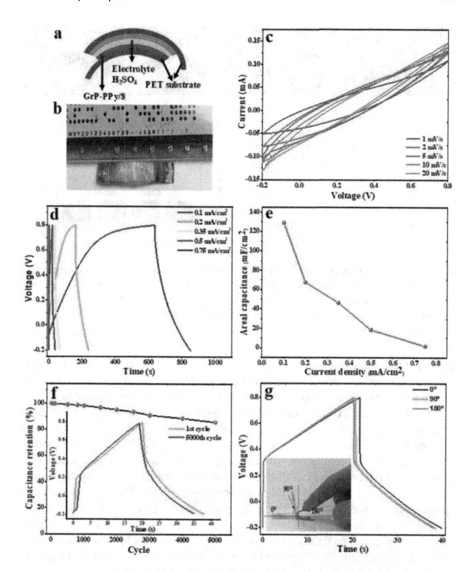

FIGURE 13.4 Adaptable Airplane seat device (AASD) fabrication using graphene-polypyrrole: (a) schematic of AASD, (b) photograph of AASD, (c) CV curves at different scan rates, (d) charge–discharge at various scan rates, (e) variation of aerial capacitance with current density, (f) stability studies after 5,000 cycles, and (g) flexibility at various bending angles. Inset in (g) is a photograph of AASD at bending condition. (Reprinted with permission [33], copyright 2020, Elsevier.)

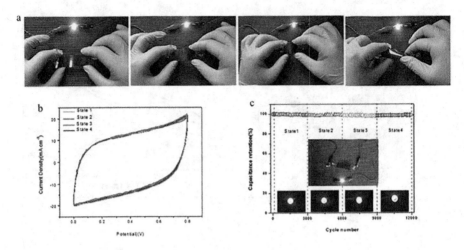

FIGURE 13.5 (a) Photograph of LED power, (b) CV curves, and (c) cyclic stability for different bending states of bent, fold, and twisted for 12,000 cycles. Inset of (c) is supercapacitor connected in series to light LED of graphene for graphene-polypyrrole hydrogel-based solid-state supercapacitor, respectively. (Reprinted with permission from [35], Copyright 2017, Elsevier.)

are found to be superior to the sandwich-type structure [38]. The nanocomposite of polypyrrole/graphene/single-walled carbon nanotubes prepared by Dhibar et al. by *in situ* polymerization method showed a maximum specific capacitance of 1,224 F g^{-1}. All solid-state flexible supercapacitor formed using this nanocomposite showed a specific capacitance of 324 F g^{-1} [39]. The rGO-polypyrrole-based electrode on PEG-modified Ni foam for wearable supercapacitor application was fabricated by Cai and group via combining different approaches. First, they electrodeposited polypyrrole-PEG on the surface of Ni foam, GO then wrapped on it using dip-coating method which was further reduced using hydrazine hydrate. The composite showed specific capacitance of 415 F g^{-1} with 96% capacitance retention over 8,000 cycles. A solid-state symmetric device was fabricated, and further three symmetric devices connected in series and fixed on the wrist can light up 5 LEDs for more than 15 minutes without change in the brightness when the wrist is moved to different positions confirming the excellent performance and flexibility of the device [40].

To avoid the problem of restacking in rGO, Khasim et al. proposed formation of rGO with PEDOT:PSS as an additive via secondary doping. A maximum capacitance of 174 F g^{-1} was observed with 90% cyclic stability over 5,000 cycles with 10 wt% of rGO [41]. Ternary composite of graphene/activated carbon/polypyrrole is studied for supercapacitor application. The AC was sandwiched within graphene nanosheets and used as a flexible substrate on which polypyrrole was electrodeposited which formed coral-like morphology. A flexible symmetric supercapacitor device fabricated using the nanocomposite showed 27% degradation in specific capacitance after 180° bending confirming its flexibility [42]. A binder-free approach is used for deposition of graphene and PANI on the carbon cloth. The electrode showed a specific capacitance of 793 F g^{-1} at 0.5 A g^{-1} with 81% cyclic stability after 10,000 cycles. The synthesized

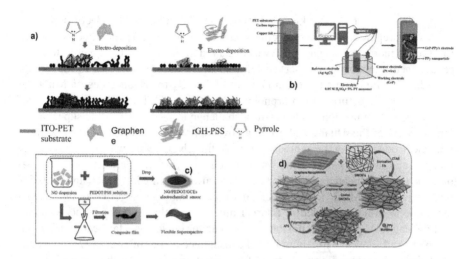

FIGURE 13.6 Schematic illustration of synthesis strategies for (a) electro co-deposition of polypyrrole-graphene with and without PSS surfactant, (b) graphene paper-polypyrrole using electrodeposition, and (c) polypyrrole/graphene nanoplatelets/single-walled carbon nanotubes. (Reprinted with permission from [32] Copyright 2017, [33] Copyright 2020, [36] copyright 2020, [39] Copyright 2019, respectively.)

material was used for the fabrication of symmetric supercapacitor which showed excellent stability and capacitance even after bending up to 150°, making it promising candidate for flexible supercapacitors [43].

The synthesis strategies for some graphene-polymer-based electrodes for flexible supercapacitor are shown in Figure 13.6.

13.4 GRAPHENE WITH METAL OXIDE

Transition metal oxides such as RuO_2, MnO_2, Co_3O_4, SnO_2, NiO, and TiO_2 are typical supercapacitor electrode materials. The pseudo-capacitance mechanism in these materials provide a higher specific capacitance value. The composite of graphene and metal oxide can give rise to synergistic effects proving better electrochemical properties. Amir et al. used electrophoretic deposition mode to synthesize electrodes of $rGO-MnO_2$ nanosheets for solid-state symmetric supercapacitor device. A voltage range of 1.6 V was achieved by this supercapacitor using $PVA-LiClO_4$ electrolyte. The supercapacitor showed an aerial capacitance of 46 mF cm^{-2}. The CV of the device when bending above 180° was applied showed an increase in the current density proving its excellent flexibility [44]. The 3D architecture of MnO_2 over rGO surface was synthesized by Jadhav et al. using hydrothermal oxidation of Mn precursor. The 3D architecture provides easy access to electrolyte ions due to improved pore size because of the incorporation of rGO. The material showed a high energy density of 42.7 Wh kg^{-1} with a specific capacitance of 759 F g^{-1}. A symmetric supercapacitor device fabricated using gel electrolyte showed good charge storage stability even after 60 days. An increased potential window of 3V was achieved with this supercapacitor which is three times larger than the pristine electrode resulting in an energy density of 64.6 Wh kg^{-1} [45]. Beyazay et al. adopted the electrodeposition method to prepare MnO_2 nanostructures on self-standing rGO paper. In cyclic stability studies, increase in capacitance was observed after 2,500 cycles; thereafter, 97% capacitance retention was observed for 10,000 cycles, which is due to the change from hausmannite-structured nanoflowers of Mn_3O_4 to nanoneedles of MnOx as confirmed by structural and morphological studies [46]. Flexibility and electrical conductivity of transition oxide is still a challenge to fabricate flexible supercapacitors. In view of this Huang and group a novel approach to fabricate Cu-rGO-MnO_2 fiber electrode. During the synthesis process, first the Cu wire was placed in a glass pipeline and then GO suspension was injected on it into the glass pipeline. The gas pipeline then was heated at 230° for 2.5 hours, after soaking in $kMnO_4$ and refluxing the formation of MnO_2 takes place. This type of fabrication provides two advantages one is MnO_2 deposited in a larger amount and the second Cu acts as a current collector. A symmetric flexible supercapacitor device was fabricated using this assembly and showed a maximum areal capacitance of 140 mF cm^{-2} [47]. Jeon et al. used simple microwave-assisted route for deposition of rGO on the carbon for flexible symmetric supercapacitor. The synthesis route is quick resulting into porous structure with a specific surface area of 285.4 m^{2}g^{-1}. The asymmetric supercapacitor comprising MnO_2 and rGO on carbon cloth as positive and negative electrode showed a maximum energy density of 64 μWh cm^{-2} and a power density of 1 mW cm^{-2} [48]. Jia et al. prepared fiber-shaped supercapacitor containing core-shell flower like MnO_2 wrapped on graphene using microfluidic spinning technique. Graphene fibers of about length 40 cm were generated by this method on which MnO_2 flowers were grown using *in situ* chemical deposition. The schematic illustration of the methodology used for fabrication is shown in Figure 13.7. The supercapacitor showed a maximum power density of 662 mW cm^{-3} and was able to lighten 10 LEDs as well as electronic timer confirming its excellent electrochemical performance [49].

Ao et al. used chemical vapor–deposited CNT as a substrate to deposit Fe_3O_4 by solvothermal method on which electrodeposition of PANI was carried out. On the surface of CNT-Fe_3O_4-PANI surface, rGO was deposited using the electrophoresis method. A symmetric supercapacitor fabricated using deposited material showed an operating potential of 2 V at 10 A g^{-1}. The electrode showed good flexibility with 95% capacitance retention with bending through 180° after 10,000 cycles. An energy density of 60.8 Wh kg^{-1} with the power density of 45.2 kW kg^{-1} was reported. Combination of different kinds of materials brings synergistic effects to increase the conductivity and pseudo-capacitance of the material [50]. A PANI-Fe_2O_3-graphene composite hydrogel on carbon cloth surface is used in a flexible supercapacitor that showed a high specific capacitance of 1,124 F g^{-1} and a high rate capacity of 82.2% owning to synergistic effects due to combination of different materials [51]. Huang and group prepared transparent graphene wrapped Fe_2O_3 nanowire using bio-inspired gas diffusion method. Due to graphene wrapping, the capacitance retention was improved from 16.1% to 92% after 10,000 cycles. The symmetric supercapacitor showed good flexibility and electrochemical performance. Two symmetric supercapacitors connected in series widened the voltage window to 2.5 V. The supercapacitor showed higher energy density of 8 mWh cm^{-3} [52]. Liu et al. used one-step laser deposition for preparation of Fe_3O_4 nanoparticle on laser-induced graphene

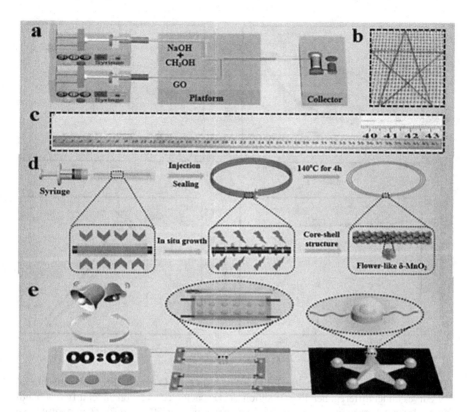

FIGURE 13.7 Microfluidic spinning assembly process and the as-prepared graphene fibers: (a) schematic illustration and (b) photograph of fibers being woven in cotton fabric. (c) Photographs of a long MnO_2-graphene fiber with a length of 40 cm. (d) Schematic illustration of the formation of MnO_2-graphene fiber. (e) Schematic illustration of fiber used to power electronics. (Reprinted with permission from Ref. [49], Copyright 2021, Elsevier.)

(LIG). The supercapacitor showed high energy density of 60.20 µwh cm^{-2} [53]. MnO_2 nanowires were grown on rGO coated carbon cloth for flexible supercapacitor using hydrothermal method. From the SEM studies, growth of very thin nanowires of nearly 5 nm diameter was observed forming 3D open porous structure. The ultrafine structure of the nanowires helps to ease electrolyte diffusion in the material to reduce the internal resistance. The electrode showed good flexibility and a rate capability of 75.5% [54]. A simple wet spinning method was employed by Chen et al. to synthesize fiber-shaped electrode consisting of rGO and Mn_3O_4 nanocrystals. A symmetric solid-state supercapacitor fabricated with fiber electrode showed excellent flexibility and electrochemical properties as shown in Figure 13.8 [55].

Cho et al. used cathodic electroplating to deposit RuO_2-Graphene (G) film on Cu foil. The electrochemical performance of RuO_2/Cu and RuO_2-G/Cu was studied. Interestingly, it was observed that although RuO_2 on Cu foil showed maximum specific capacitance of 1,891 F g^{-1} compared to 1,494 F g^{-1} for RuO_2-G on Cu foil but in bent conditions, the later showed improved capacitance due to the stable adhesion of RuO_2 on Cu foil [56]. Cellulose fiber is a good flexible substrate for the deposition of

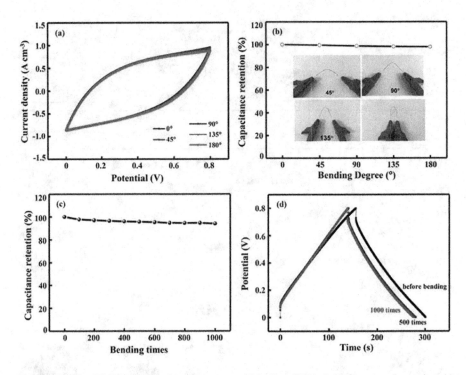

FIGURE 13.8 Electrochemical performance of Mn_3O_4–rGO hybrid fiber supercapacitor (a) CV curves under different bending conditions, (b) capacitance retention under different bending angles, (c) capacitance retention and bending times for 900, and (d) charge–discharge at 500 A cm^{-3} before and after bending. (Reprinted with permission [55], Copyright 2017, Elsevier.)

active material in supercapacitors however it suffers the drawback of low capacitance therefore increase in the active material is required to increase the capacitance. Song et al. developed a novel layer-by-layer growth approach to overcome this problem. They first did *in situ* polymerization of PANI on the printing paper after that graphene sheets were added to form PANI-graphene composite paper. The paper was then dipped in $kMnO_4$ solution where MnO_2 is generated and deposited on the paper and hence prepared PANI-graphene-MnO_2 composite paper. The areal capacitance of 3.5 F cm^{-2} with 82% capacitance retention was observed for the composite [57]. A self-supported solid-state supercapacitor electrode comprising N-doped rGO with V_2O_3 nanoflakes was synthesized using hydrothermal and vacuum filtration method. A specific capacitance of 206 F g^{-1} was observed at 1 mA cm^{-2}. A symmetrical supercapacitor device showed a capacitance of 8.1 F g^{-1} at 0.1 A cm^{-3} with 76% rate capability and good flexibility [58]. A flexible hybrid solid-state supercapacitor comprising SnO_2-rGO-PEDOT:PSS on bacterial nanocellulose was fabricated. Incorporation of SnO_2 leads to improved capacitance of the electrode with better cycling stability [59]. A solid-state flexible supercapacitor device was fabricated using rGO, silver nanowire, and nickel–aluminum-layered double hydroxide using vacuum filtration and hydrothermal method. The Ni–Al double-layered hydroxide first grown

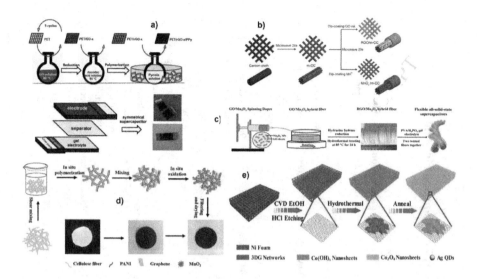

FIGURE 13.9 Schematic illustration of different strategies for preparation of (a) rGO-manganese oxide using electrodeposition, (b) graphene–MnO_2 using microwave-assisted approach, (c) rGO–Mn_3O_4 fiber, (d) PANI-graphene paper wrapped by MnO_2 and (e) Ag-Co_3O_4-graphene. (Reprinted with permission from [46] Copyright 2019, [48] Copyright 2018, [55] Copyright 2017, [57] Copyright 2019, [62] Copyright 2017, Elsevier, respectively.)

on Ag nanowire using hydrothermal method resulting into core-shell composite which is then combined with rGO. The capacitance performance of the hybrid film was increased from 762.5 to 1,148 F g^{-1} after incorporation on Ag nanowires with a cycle life of 77.2% after 10,000 cycles [60]. A sandwich-type supercapacitor was fabricated by Li et al. using ultrafine CuO in 3D graphene network using thermal decomposition. A maximum specific capacitance of 1,539.8 F g^{-1} was observed for the electrode at 6 mA cm^{-2}, with good flexibility [61].

Ag doped hydrothermally grown nanosheets of Cu_3O_4 was deposited on a 3D graphene structure for flexible supercapacitor by Wang et al. The hybrid material showed a high specific surface area of 242.8 $m^2 g^{-1}$. The maximum specific capacitance of 1052.5 F g^{-1} was observed after 4,000 cycles [62].

Figure 13.9 shows the schematic illustration of different techniques used for the preparation of graphene-metal oxides for flexible supercapacitors.

The comparison of electrochemical performance of graphene-based materials described in this book chapter is given in Table 13.1.

13.5 CONCLUSION

Flexible supercapacitors are attracting a lot of attention and study because of the growing demand for flexible/wearable energy storage devices. In this book chapter, we have tried to summarize the latest progress on graphene-based materials for flexible supercapacitor electrode material, with a special emphasis on fabrication methods, different types of substrates, types of structures required, and its supercapacitor

TABLE 13.1

Comparison of Preparation Methods, Capacitances, Energy Density, Power Density, Retention and Flexibility of the Graphene-Based Supercapacitor Electrodes

Sr. No.	Active Materials	Template Materials	Preparation Method	Capacitance	Energy Density	Power Density	Retention	Flexibility	Ref
	Graphene		Electrochemical exfoliation	3.6 F cm^{-3} at 200 mV s^{-1}	–	–	94% (20,000 cycles)	No significant change in CV after bending by 90°, and 180°	[21]
	Graphene	Carbon thread	Microwave-assisted method	38.1 mF cm^{-1} at 10 mV s^{-1}	56.5 mWh cm^{-1}	2.29 W cm^{-1}	99% (10,000 cycles)	96.5% capacitance retention in bent state for 10,000 cycles	[22]
	rGO	Scotch tape	Co reduction and drying	6 mF cm^{-2} at 80 mV s^{-1}	0.137 mWh cm^{-3}	164.06 mew cm^{-3}		96% capacitance retention for 1,000 bending cycles	[23]
	Graphene	Graphite paper	Chemical vapor deposition	260 F g^{-1}	8.87 Wh kg^{-1}	178.5 W kg^{-1}	112% (10,000 cycles)	Not much distortion in CV curves for bending, twisting and rolling conditions	[24]
	Graphene	Ni foam	Electrochemical exfoliation Freeze dry technique	45.4 F g^{-1} at 5 mV s^{-1}	13.28 Wh kg^{-1}	206.62 W kg^{-1}	96.34% (500 cycles)	94% capacitance retention after 180° bending	[25]
	Graphene	Polyimide	Chemical vapor deposition	1.5 mF cm^{-2} at 10 V s^{-1}	0.38 μWh cm^{-2}	14.4 mW cm^{-2}	95% (20,000 cycles)	85% capacitance retention after bending through 90° for 5,000 cycles	[26]

(Continued)

TABLE 13.1 (Continued)

Comparison of Preparation Methods, Capacitances, Energy Density, Power Density, Retention and Flexibility of the Graphene-Based Supercapacitor Electrodes

Sr. No.	Active Materials	Template Materials	Preparation Method	Capacitance	Energy Density	Power Density	Retention	Flexibility	Ref
	Boron-doped rGO		Hydrothermal and pyrolysis	122 F g⁻¹	67.8 Wh kg⁻¹	1,000 W kg⁻¹	98% (5,000 cycles)		[27]
	Graphene	Stainless steel	Chemical approach	55 F g⁻¹ at 3 A g⁻¹	17.36 Wh kg⁻¹	191.7 W kg⁻¹	91% (1,000 cycles)	Not much difference in CV curves for 45°, 90°, 135° bending angles as well as twisting	[28]
	Graphene	PET	Wet spinning technique	145.6 mF cm⁻² at 0.086 mA cm⁻²	3.23 μWh cm⁻²	0.017 mW cm⁻²		95.3% capacitance retention after bending through 180° for 1,000 cycles	[29]
	rGO-Carbon nanoparticles-PANI	PTFE	Hydrothermal	787.3 F g⁻¹ at 1 A g⁻¹	–		92.29 % (2,000 cycles)	Low loss in cv after bending through 180° for 200 cycles	[31]
	rGO-Polypyrrole-PSS	ITO-PET	Cyclic Voltammetry	640.8 F g⁻¹ at 1 A g⁻¹	68.2 Wh kg⁻¹	438.5 W kg⁻¹	90 % (2,000 cycles)		[32]

(Continued)

TABLE 13.1 (Continued)

Comparison of Preparation Methods, Capacitances, Energy Density, Power Density, Retention and Flexibility of the Graphene-Based Supercapacitor Electrodes

Sr. No.	Active Materials	Template Materials	Preparation Method	Capacitance	Energy Density	Power Density	Retention	Flexibility	Ref
	Graphene-Polypyrrole	PET (polyethylene Terephthalate)	Electrodeposition	128.9 mF cm^{-2}	$16.1 \text{ mWh cm}^{-12}$	180 mW cm^{-2}	85% (5,000 cycles)	Withstand 90° and 180° bending without much change in capacitance	[33]
	Graphene-polyaniline	Stainless steel fabric	Electrodeposition	$1,506.6 \text{ mF cm}^{-2}$	–	–	92% (5,000 cycles)	95.8% retention of capacitance after 1,000 bending cycles	[34]
	Graphene-polypyrrole hydrogel	Au-coated Polyamide		363 F cm^{-2} at 1 mA cm^{-3}				98.6% capacity retention (bent, fold, and twisted modes) after 12,000 cycles	[35]
	Nitrogen doped graphene-PEDOT	Carbon cloth	Flow directed self-assembly	206 F g^{-1} at 0.25 A g^{-1}			95% (1,000 cycles)	No significant change in CV curve for 90° and 180° bending angles	[36]
	rGO-PEDOT:PSS	Carbon cloth	Spray deposition	170 F g^{-1} at 10 mV s^{-1}	11.0 Wh kg^{-1}	$4,460 \text{ W kg}^{-1}$	100% (2,000 cycles)	Capacitance retention of 95% under bending	[37]

(Continued)

TABLE 13.1 (Continued)

Comparison of Preparation Methods, Capacitances, Energy Density, Power Density, Retention and Flexibility of the Graphene-Based Supercapacitor Electrodes

Sr. No.	Active Materials	Template Materials	Preparation Method	Capacitance	Energy Density	Power Density	Retention	Flexibility	Ref
	Graphene-PANI	Gold-coated PET	Inkjet printing	554 F cm^{-3} at 1 mV s^{-1}	76.94 Wh cm^{-3}	5,593.7 W cm^{-3}	>90% (2,000 cycles)	No change in capacitance after bending	[38]
	Polypyrrole-graphene-CNT		*In situ* polymerization	324 F g^{-1} at 0.5 A g^{-1}	28.8 Wh kg^{-1}	1,975.22 W kg^{-1}	91% (5,000 cycles)		[39]
	rGO-Polypyrrole-Polyethylene glycol foam	Ni foam	Electrodeposition and chemical reduction	1,019 mF cm^{-2} at 2 mV s^{-1}			95% (4,000 cycles)	Negligible capacitance loss after 1,000 bending cycles	40
	PEDOT-PSS-Graphene		Secondary doping	174 F g^{-1}	810 Wh kg^{-1}		90% (5,000 cycles)		[41]
	Graphene-Activated Carbon-Polypyrrole		Vacuum filtration and electrodeposition	178 F g^{-1}			64.4 % (5,000 cycles)	83.6% capacitance retention after 500 stretching-bending cycles	[42]
	Graphene-PANI	Carbon cloth	Electrochemical deposition	512 F g^{-1} at 0.5 A g^{-1}	11.38 Wh kg^{-1}	199.80 W kg^{-1}	85% (10,000 cycles)		[43]
	rGO-MnO2	Au-coated PET	Electrophoretic deposition	182 F cm^{-3}	48.8 mWh cm^{-3}	8.34 W cm^{-3}	80% (2,000 cycles)	No change in CV for bending upto 180°	[44]
	rGO-MnO$_2$		Hydrothermal	759 F g^{-1} at 2 A g^{-1}	64.6 Wh kg^{-1}	15 kW kg^{-1}	88% (3,500 cycles)		[45]
	rGO-MnO$_2$		Electrodeposition	546 F g^{-1} at 0.5 A g^{-1}	6.9 Wh kg^{-1}	143 W kg^{-1}	97% (10,000 cycles)		[46]

(Continued)

TABLE 13.1 (Continued)

Comparison of Preparation Methods, Capacitances, Energy Density, Power Density, Retention and Flexibility of the Graphene-Based Supercapacitor Electrodes

Sr. No.	Active Materials	Template Materials	Preparation Method	Capacitance	Energy Density	Power Density	Retention	Flexibility	Ref
	Cu-rGO-MnO_2		Hydrothermal	140 mF cm^{-2} at 0.1 mA cm^{-2}			88% (5,000 cycles)	97% (after 500 bending cycles with bending angle 120°)	[47]
	rgO-MnO_2	Carbon cloth	Microwave-assisted method	264 mF cm^{-2} at 1 mA cm^{-2}	64 μWh cm^{-2}	15 mW cm^{-2}	95% (1,000 cycles)	97% capacitance retention for 2,000 cycles after bending and twisting conditions	[48]
	MnO_2-Graphene		Microfluidic spinning technique	164.2 F cm^{-3}	4.3 mWh cm^{-3}	662 mW cm^{-3}	93.6% (10,000 cycles)	Negligible loss in capacitance after bending to 45°, 90°, 135°, and 180°	[49]
	rGO-CNT-Fe_3O_4-PANI		Solvothermal, electrodeposition, electrophoresis	414.5 F g^{-1} at 1 A g^{-1}	60.8 Wh kg^{-1}	45.2 kW kg^{-1}		95.1% capacitance retention after bending to 180° for 10,000 cycles	[50]
	PANI-Graphene-FeO	Carbon cloth	In situ polymerization	1,124 F g^{-1} at 0.25 A g^{-1}	14.4 Wh kg^{-1}	58 W kg^{-1}			[51]
	Graphene-Fe_2O_3	ITO/PET		3.3 mF cm^{-2}	8 mWh cm^{-3}	191.3 mW cm^{-3}	85.6% (5,000 cycles)		[52]

(Continued)

TABLE 13.1 (Continued)

Comparison of Preparation Methods, Capacitances, Energy Density, Power Density, Retention and Flexibility of the Graphene-Based Supercapacitor Electrodes

Sr. No.	Active Materials	Template Materials	Preparation Method	Capacitance	Energy Density	Power Density	Retention	Flexibility	Ref
	Fe_3O_4-graphene	Ferric chloride-coated PI film	Laser-assisted deposition	719.28 F cm^{-2}	60.20 µWh cm^{-2}			90% capacitance retention for 45°, 90°, 135°, and 180° bending after 2,000 cycles	[53]
	MnO_2-rGO	Carbon cloth	Hydrothermal	506.8 mF cm^{-2} at 0.128 mA cm^{-2}			98.6% (10,000 cycles)		[54]
	rGO-Mn_3O_4			45.5 F cm^{-3} at 50 mA cm^{-3}	4.05 mWh cm^{-3}	268 mW cm^{-3}		97% capacitance retention (for 90° bending angle after 100 cycles)	[55]
	RuO_2-Graphene	Cu foil	Cathodic electroplating	1,561 F g^{-1} at 5 mV s^{-1}	~13 Wh kg^{-1}	~21 kW kg^{-1}	98%		[56]
	PANI-Graphene-MnO_2	Printing paper	Layer-by-layer in situ growth		5.2 mWh cm^{-3}			90% capacitance retention after 1,000 bending cycles	[57]
	N-doped rGO-V_2O_3		Vacuum filtration	8.1 F cm^{-3} at 0.1 A cm^{-3}	0.55 mWh cm^{-3}	0.035 W cm^{-3}	81% (10,000 cycles)	Negligible loss in capacitance for 45°, 90°, 135°, and 180° bending	[58]

(Continued)

TABLE 13.1 (*Continued*)
Comparison of Preparation Methods, Capacitances, Energy Density, Power Density, Retention and Flexibility of the Graphene-Based Supercapacitor Electrodes

Sr. No.	Active Materials	Template Materials	Preparation Method	Capacitance	Energy Density	Power Density	Retention	Flexibility	Ref
	Nanocellulose-rGO-SnO$_2$-PEDOT:PSS		Bacteria-meditated growth	445 F g^{-1} at 2 A g^{-1}			84.1% (2,500 cycles)	89.8%	[59]
	rGO-Ag-NiAl hydroxide	Au-PET	Hydrothermal and vacuum filtration	127.2 F g^{-1} at 1A g^{-1}	35.75 mWh cm^{-3}	1.01 W cm^{-3}	83.2 % (10,000 cycles)	93.1% capacitance retention (for 90° bending angle after 100 cycles)	[60]
	CuO-Graphene	Carbon cloth	Thermal decomposition	1,539.8 F g^{-1} at 6 mA cm^{-2}	248 mWh cm^{-2}	5.6 W cm^{-2}	82.6% (1,000 cycles)	No change in CV curves after bending by 45°, 90°, and 180°	[61]
	Ag-Co$_3$O$_4$-Graphene	Ni foam	Chemical vapor deposition and hydrothermal	421 mF cm^{-2} at 1 mA cm^{-2}	26.7 Wh kg^{-1}	15,000 W kg^{-1}	91.6% (10,000 cycles)		[62]

performance in the form of device. Recent work which is going on micro, fiber, and cable types of flexible supercapacitors is also discussed. Different parameters such as active material, synthesis strategies, and type of supercapacitor device have major effects on the electrochemical performance of the flexible supercapacitors. It is expected that the flexible supercapacitor should show excellent electrochemical performance under the bending conditions, but the flexibility measurement standards are not the same in every literature which can be a concern considering its practical application. If this challenge is tackled, then we believe graphene-based materials will be one of the promising electrodes for the flexible supercapacitor.

REFERENCES

1. Supriya J. Marje, Vinod V. Patil, Vinayak G. Parale, Hyung-Ho Park, Pragati A. Shinde, Jayavant L. Gunjakar, Chandrakant D. Lokhande, Umakant M. Patil, (2022). "Microsheets like nickel cobalt phosphate thin films as cathode for hybrid asymmetric solid-state supercapacitor: Influence of nickel and cobalt ratio variation", *Chem. Eng. J.*, Vol. 429:132184, https://doi.org/10.1016/j.cej.2021.132184
2. Antonino Salvatore Arico, Peter Bruce, Bruno Scrosati, Jean-Marie Tarascon, and Walter Van Schalkwijk (2005). "Nanostructured materials for advanced energy conversion and storage devices", *Nat. Mater.* Vol. 4:366–377. https://doi.org/10.1038/nmat1368.
3. John Chmiola, G. Yushin, Yury Gogotsi, Christele Portet, Patrice Simon, and Pierre-Louis Taberna (2006). "Anomalous increase in carbon capacitance at pore sizes less than 1 nanometer", *Science*, Vol. 313:1760–1763. https://doi.org/10.1126/science.1132195
4. Tian Lv, Yao Yao, Ning Li, and Tao Chen. (2016). "Highly stretchable supercapacitors based on aligned carbon nanotube/molybdenum disulfide composites", *Angew. Chem. Int. Ed.* Vol. 55:9191–9195. https://doi.org/10.1002/anie.201603356
5. Cheng Zhong, Yida Deng, Wenbin Hu, Jinli Qiao, Lei Zhang, and Jiujun Zhang (2015). "A review of electrolyte materials and compositions for electrochemical supercapacitors", *Chem. Soc. Rev.* Vol. 44:7484–7539. https://doi.org/10.1039/C5CS00303B
6. Peihua Yang, and Wenjie Mai (2014), "Flexible solid-state electrochemical supercapacitors", *Nano Energy*, Vol. 8:274–290. https://doi.org/10.1016/j.nanoen.2014.05.022
7. Shahram Ghasemi, Majid Jafari, and Fatemeh Ahmadi. (2016). "Cu$_2$O-Cu(OH)$_2$-graphene nanohybrid as new capacitive material for high performance supercapacitor", *Electrochim. Acta*, Vol. 210:225–235. https://doi.org/10.1016/j.electacta.2016.05.155
8. Fan Bu, Weiwei Zhou, Yihan Xu, Yu Du, Cao Guan, and Wei Huang (2020). "Recent developments of advanced micro-supercapacitors: Design, fabrication and applications" *NPJ Flex. Electron.*, Vol. 31. https://doi.org/10.1038/s41528-020-00093-6
9. Dao-Yi Wu, Jiao-Jing Shao (2021). "Graphene-based flexible all-solid-state supercapacitors", *Mater. Chem. Front.*, Vol. 5:557–583 https://doi.org/10.1039/D0QM00291G
10. Dingshan Yu, Qihui Qian, Li Wei, Wenchao Jiang, Kunli Goh, Jun Wei, Jie Zhang, Yuan Chen (2015). "Emergence of fiber supercapacitors", *Chem. Soc. Rev.*, Vol. 44:647–662. https://doi.org/10.1039/C4CS00286E
11. Xu Peng, Lele Peng, Changzheng Wu, and Yi Xie (2014). "Two dimensional nanomaterials for flexible supercapacitors", *Chem. Soc. Rev.*, Vol. 43:33033323. https://doi.org/10.1039/C3CS60407A
12. Francesco Bonaccorso, Luigi Colombo, Guihua Yu, Meryl Stoller, Valentina Tozzini, Andrea C. Ferrari, Rodney S. Ruoff, and Vittorio Pellegrini. (2015). "Graphene, related two-dimensional crystals, and hybrid systems for energy conversion and storage", *Science*, Vol. 347:1246501. https://doi.org/10.1126/science.1246501

13. Jilin Xia, Fang Chen, Jinghong Li, and Nongjian Tao. (2009). "Measurement of the quantum capacitance of graphene", *Nat. Nanotechnol.*, Vol. 4(8):505–509. https://doi.org/10.1038/nnano.2009.177

14. Yuanlong Shao, Maher F. El-Kady, Lisa J. Wang, Qinghong Zhang, Yaogang Li, Hongzhi Wang, Mir F. Mousaviae, Richard B. Kaner (2015). "Graphene-based materials for flexible supercapacitors", *Chem. Soc. Rev.*, Vol. 44:3639–3665.

15. Rinaldo Raccichini, Alberto Varzi, Stefano Passerini, and Bruno Scrosati. (2015). "The role of graphene for electrochemical energy storage". *Nat Mater.*, Vol. 14:271–279. https://doi.org/10.1038/nmat4170

16. Yanwu Zhu, Shanthi Murali, Weiwei Cai, Xuesong Li, Ji Won Suk, Jeffrey R. Potts, and Rodney S. Ruoff (2010), "Graphene and graphene oxide: Synthesis, properties, and applications", *Adv. Mater.*, Vol. 22:3906–3924. https://doi.org/10.1002/adma.201001068

17. Daniel R. Dreyer, Sungjin Park, Christopher W. Bielawski, Rodney S. Ruoff (2010). "The chemistry of graphene oxide", *Chem. Soc. Rev.*, Vol. 39:228–240. https://doi.org/10.1039/B917103G

18. Yu Zhu, Dustin K. James, and James M. Tour. (2012), "New routes to graphene, graphene oxide and their related applications", *Adv. Mater.*, Vol. 24:4924–4955. https://doi.org/10.1002/adma.201202321

19. Xihong Lu, Minghao Yu, Gongming Wang, Yexiang Tong, Yat Li (2014). "Flexible solid-state supercapacitors: Design, fabrication and applications", *Energy Environ. Sci.*, Vol. 4:2160–2181. https://doi.org/10.1039/C4EE00960F

20. Yi-Zhou Zhang, Yang Wang, Tao Cheng, Wen-Yong Lai, Huan Pang, Wei Huang (2015), "Flexible supercapacitors based on paper substrates: A new paradigm for low-cost energy storage", *Chem. Soc. Rev.*, Vol. 44:5181–5199. https://doi.org/10.1039/C5CS00174A

21. Dongxu He, Alexander J. Marsden, Sheeling Li, Rui Zhao, Weidong Xue, Mark A. Bissett (2018). "Fabrication of a graphene-based paper-like electrode for flexible solid-state supercapacitor devices", *J. Electrochem. Soc.*, Vol. 165:A3481. https://doi.org/10.1149/2.1041814jes

22. Seung Hwa Park, Bong Gill Choi (2019). "High-performance of flexible supercapacitor cable based on microwave-activated 3d porous graphene/carbon thread", *Appl. Chem.*, Vol. 30:23–28. https://doi.org/10.14478/ace.2018.1093

23. Himadri Raha, Bibhas Manna, Debabrata Pradhan, Prasanta Kumar Guha (2019). "Quantum capacitance tuned flexible supercapacitor by UV-ozone treated defect engineered reduced graphene oxide forest", *Nanotechnology*, Vol. 30: 435404. https://doi.org/10.1088/1361-6528/ab331a

24. A. Ramadoss, Ki-Yong Yoon, Myung-Jun Kwak, Sun-I. Kim, Seung-Tak Ryu, Ji-Hyun Jang (2017). "Fully flexible, lightweight, high performance all-solid-state supercapacitor based on 3-dimensional-graphene/graphite-paper", *J. Power Sources*, Vol. 337: 159–165 https://doi.org/10.1016/j.jpowsour.2016.10.091

25. N. P. Sari, D. Dutta, A. Jamaluddin, Jeng-Kuei Chang, Ching-Yuan Su (2017). "Controlled multimodal hierarchically porous electrode self-assembly of electrochemically exfoliated graphene for fully solid-state flexible supercapacitor", *Phys. Chem. Chem. Phys.*, Vol. 19:30381–30392. https://doi.org/10.1039/C7CP05799G

26. Lu Zhang, Derek De Armond, Noe T. Alvarez, Rachit Malik, Nicholas Oslin, Colin McConnell, Paa Kwasi Adusei, Yu-Yun Hsieh, Vesselin Shanov(2017). "Flexible micro-supercapacitor based on graphene with 3D structure", Vol. 10:1603114. https://doi.org/10.1002/smll.201603114

27. P. Muthu Pandian, A. Pandurangan (2021). "Flexible asymmetric solid-state supercapacitor of boron doped reduced graphene for high energy density and power density in energy storage device", *Diamond Related Mater.*, Vol. 118:108495. https://doi.org/10.1016/j.diamond.2021.108495

28. Arvind Singh, Sumeet Kumar, Animesh K. Ojha (2020). "Charcoal derived graphene quantum dots for flexible supercapacitor oriented applications", *New J. Chem.*, Vol. 44: 11085–11091. https://doi.org/10.1039/D0NJ00899K

29. Tuxiang Guan, Liming Shen, Ningzhong Bao (2019). "Hydrophilicity improvement of graphene fibers for high-performance flexible supercapacitor", *Ind. Eng. Chem. Res.*, Vol. 58:17338–17345. https://doi.org/10.1021/acs.iecr.9b02504

30. P.R. Deshmukh, N.M. Shinde, S.V. Patil, R.N. Bulakhe, C.D. Lokhande (2013). "Supercapacitive behavior of polyaniline thin films deposited on fluorine doped tin oxide (FTO) substrates by microwave-assisted chemical route", *Chem. Eng. J.*, Vol. 223:572–577. https://doi.org/10.1016/j.cej.2013.03.056

31. Dong Liu, Hongxing Wang, Pengcheng Du, Wenli Wei, Qi Wang, Peng Liu (2018). "Flexible and robust reduced graphene oxide/carbon nanoparticles/polyaniline (RGO/CNs/PANI) composite films: Excellent candidates as free-standing electrodes for high-performance supercapacitors", *Electrochim. Acta*, Vol. 259:161–169. http://dx.doi.org/10.1016/j.electacta.2017.10.165

32. Prasit Pattananuwat, Duangdao Aht-ong (2017). "Controllable morphology of polypyrrole wrapped graphene hydrogel framework composites via cyclic voltammetry with aiding of poly (sodium 4-styrene sulfonate) for the flexible supercapacitor electrode", *Electrochim. Acta*, Vol. 224:149–160. https://doi.org/10.1016/j.electacta.2016.12.036

33. W. Wang, Omer Sadak, Jiehao Guana, Sundaram Gunasekaran (2020). "Facile synthesis of graphene paper/polypyrrole nanocomposite as electrode for flexible solid-state supercapacitor", *J. Energy Storage*, Vol. 30:101533. http://dx.doi.org/10.1016/j.est.2020.101533

34. Jianhui Yu, Feifei Xie, Zhengchen Wu, Ting Huang, Jifeng Wu, Dandan Yan, Chaoqiang Huang, Lei Li (2018). "Flexible metallic fabric supercapacitor based on graphene/polyaniline composites", *Electrochim. Acta*, Vol. 259:968–974. http://dx.doi.org/10.1016/j.electacta.2017.11.008

35. Xinming Wu, Meng Lian (2017). "Highly flexible solid-state supercapacitor based on graphene/polypyrrole hydrogel", *J. Power Sources*, Vol. 362: 184–191, https://doi.org/10.1016/j.jpowsour.2017.07.042

36. He Teng, Jingyao Song, Guiyun Xu, Fengxian Gao, Xiliang Luo (2020). "Nitrogen-doped graphene and conducting polymer PEDOT hybrids for flexible supercapacitor and electrochemical sensor", *Electrochim. Acta*, Vol. 355:136772, https://doi.org/10.1016/j.electacta.2020.136772.

37. Neetesh Kumar, Riski Titian Ginting, Jae-Wook Kang (2018). "Flexible, large-area, all-solid-state supercapacitors using spray deposited PEDOT:PSS/reduced-graphene oxide", *Electrochim. Acta*, Vol. 270:37–47, https://doi.org/10.1016/j.electacta.2018.03.069.

38. Jianglin Diao, Jia Yuan, Ailing Ding, Jiushang Zheng, Zhisong Lu (2018), "Flexible Supercapacitor based on inkjet-printed graphene@polyaniline nanocomposites with ultrahigh capacitance", *Macromol. Mater. Eng.* Vol. 2018:1800092. https://doi.org/10.1002/mame.201800092

39. Saptarshi Dhibar, Arkapal Roy, Sudip Malik (2019). "Nanocomposites of polypyrrole/graphene nanoplatelets/single walled carbon nanotubes for flexible solid-state symmetric supercapacitor", *Eur. Polym. J.*, Vol. 120:109203, https://doi.org/10.1016/j.eurpolymj.2019.08.030.

40. C. Cai, Jialong Fu, Chengyan Zhang, Cheng Wang, Rui Sun, Shufang Guo, Fan Zhang, Mingyan Wang, Yuqing Liu, Jun Chen (2020). "Highly flexible reduced graphene oxide@polypyrrole–polyethylene glycol foam for supercapacitors", *RSC Adv.*, Vol. 10:29090. https://doi.org/10.1039/D0RA05199C.

41. Syed Khasim, Apsar Pasha, Nacer Badi, Mohana Lakshmi, Yogendra Kumar Mishra (2020). "High performance flexible supercapacitors based on secondary doped PEDOT–PSS–graphene nanocomposite films for large area solid state devices" *RSC Adv.*, Vol. 10:10526 https://doi.org/10.1039/D0RA01116A.

42. Lanshu Xu, Mengying Jia, Yue Li, Shifeng Zhang, Xiaojuan Jin (2017). "Design and synthesis of graphene/activated carbon/polypyrrole flexible supercapacitor electrodes, *RSC Adv.*, Vol. 7:31342. https://doi.org/10.1039/C7RA04566B.

43. Lele Wen, Ke Li, Jingjing Liu, Yanshan Huang, Fanxing Bu, Bin Zhao, Yuxi Xu (2017). "Graphene/polyaniline@carbon cloth composite as a high-performance flexible super-capacitor electrode prepared by a one-step electrochemical co-deposition method", *RSC Adv.*, Vol. 7:7688 https://doi.org/10.1039/C6RA27545A

44. F. Z. Amir, V. H. Pham, E. M. Schultheis, J. H. Dickerson (2018). "Flexible, all-solid-state, high-cell potential supercapacitors based on holey reduced graphene oxide/manganese dioxide nanosheets", *Electrochim. Acta*, Vol. 260:944–951. https://doi.org/10.1016/j.electacta.2017.12.071

45. Sarika Jadhav, Ramchandra S. Kalubarme, Chiaki Terashima, Bharat B. Kale, Vijay Goadbole, Akira Fujishima, Suresh W. Gosavi (2019). "Manganese dioxide/reduced gra-phene oxide composite an electrode material for high-performance solid state supercapac-itor", *Electrochim. Acta*, Vol. 299:34–44. https://doi.org/10.1016/j.electacta.2018.12.182.

46. Tugce Beyazay, F. Eylul Sarac Oztuna, Ugur Unal (2019), "Self-standing reduced gra-phene oxide papers electrodeposited with manganese oxide nanostructures as electrodes for electrochemical capacitors", *Electrochim. Acta*, Vol. 296:916–924, https://doi.org/10.1016/j.electacta.2018.11.033

47. Miaomiao Huang, Lu Wang, Shuangbao Chen, Liping Kang, Zhibin Lei, Feng Shi, Hua Xu, Zong-Huai Liu (2017). "Highly flexible all-solid-state cable-type supercapaci-tors based on Cu/reduced graphene oxide/manganese dioxide fibers", *RSC Adv.*, Vol. 7:10092 https://doi.org/10.1039/C6RA28117F.

48. Hyeonyeol Jeon, Jae-Min Jeong, Seok Bok Hong, MinHo Yang, Jeyoung Park, Do Hyun Kim, Sung Yeon Hwang, Bong Gill Choi (2018). "Facile and fast microwave-assisted fabrication of activated and porous carbon cloth composites with graphene and MnO_2 for flexible asymmetric supercapacitors", *Electrochim. Acta*, Vol. 280:9–16, https://doi.org/10.1016/j.electacta.2018.05.108.

49. Yunming Jia, Xiaying Jiang, Arsalan Ahmed, Lan Zhou, Qinguo Fan, Jianzhong Shao (2021). "Microfluidic-architected core–shell flower-like δ-MnO_2@graphene fibers for high energy-storage wearable supercapacitors", *Electrochim. Acta*, Vol. 372:137827, https://doi.org/10.1016/j.electacta.2021.137827.

50. Jing Ao, Ran Miao, Jinsong Li (2019). "Flexible solid-state supercapacitor based on reduced graphene oxide-enhanced electrode materials", *J. Alloys Compd*, Vol. 802:355–363, https://doi.org/10.1016/j.jallcom.2019.06.203.

51. Anjli Gupta, Silki Sardana, Jasvir Dalal, Sushma Lather, Anup S. Maan, Rahul Tripathi, Rajesh Punia, Kuldeep Singh, Anil Ohlan (2020). "Nanostructured polyaniline/graphene/Fe_2O_3 composites hydrogel as a high-performance flexible super-capacitor electrode material", *ACS Appl. Energy Mater.*, Vol. 3:6434–6446 https://doi.org/10.1021/acsaem.0c00684

52. Xuankai Huang, Haiyan Zhang, Na Li (2017). "Symmetric transparent and flexible supercapacitor based on bio-inspired graphene-wrapped Fe_2O_3 nanowire networks", *Nanotechnology*, Vol. 28:075402. https://doi.org/10.1088/1361-6528/aa542a

53. Huilong Liu, Kyoung-sik Moon, Jiaxiong Li, Yingxi Xie, Junbo Liu, Zhijian Sun, Longsheng Lu, Yong Tang, Ching-Ping Wong (2020), "Laser-oxidized Fe_3O_4 nanopar-ticles anchored on 3D macroporous graphene flexible electrodes for ultrahigh-energy in-plane hybrid micro-supercapacitors", *Nano Energy*, Vol. 77:105058, https://doi.org/10.1016/j.nanoen.2020.105058.

54. Zhihui Xu, Shishuai Sun, Wen Cui, Dan Yua, Jiachun Deng (2018). "Ultrafine MnO_2 nanowires grown on RGO-coated carbon cloth as a binder-free and flexible super-capacitor electrode with high performance", *RSC Adv.*, Vol. 8:38631, https://doi.org/10.1039/c8ra05890c

55. Shuangbao Chen, Lu Wang, Miaomiao Huang, Liping Kang, Zhibin Lei, Hua Xu, Feng Shi, Zong-Huai Liu (2017). "Reduced graphene oxide/Mn$_3$O$_4$ nanocrystals hybrid fiber for flexible all-solid-state supercapacitor with excellent volumetric energy density", *Electrochim. Acta*, Vol. 242:10–18 https://doi.org/10.1016/j.electacta.2017.05.013

56. Sangeun Cho, Jongmin Kim, Yongcheol Jo, Abu Talha Aqueel Ahmed, H.S. Chavan, Hyeonseok Woo, A. I. Inamdar, J. L. Gunjakar, S. M. Pawar, Youngsin Park, Hyungsang Kim, Hyunsik Im (2017). "Bendable RuO$_2$/graphene thin film for fully flexible supercapacitor electrodes with superior stability", *J. Alloys Compd.*, Vol. 725:108–114 http://dx.doi.org/10.1016/j.jallcom.2017.07.135

57. Ningning Song, Yue Wu, Wucong Wang, Ding Xiao, Huijun Tan, Yaping Zhao (2019), "Layer-by-layer in situ growth flexible polyaniline/graphene paper wrapped by MnO$_2$ nanoflowers for all-solid-state supercapacitor", *Mater. Res. Bull.*, Vol. 111:267–276 https://doi.org/10.1016/j.materresbull.2018.11.024

58. Z. Q. Hou, Z. Y. Wang, L. X. Yang, Z. G. Yang (2017). "Nitrogen-doped reduced graphene oxide intertwined with V$_2$O$_3$ nanoflakes as self-supported electrodes for flexible all-solid-state supercapacitors", *RSC Adv.*, Vol. 7:25732–25739. https://doi.org/10.1039/c7ra02899g

59. Keng-Ku Liu, Qisheng Jiang, Clayton Kacica, Hamed Gholami Derami, Pratim Biswas Srikanth Singamaneni (2018). "Flexible solid-state supercapacitor based on tin oxide/reduced graphene oxide/bacterial nanocellulose", *RSC Adv.*, Vol. 8:31296, https://doi.org/10.1039/c8ra05270k

60. Lei Li, Kwan San Hui, Kwun Nam Hui, Tengfei Zhang, Jianjian Fu, Young Rae Cho (2018). "High performance solid state flexible supercapacitor based on reduced graphene oxide/hierarchical core shell Ag nanowire@NiAl layered double hydroxide film electrode", *Chem. Eng. J.*, Vol. 348:338–349. https://doi.org/10.1016/j.cej.2018.04.164

61. Yanrong Li, Xue Wang, Qi Yang, Muhammad Sufyan Javed, Qipeng Liu, Weina Xu, Chenguo Hu, and Dapeng Wei (2017). "Ultra-fine CuO nanoparticles embedded in three-dimensional graphene network nano-structure for high-performance flexible supercapacitors", *Electrochim. Acta*, Vol. 234:63–70. https://doi.org/10.1016/j.electacta.2017.02.167

62. Junya Wang, Wei Dou, Xuetao Zhang, Weihua Han, Xuemei Mu, Yue Zhang, Xiaohua Zhao, Youxin Chen, Zhiwei Yang, Qing Su, Erqing Xie, Wei Lan, Xinran Wang (2017). "Embedded Ag quantum dots into interconnected Co$_3$O$_4$ nanosheets grown on 3D graphene networks for high stable and flexible supercapacitors", *Electrochim. Acta*, Vol. 224:260–268. http://dx.doi.org/10.1016/j.electacta.2016.12.073

14 Graphene Oxide Application for Flexible Energy Storage Devices

Chinemerem Jerry Ozoude
University of Nigeria

Raphael M. Obodo
University of Nigeria
University of Agriculture and Environmental
Sciences, Umuagwo
Quaid-i-Azam University
Northwestern Polytechnical University

Ikechukwu Ifeanyi Timothy
University of Nigeria

Fabian I. Ezema
University of Nigeria
Northwestern Polytechnical University
iThemba LABS-National Research Foundation
University of South Africa (UNISA)

CONTENTS

DOI: 10.1201/9781003215196-14

14.1 INTRODUCTION

There is no denying that the expected depletion of fossil fuel reserves, as well as the rapid rate of worsening environmental pollution, and climate challenges caused by massive fossil fuel consumption, are causing a rapid shift in the way energy is produced, stored, and consumed. This shift aided the major purpose of the Paris Agreement, which was signed in 2015 and intends to investigate practical ways to address climate change by exploring alternative energy sources [1]. Renewable energy, which is a plentiful and environmentally acceptable form of energy, has proven to be ideal for replacing fossil fuels. However, the fundamental limits of energy created through this medium are intermittent in nature. Solar and wind energy, for example, are now prospering in the energy market, but their inconstancy needs the inclusion of an energy storage or conversion device in order to improve the efficiency and cost-effectiveness of the system as a result [2]. Fuel cells are a new type of energy conversion technology that has little or no environmental impact because of their high theoretical density and efficiency. When it comes to energy storage technologies, batteries and supercapacitors (SCs) are two that are now undergoing extensive study to boost their energy density in order to keep up with the rapid development in renewable energy sources [3]. However vital the need for an alternative energy source and storage devices are, the need for better storage devices is which are flexible and can withstand high stresses is becoming pressing as these devices now take in new technologies where they are subjected under high, cyclic, and creep loading. To accommodate this necessity, graphene-based materials are continuously in research for use in flexible, portable, and wearable electronics [4].

14.2 BRIEF HISTORY

The discovery of graphene by Novoselov and Geim in 2004 spurred a flurry of scientific investigation. In recent years, graphene-based materials have been intensively investigated as field-effect semiconductor devices (FESD) electrode materials due to their remarkable physical characteristics, which include high strength (1 TPa), electrical conductivity, and specific surface area (SSA) ($2,630\,m^2g^{-1}$) [5]. Graphene single-layer capacitance measured at 21 F cm^2 was shown to have intrinsic electrochemical double-layer capacitance. In all carbon materials, including carbon nanotubes (CNTs), carbon fiber (CF), activated carbon, and graphene, this anticipated electrochemical double-layer capacitance of 550 F g^{-1} is the greatest limit capacitance yet measured. Van der Waals' interactions between sheets, on the other hand, can result in the restacking of graphene sheets (GSs), for instance. With increasing graphene layers, the effective SSA of graphene drops, resulting in a fast reduction in ion transport during charging and discharging. In order to resist layer restacking, researchers have created curved and crumpled graphene with one-dimensional (1D) graphene fiber morphologies, two-dimensional (2D) graphene films, and three-dimensional (3D) graphene foams. The use of graphene-based hybrid nanostructures with high power and energy density as a building block for adjustable functionalization with other electroactive components, such as conducting polymers (CPs), has been shown for field-effect transistors (FESDs). It is important to note that graphene is flexible in addition to the electrochemical characteristics listed above. According to the structural description of nanostructured graphene materials, they may be organized into 1D fibers, 2D films, and 3D graphene foams. Among them, 1D fibers consisting of 2D microscopic rGO sheets stacked in the axial direction, 2D films with rGO sheet alignment, and 3D foams with porous structure all have high mechanical strength, which results in superior FESD flexibility in a variety of applications. In recent years, there has been a growth in the number of publications on graphene-based materials for FESDs to meet the needs of industry. Example: Lightweight graphene-based FESDs may be bent, rolled, twisted, and stretched to form wearable displays and touch-screens, such as popular curved screen mobile phones and LCD televisions with curved screens.

14.3 DEFINITION OF TERMS

14.3.1 GRAPHENE

Graphene is a carbon-based substance made up of carbon atoms linked together in a hexagonal arrangement. Because graphene is extremely thin, it is classified as a 2D material. Graphene is the world's strongest substance, as well as one of the most electrically and thermally conductive. In practically every business, graphene has limitless potential applications.

14.3.2 GRAPHENE OXIDE (GO)

One of the most important discoveries in recent years has been the discovery of graphene oxide (GO), a single-atomic-layered substance generated by a strong oxidation of graphite, a very inexpensive and plentiful material. GO is a kind of graphene that has been oxidized and laced with oxygen-containing groups, and it is a material that is used in electronics. For this reason, it is regarded straightforward to handle, and it may even be utilized in the production of graphene because it disperses in water (and other solvents). Despite the fact that GO is not a good conductor, there are methods to increase its properties in the future. It is typically marketed as a powder, as a dispersion, or as a coating for a variety of substrates.

14.3.3 REDUCED GRAPHENE OXIDE (rGO)

Reduced graphene oxide (rGO) is a type of GO in which the oxygen content has been reduced through chemical, thermal, and other methods, whereas graphite oxide is a material formed by oxidizing graphite, resulting in increased interlayer spacing and functionalization of the graphite basal planes.

14.4 FLEXIBLE GRAPHENE MATERIALS FOR ENERGY STORAGE

Depending on how they store energy, SCs (also known as electrochemical capacitors) are classified into two types: electrical double-layer capacitors (EDLCs) and pseudo-SCs [6]. A large portion of the capacitance of an electrolyte-dispersed liquid crystal (EDLC) is derived from charge partitioning and collecting at the anode/electrolyte interface (Figure 14.2a) [6]. As a result, the concave surface components of anode materials, as well as the pore-size distribution of anode materials, have an influence on the limit of the EDLC. Carbon materials such as activated carbon, CNTs, and graphene are often utilized as cathode materials in electrochemically discharged liquid crystal (EDLC) batteries due to their convenience of use, ease of handling, and high compound strength, among other reasons [7]. The actual charging/releasing contact favors EDLCs with a high force density because of the nature of the interaction. In contrast, because of the limited explicit surface space available on cathode materials during electrochemical contact with the electrolyte, the energy density of EDLCs is frequently insufficient [8]. Materials that create pseudo-capacitance as a result of reversible faradaic reactions, such as CPs and changeable metal oxides, do so in an unexpected way, resulting in a comparatively high energy density compared to other materials. Because of their low conductivity, they have a very weak force on the ground. In a similar line, the consolidation of pseudo-capacitive materials into carbon structures is crucial for the performance of SC cathodes. Carbon and lithium-metal oxide are utilized as the anode and cathode in a conventional lithium-ion battery (LIB), with carbon serving as the anode. (i) Lithium particles concentrate and supplement between two terminals in a reversible manner, while electrons are evacuated and expanded (Figure 14.2b). Because of sodium's inexpensiveness and several beneficial properties, sodium-particle batteries (SIBs) are gaining a lot of interest these days. They have a response component that is comparable to that of LIBs, and they have been identified as one of the most anticipated energy storage options [9] (Figure 14.1).

Although the energy density of the two batteries is far lower than that required by flexible electronic devices, the energy density of addition-type LIBs is approaching its theoretical maximum. When using a single particle intercalation response instrument, LIB or SIB cathode with a solitary particle intercalation response instrument has limitations that are fewer than $250 \, \text{mA h g}^{-1}$ [11], S and O_2 dependent on the multi-particle change chemistry might have an uncommonly high hypothetical limit of $1,675 \, \text{mA h g}^{-1}$ [11]. As a result of this fact, Li-S and Li-O_2 batteries are rapidly improving, with possible energy densities of up to 3,505 and $2,567 \, \text{Wh kg}^{-1}$, respectively. Cathode material, current authority, separator, electrolyte, and bundling are all included in the two batteries and SCs [12]. This vast number of pieces, regardless of

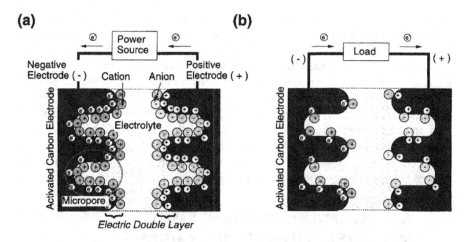

FIGURE 14.1 Schematic illustration of electric double-layer capacitor: (a) charge state and (b) discharge state. *Copyright* permission, carbon alloys, 2022 [10].

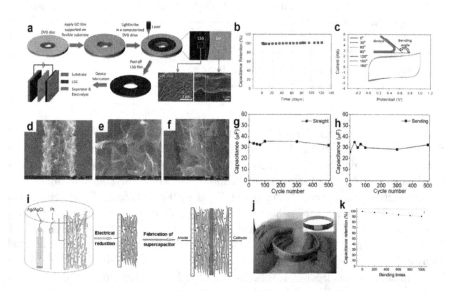

FIGURE 14.2 (a) Schematic diagram of the fabrication of LSG-SCs; (b) a shelf-life test showing good stability for over 4 months; (c) CVs when the LSG-SCs were bent at 1,000 mV s⁻¹. Reproduced with permission. An SEM image of a GF@3D-G: (e) an enlarged view of GF@3D-G (d); (f) the edge view of a GF@3D-G; (g, h) the capacitance stability of GF@3D-G fiber SC when bent and straight, respectively. (i) Schematic illustration for the synthesis of the rGO hydrogel film and the fabrication of an SC; (j) a photo of the belt-like flexible SC; and (k) capacitance retention after 1,000 bending tests [19]. (Copyright 2022, Energy Storage Materials.)

whether they are packed together, must be adjustable for adaptable SCs or batteries. Overall, the electrolytes and separators are adaptable. In contrast, current gatherer bundling materials and dynamic anode materials have weak mechanical properties and are incapable of withstanding significant strain. Standard SCs and batteries with jumbled configurations that are heavy, large, and unresponsive would fall short of addressing the specific requirements of adaptive and wearable devices. The devices themselves must be built to be tiny and integrated, as well as flexible and adaptable with an updated structure, in order to function with a variety of elastic electronic devices and, as a result, produce completely flexible electronic frameworks. As a result, continuing research into the fabrication of flexible electrodes as well as the design of flexible energy storage device topologies should be given top priority.

14.5 NANOSTRUCTURED GRAPHENE-BASED MATERIALS FOR FLEXIBLE ENERGY STORAGE

14.5.1 Flexible Supercapacitors

Traditionally utilized energy storage devices are not suitable for usage in FESDs because of their stiffness, heavyweight, and large volume characteristics [13]. As a result, in recent years, researchers have focused their efforts on developing light-weight, flexible, and portable energy storage systems [14]. To a certain extent, FESDs are comprised mostly of:

- Stretchable batteries and SCs
- Micro-scale batteries and micro-scale SCs
- LIBs that are both flexible and lightweight
- LIBs

In recent years, however, the development of flexible SCs and batteries has overtaken the development of new $Li-O_2$ and Al-ion batteries, which has resulted in a significant price premium. SCs (also known as electrochemical capacitors) are a form of FESD that is widely used in many applications [10]. High SSA, outstanding electrical conductivity, and great chemical stability are all advantages of graphene-based materials [15]. As a result, it has a lot of potential for versatile SC applications. EDLCs, pseudo-capacitors, and hybrid capacitors are the three types of SCs [16]. Electrostatic charges accumulate in EDLCs, which allows them to store electrical energy. Electrical conductivity, pore structures, and SSA are all important factors that influence the electrochemical performance of EDLC electrodes.

Meanwhile, reversible and rapid redox processes are used to store energy in pseudo-capacitors.

Therefore, the capacitance and energy density are lower than those of a pseudo-capacitor. Such elements as theoretical capacitance and the electrical behavior of electrode materials have a significant impact on the electrochemical performance of pseudo-capacitors [17]. As a result of the belief that electrodes play a critical role in SCs, various research has been undertaken in this field. To fabricate high-performance SCs with flexibility, the fundamental challenge is to design and synthesize electrode materials that have superior mechanical flexibility, high energy density, high power density, and excellent cycle stability, and then combine these materials with flexible current collectors and compatible electrolytes in an adjustable assembly [18]. High mechanical strength in the microstructures of microstructural foam films with 1D, 2D, and 3D microstructures results in exceptional flexibility in the films, which may be ascribed to the high mechanical strength of the microstructures. Furthermore, solid electrolytes such as polyvinyl alcohol (PVA)/H_3PO_4, H_2SO_4-PVA, and PVA/LiCl, which can protect FESDs from current leakage and endure mechanical stresses, can be used to safeguard FESDs [13].

14.5.2 Pure Graphene

In the field of static charge, graphene-based materials such as GO and rGO are commonly used as SC electrodes; however, pure graphene electrodes are required in the field of static charge. Due to the fact that it is a carbon substance, graphene may be used as an EDLC electrode material, on the one hand. As a result, they have a higher capacitance and higher energy density than EDLCs, which is advantageous.

The effective SSA of graphene is related to the electrochemical performance of graphene materials. Graphene structures with curved and crumpled surfaces, such as 1D fibers, 2D films, and 3D foams, must be created in order to reduce graphene restacking. 1D graphene fibers have a high SSA, excellent electrical conductivity, high mechanical strength, amazing flexibility, and excellent electrochemical performance, to name a few properties. Meng et al., for example, developed a GF@3D-graphene (GF@3DG-G) system to demonstrate its capabilities. The graphene layers that had been deposited were evenly distributed throughout the fiber (Figure 14.2d–f). Although the fiber SC was bent 500 times, the capacitance remained constant at roughly 30–40 F, demonstrating the fiber SC's excellent stability when twisted (Figure 14.2g and h).

rGO-GO-rGO fibers are easily created by reducing GO fibers in specified regions under laser irradiation [20]. After 160 bending tests, the SC based on fibers as electrodes and GO as a separator demonstrated excellent structural stability. Researchers have created a fiber-shaped solid SC based on electrochemically rGO that exhibits enhanced electrochemical characteristics as well as more flexibility than previous solid SC materials. A flexible cable-type SC based on rGO nanosheets was developed, and when used in conjunction with a redox additive electrolyte, it demonstrated improved electrochemical performance [21] is also developed.

When bent, the device also demonstrated good flexibility. This substantial mechanical flexibility is provided by the structure of the 2D freestanding GS, which may be used to fabricate graphene electrodes that can be used as self-standing or substrate-supporting electrodes for SCs [22]. El-Kady and colleagues synthesized graphene directly from GO layers using a LightScribe DVD optical drive and laser reduction (Figure 14.2a) [23]. Laser-scribed graphene (LSG) sheets are produced, which exhibit superior electrochemical performance over GO, with a cyclic voltammetry (CV) shape that is almost rectangular at $1,000\,mV\ s^{-1}$ when compared to GO. In addition, a lightweight FESD based on LSG films was created, and it displayed outstanding cycling stability (>97% after 10,000 cycle tests) as well as robust stability (during 4 months of testing) (Figure 14.2b). According to Figure 14.3c, when the device was bent from 0° to 180°, there was no visible change in CVs, and after more than 1,000 cycles of bending, there was a 5% decrease in CVs, showing that the device was flexible and stable. Through the use of direct printing and photo-thermal reduction of GO, Jung et al. developed a novel approach for synthesizing a highly porous pattern of interdigitated SC with interdigitation [24]. When bent at different angles, the performance of the printed-rGO-based SC was only marginally degraded. It has been demonstrated that by applying a simple irradiation and selective reduction of GO, Xue and colleagues have effectively generated multiscale, planar patterning of GO-rGO. The capacitance of the produced SCs based on the GO-rGO composite electrodes was 141.2 F g^{-1}, whereas Ramadoss et al. employed a straightforward chemical vapor deposition (CVD) approach to manufacture a 3D graphene/graphite paper in their research [25]. When tested as part of a three-electrode system, the SC demonstrated maximum capacitance of 260 F g^{-1} and maximum capacitance of 80 F g^{-1} in a whole cell, with excellent capacitance retention. The SC also maintained acceptable electrochemical performance when deformed in any way (bent, rolled, or twisted), which further demonstrates the versatility of the integrated device. The arbitrary-shaped SCs developed by Zheng et al. differ from the 2D films-based SCs previously described because they were created using a novel printing technique that employed electrochemically exfoliated graphene as both anode and cathode, along with GO as a separator on a polyethylene terephthalate (PET) substrate [18]. The as-fabricated SC had a volumetric capacitance of 280 F cm^{-3} when tested with a redox electrolyte and exhibited no noticeable degradation for rectangle-shaped SCs after 10,000-cycle testing, demonstrating the exceptional flexibility of the graphene-based planar sandwich SC. Comparatively to 1D graphene fibers and 2D graphene films, 3D graphene foams with an organized porous structure and great flexibility have an increased effective SSA, as well as rapid electron and ion transmission, and have been extensively investigated for use in SC applications [26]. Graphene foams can be produced from GO by a thermal and chemical reduction process, or they can be produced directly using a CVD method on a template. Researchers Feng and colleagues employed electrochemical reductions in situ to generate a non-stacked rGO hydrogel layer (Figure 14.2i). Due to the high SSA of the porous films made as-is, they may be employed as SC electrodes without the need for drying. Electrochemically generated rGO (ERGO) film possesses specific gravimetric and volumetric capacities of 206 F g^{-1} and 231 F m^{-3}, respectively, when measured electrochemically. As shown in Figure 14.2j, a belt-like packed flexible capacitor made of two ERGO electrodes and titanium mesh as a current collector was also developed and tested. It was found that even after being bent 1,000 times, 91.5% of the capacitance was intact, demonstrating that the device was adaptable (Figure 14.2k). Furthermore, the capacitor gadget has the capability of powering a light-emitting diode (LED) bulb for more than 2 minutes at a time without requiring a power source. Chemically converted graphene (CCG), also known as rGO, may be dispersed in water and self-assemble into oriented 3D graphene hydrogels, according to the researchers. On the basis of hydrogels, Yang et al. employed (CCG) with corrugated 2D structures and self-assembly property to create an SC with high capacitance and maximum energy density of 59.9 W h L^{-1} that has a maximum energy density of 50.9 W h L^{-1} [27]. Yu et al.

created a flexible SC using CCG as the active material and stainless-steel fabrics as the current collector, which differs from the previous work [28]. After 7,500-cycle testing, the capacitor was found to have a high capacitance of 180.4 mF cm^{-2} and a capacitance retention of 96.8%, indicating that it was well-designed. Furthermore, the SC's flexibility was tested, and it was shown to have a capacitance of 96.4% even after 800 bending tests. A simple hydrothermal reduction procedure was employed by Shi et al. to produce ordered 3D graphene with a superior gravimetric capacitance of 220 F g^{-1} at 2 A g^{-1} and outstanding cycle stability, according to their findings [29]. Although the capacitor was bent at a scan rate of 200 mV s^{-1}, the capacitance retention was greater than 80% after 10,000 cycle tests.

14.5.3 GRAPHENE/METAL OXIDE

Metal oxides and CPs are among the pseudo-capacitor electrode materials that have been produced so far [30]. Electrode materials made of electrolyte-dielectric liquid (EDLC) offer greater rate capability, higher power density, longer operating lives, lower capacitance, and lower energy densities than electrode materials made of pseudo-capacitor. As a result, combining EDLC and pseudo-capacitive materials is an option worth studying [31,32]. Pseudo-capacitors made of metal oxides such as MnO_2, Fe_2O_3, and Fe_3O_4 nanoparticles (NPs) are the most commonly studied pseudo-capacitor materials. Furthermore, microstructures such as 1D hybrid fibers, 2D films, and 3D foams are used to alter the flexible properties. Wet-spinning may be used to create MnO_2 nanorods/graphene hybrid fibers and MoO_3 nanorods/rGO hybrid fibers, which are both useful materials. Using a straightforward wet-spinning technique, Ma et al. were able to fabricate a MnO_2 nanowire/graphene hybrid fiber with a hierarchical structure [33,34].

The X-ray diffraction patterns revealed that the -MnO_2 had a low crystallinity, which made it a better choice for SCs than the other two compounds. SEM scans of nanowires with diameters ranging between 5 and 20 nm and lengths ranging between 5 and 10 μm (Figure 14.3a–c) revealed a 3D connected porous network structure composed of folded GSs (Figure 14.3a–c). Assembly of a high-performance fiber-optic power supply resulted in high volumetric capacitance of 66.1 F cm^3 and capacitance retention of 96% after 10,000 cycles, both of which were accomplished (flexible all-solid-state SC (FSSC)). It was possible to reach high energy densities of 5.8 mWh cm^{-3} as well as power densities of 0.51 W cm^{-3} (Figure 14.3d). The CV curves under various bending degrees and galvanostatic charge–discharge (GCD) curves after changing bending periods were tested, and the results revealed that the FSSC's flexibility did not appear to alter much (Figure 14.3h and i). Even when the device is linked in series, a red LED may be observed, suggesting that it has the potential to be used in FESDs. In hybrid fibers, rGO sheets can be used to support and convey fibers that are electrically and mechanically coupled to one another. The authors employed MnO_2 nanorods/rGO hybrid fibers as the positive electrode, MoO_3/rGO as the negative electrode, and H_3PO_4/PVA as the electrolyte to construct a flexible asymmetric super Capacitor (a-SC) in the same manner as Ma et al. [34]. The improved a-SC had a good volumetric energy density of 18.2 mW cm^{-3} and a good power density of 76.4 mW cm^{-3}, both of which were excellent. Aside from that,

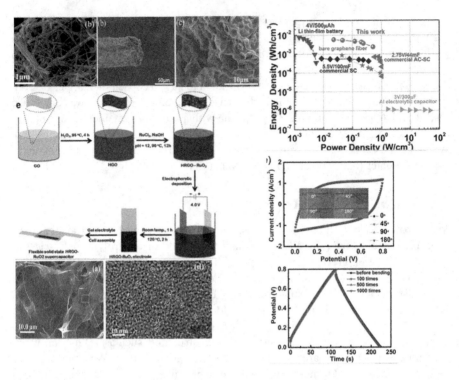

FIGURE 14.3 (a) SEM image of the MnO₂ nanowires; (b, c) cross-sectional SEM images of the MnO₂/GO fibers; (d) energy and power densities of the SC compared with commercially available energy storage systems; (e) schematic illustration of the fabrication of the flexible solid-state holey rGO-RuO₂ SC; (f) SEM image of freeze-dried holey rGO-RuO₂; (g) TEM images of holey rGO-RuO₂ (gray and dark gray circles represents RuO₂ NPs and the in-plane nanopores, respectively); (h) CV curves under bending at with different angles; and (i) GCD curves before and after bending for 100, 500, and 1,000 cycles (a–d, h, i). (Reproduced with permission, Copyright 2022, Energy Storage Materials [19].)

the a-SC demonstrated outstanding cyclability as well as flexibility and mechanical stability. Guo et al. employed polyester as the flexible substrate in comparison to the above MnO₂/rGO, a wet-spinning approach to the synthesis of GO/polyester composites, NaBH₄ solution immersion to make graphene/polyester hybrid fibers, and a hydrothermal procedure for MnO₂/graphene/polyester. Material shape, electrode structure, and mechanical bending or stretching all have an influence on the electrochemical performance of MnO₂/graphene/polyester composite electrode materials. Another way of putting it is that an FSSC based on the MnO₂/graphene/polyester composite textiles electrode exhibited an overall specific capacitance of 332 F g⁻¹ at 2 mV s⁻¹, but the specific capacitance of the composite fabric's electrode exhibited a specific capacitance of 265.8 F g⁻¹. The electrochemical performance of FSSC, on the other hand, remained steady despite mechanical bending and stretching.

A three-step procedure can be used to create 2D metal oxide/graphene films. There are three steps:

1. The modified Hummers process is used to synthesize GO
2. Second, metal oxides and GO are mixed by simple solution mixing
3. Metal oxides-GO are reduced to metal oxides-rGO.

After the GO reduction process, metal oxides can be combined with rGO to create new materials. In addition, FESDs benefit from flexible substrates and electrolytes, as well as 2D film microstructure. Porous RuO_2-rGO composites (Figure 14.3e) were generated using basic sol–gel and electrophoretic deposition techniques [35]. Big-holey rGO-RuO_2 composite sheets were decorated with tiny RuO_2 NPs in Figure 14.3f, while the composite sheets themselves had large rGO holes up to tens of microns across, as shown. It was discovered that the composites' basal flat surface included an excessive number of in-plane holes with sizes ranging from 1.0 to 4.0 nm in diameter (Figure 14.3g). A flexible SC constructed using rGO-RuO_2 films, Au-coated PET substrates, and a PVA-H_2SO_4 gel electrolyte-maintained 88.5% of its capacitance after 10,000 cycles. When the bent was turned 180°, just 4.9% of mechanical flexibility was lost. The integration of pseudo-capacitive RuO_2 NPs, the distinct features of rGO nanopores, and the layered structure of rGO-RuO_2 sheets are all aspects that contribute to the excellent performance and flexibility of the material.

Choi et al. employed a straightforward approach to create RuO_2-liquid functionalized-chemically modified graphene (IL-CMG) films for use as a positive electrode for SCs.

RuO_2 NPs were found to be uniformly dispersed on the surface of the IL-CMG thin films. Using RuO_2-IL-CMG films, specific capacitances of 175, 166, 144, 141, and 139 F g^{-1} were achieved at varied current densities by using different current densities. After more than 2,000 cycles, the specific capacitance of this device maintained a value of 95% of its initial value. In FESD applications, the RuO_2-IL-mechanical CMG's ruggedness, interfacial contact, and electron transmission among the components emphasize the material's usefulness. A 2D V_2O_5 is what it is. Solvent mixing and hydrothermal reduction are the two most often used techniques for fabricating graphene-based hybrid films, according to the literature. When used in this manner, the V_2O_5 NPs inhibited GS stacking and provided channels for rapid ionic transport, which in turn assisted the EDLCs in achieving their maximum potential surface-specific absorption when applied in this manner (SSA). It was developed by Foo and his colleagues in both the flat and bending states using rGO/V_2O_5-rGO, which showed good cyclability, real power, and energy density [36]. When measured in a flat state, the maximum energy density was determined to be 13.3 W h kg^{-1}, with the power density declining to 12.5 W kg^{-1}. In a bent state, the SC's energy density was 13.6 W h kg^{-1}. A flexible graphene/V_2O_5/H_2O ultrathin-film SC developed by Bao et al. showed a capacitance of 11,718 F cm^{-2} at 0.25 A m^{-2} and an excellent cycle life [37]. The as-obtained FSSC also provided a good energy density of 1.13 mW h cm^{-2} at a power density of 10.0 mW. GCD curves and cycling stability tests in a bent shape were employed to evaluate the SC's flexibility. The capacitance of the SC did not decrease after 500 cycles because of the 2D thin sheets' incredibly thin nature. To generate oxygen-deficient Bi_2O_3/graphene (r-Bi_2O_3), Liu and his colleagues used simple solvothermal and solution reduction procedures. The solvothermal Bi_2O_3/GN (graphene nanocoating) and the subsequent reduction treatment of the r-Bi$_2$O$_3$/GN had distinct transmission electron microscopy structure. Preliminary tests demonstrate that GSs are covered with ultrathin Bi$_2$O$_3$ nanosheets (GN). Bi$_2$O$_3$/GN was found to have less standing Bi$_2$O$_3$ nanosheets following reduction treatment, which allowed for excellent charge transfer through the highly conductive GN after the reduction process. Flexibility and 55.1 MPa tensile strength of the flexible substrate bacterial cellulose (BC) played a critical role in the development of flexible SCs. With great flexibility and no discernible capacitance decay under varying bending conditions, the flexible a-SC made of r-Bi$_2$O$_3$/GN/BC and Co$_3$O$_4$/GN/BC paper achieved high energy density, maximum power density, and excellent flexibility. It is also vital to take into account 3D hybrid networks, which may effectively separate GSs and keep high SSA, as well as resulting in high-performance electrochemical and flexible devices attributable to their structure and effects on mechanical flexibility [38].

It is possible to generate 3D hybrid networks by using vacuum filtering, heat reduction, or electrochemistry to create them. Graphene/MnO$_2$ composite electrodes have been successfully fabricated for use in SCs [39]. Optimizing the MnO$_2$ content of the electrode allowed for a maximum specific capacitance of 130 F g^{-1}. Graphene/MnO$_2$ networks, PET membranes, and polymer separators have also been studied in terms of their electrochemical performance and flexibility. After 5,000 cycles, the lightweight and slim SC achieved 24.2 F g^{-1} capacitance, which is an outstanding performance. The graphene/MnO$_2$-based SC's specific capacitance remained at 92% even after 200 bending tests with a bending angle of 90°, confirming its exceptional flexibility.

Fe$_2$O$_3$/rGO paper had a capacitance of 178.3 F cm^{-3} when compared to the capacitance of the pure Fe$_2$O$_3$/rGO paper, which was 106.2 F cm^3. Using an innovative three-step process, Liu et al. created a Fe$_3$O$_4$/GS composite paper [40]. As the NPs were synthesized in situ and fixed on GSs, their diameter was less than 5 nm. NPs of Fe$_3$O$_4$ are better dispersed in 3D paper because of their increased electrical conductivity and better dispersion. Specific capacitance values were observed to range from 368 F g^{-1} at 1 A to 245 F g^{-1} at 5 A, suggesting that the Fe$_3$O$_4$/GS paper could be used in future high-energy scavenging devices (FESDs).

14.5.4 POLYMER/GRAPHENE CONDUCTIVE POLYMER

CPs have a variety of benefits over other electric active materials, including low cost, wide voltage ranges, environmental friendliness, excellent reverse reversibility, and adjustable electrochemical activity, to name a few. Because graphene and carbon nanotubes (CPs) are both capable of delivering high energy density, power density, and cyclability, attempts to combine the two materials have been pushed up in recent years [1–6]. Extensive research has been done on graphene composite sites that have been combined with chemical compounds such as polyaniline (PANI), polypyrrole (PPy), and poly (3,4-ethylene dioxythiophene) (PEDOT) for the purpose of using SCs [41]. PANI is one of these promising materials because of its good capacitive properties, low cost, and ease of preparation. Paper-like film-based flexible SCs were developed by Wu et al., and are now commercially available.

Newly developed in the lab is G-PNF30, a high-quality and flexible film with a CCG weight of 30% and excellent flexibility. It was also found that films prepared with G-PNF30 had a homogeneous distribution of PANI-nanofibers (NFs) across the CCG layer structure. In the symmetric SC device, utilizing composite materials, an average capacitance of 155 F g^{-1} was achieved during 800 cycles at 3 A g^{-1} after using composite materials 44]. A self-standing graphene/PANI electrode was developed by Cong et al. that demonstrated high electrical conductivity and excellent flexibility. At 1 A g^{-1}, the graphene/PANI [43] composite paper had a larger specific capacitance of 763 F g^{-1} than the pure PANI sheet, which had 520 F g^{-1} at the same voltage. It is remarkable that the composite electrode retained an astonishing 82% of its capacitive capacity even after 1,000 cycles of cycling. This is a huge increase from the 51.9% of the PANI film that was used. Using an in situ polymerization approach invented by Yu et al., a novel flexible rGO foam/PANI composite with a high specific capacitance of 939 F g^{-1} has been created. Despite the fact that Yu and colleague constructed this novel flexible rGO foam/PANI using composite electrodes, the symmetric SC displayed excellent cycle stability, with capacitance retention remaining at 88.7% after 5,000-cycling tests. SC CV curves also revealed that when the SC was bent 180°, there was virtually little change, demonstrating the SC's adaptability. Using oxidative polymerization in situ to produce a multiscale rose flower-based structure for the PANI/rGO composite, Chang et al. achieved the desired results. The capacitance of 626 F g^{-1} was achieved by the as-fabricated SC, which was constructed using the composite electrode, graphene paper (GP) as a current collector, and H_2SO_4 as the electrolyte. The SC's flexibility was not significantly affected by the bending deformations, and the CV curves remained rather constant. As a consequence, even though the SC with a five-cell series was compressed into a small area, it was still able to illuminate the red LED.

It is also a low-cost and straightforward preparation method for the direct development of CPs on the collector when performed in a single step. Zhou and his colleagues created PPy/GO composites with the assistance of electrochemical deposition [46]. PPy/GO composites with a fuzzy sheet-like topology demonstrated superior capacitive characteristics due to their fuzzy sheet-like structure. The capacitance of the composite was 152 mF cm^{-2} at 10 mV s^{-1}, and the rate of discharge was 10 mV s^{-1}. A novel freestanding rGO-PPy composite paper was made by Shu et al. using electrochemical polymerization and a layer-by-layer structure with PPy incorporated in graphene layers, according to the researchers [47]. After 60 minutes of deposition, the hybrid PPy60 sheets demonstrated higher areal capacitance and a maximum energy density of 61.3 W h cm^{-2}, compared to the conventional PPy60 sheets. Because of this, the areal capacitance and maximum energy density have both increased. The capacitive behavior of this FESD was substantially the same in all bent states, showing that this FESD is stable in all bent states. We have developed a straightforward method for fabricating a FSSC out of composite papers with a layered structure, which we call rGO/PPy-NP composite papers. Attempting to bend the SC resulted in a maximum areal energy density of 132.5 W h cm^2, indicating that the field of electrostatic discharge had a tremendous deal of potential. In addition, Lehtimäki et al. employed a straightforward electro-polymerization and reduction procedure to effectively create PEDOT/rGO composite films on flexible PET substrates, which is worth mentioning [46].

14.6 OTHER FLEXIBLE GRAPHENE MATERIALS

14.6.1 GRAPHENE/CARBON NANOSPHERES

Carbon nanotubes [47,48], carbon black, CFs, graphene, and other carbon compounds are examples of electrode materials used in electrochemical deposition of liquid crystals. The combination of graphene and other carbon materials results in electrode materials with lower capacitance and energy density than pseudo-capacitor electrode materials such as metal oxide/graphene and conductive polymer/graphene composites. Due to the intrinsic features of fibers, numerous hybrid fiber-shaped SCs with excellent flexible capabilities were produced for use in flexible devices in addition to electrochemical performances. Electrophoretic fabrication of flexible fiber-shaped SC electrodes with layers has been achieved by Zhang and colleagues [49]. As you can see in this image, the carbon nanospheres were well spread. An ideal capacitance of 53.56 mF cm^{-2} was achieved with the fibrous electrode, as well as a superior reversible capacity with a decay rate of 91.2% over 4,000 cycles. SC's fiber-like shape revealed its good flexibility by remaining stable even at a 90-degree bend. Zhou et al. developed an SC fiber-shaped electrode made of extremely flexible graphene hydrogels/multi-walled CNTs cotton thread and an SC fiber-shaped electrode made of SC fibers [52]. When tested at 2 mV per second, the fiber-shaped SC maintained its capacitance at 95.51% even after being exposed to 8,000 cycles. Fiber-shaped SC's capacitance did not appear to degrade under varied bending conditions. In contrast, it showed a 7.4% increase in capacitance at a bending angle of 180°. For a material to be flexible, its fiber structure must be taken into account. rGO-wrapped Fe-doped MnO_2 composite (G-MFO) and rGO-wrapped hierarchical porous carbon microspheres composite (G-HPC) were also successfully manufactured by Tang et al. [48]. To construct a-SC devices, we used G-MFO films for the positive electrodes and G-HPC films for the negative electrodes that we had obtained. Even when bent, the asymmetrical supercapacitor (ASC) gadget maintained 91.6% of its capacitive capacity during 8,000-cycling tests. ASC's CV curves remained nearly unchanged even when deformed, showing its excellent mechanical stability for SC applications.

14.6.2 CARBON, GRAPHENE, AND METAL OXIDES

When comparing binary hybrid materials that employ metal oxide CPs or carbon particles to graphene-based ternary hybrid materials, there are several flaws in the binary hybrid materials. There are a number of flaws, including a lack of cycling behavior, tiny SSAs, and non-elastic deformation of transition metal oxides (TMOs) during charge–discharge operations, among others. As a result, graphene ternary composites with high electrochemical performance and flexibility [51] are possible. Because of their low cost and high capacity, Co_3O_4 and Mn_3O_4 are considered suitable materials. A hydrothermal technique was used by Liao et al. to create a CF/vertically aligned graphene nanosheets (VAGN)/Mn_3O_4 electrode [52]. Mn_3O_4 NPs with a grain size of 10 nm were used to cover the whole surface of the VAGN and anchor it on both sides. After 10,000 cycles, the electrodes were integrated into an FSSC that had a capacitance of 562 F g^{-1}, a high energy density of 50 W kg^{-1}, and a high power density of 64 W kg^{-1}. Excellent cycle stability, with just minimal deterioration at 150°, revealed good FSSC flexibility, which can be attributed to the flexibility of the carbon fabric, hybrid fiber architectures, and electrolyte PVA/H_3PO_4 gel. Yuan et al. have created paper electrodes based on Co_3O_4/rGO/CNTs for use in solid-state capacitors (SCs). This innovative flexible paper electrode has a capacitance of 378 and 297 F g^{-1} when measured at 2 and 8 A g^{-1}, respectively, when tested.

In addition, the flexible paper displayed steady cycling behavior under a variety of current densities and temperature conditions. Over 3,000 cycles, the retention of capacitance at 2 A g^{-1} was 96%, with no detectable loss of capacitance. The hierarchical architecture of the flexible paper electrode has been attributed to the electrode's superior electrochemical properties.

14.6.3 METAL OXIDE/CONDUCTIVE POLYMER/GRAPHENE

This is induced by the charge–discharge process, which causes the CPs to compress and increase in bulk throughout the process of charge–discharge. As a result, the electrochemical properties of CPs-metal oxide composites have been investigated. Wu et al. developed a NiO-coated graphene nanosheets/PANI composite film based on graphene nanosheets [55]. The as-obtained film has a specific capacitance of 1,409 F g^{-1} at 1.0 A g^{-1} and exceptional cycle stability with a capacitance X when operated at 1.0 A g^{-1}. According to Guo and colleagues in the journal *Energy Storage Materials* more than 92% of the capacity was preserved after 2,500 cycles. Furthermore, it demonstrated outstanding electrochemical stability at a variety of angles, as well as a long cycle life, with capacitance retention of 86% after 1,000 bending tests were performed. Sankar et al. developed composites of $CoFe_2O_4$/rGO/PANI (CCGP) and Co $(OH)_2$ as well as other materials Example: The as-fabricated SC with CCGP (negative electrodes), $CoOH_2$ (positive electrodes), and a length of 9 mF cm^{-1} at 1 mA demonstrated outstanding cycling performance when used with CCGP (negative electrodes). Further evidence of the cell's adaptability may be seen in the fact that the CV tests were virtually identical across the flat and bent states. An illuminated red LED was used for demonstration reasons, allowing the contents of the LED to be readily read through it. Lee et al. produced a 3D rGO nanostructure (carbonized PPy-coated SnO_2/Co_3O_4 (CPSC)-3rGO) with carbonized PPy-coated SnO_2/Co_3O_4 NFs that were used to adorn it. Comparing the CPSC-3rGO to other intermediated materials, its specific capacitance was 446 F g^{-1}, which was much higher than the other materials. Mechanical stability of the 3D graphene-based construction was demonstrated even under bending and twisting circumstances. The specific capacitance of the SCs was also found to be considerable even under these conditions.

14.6.4 DOPED GRAPHENE

Heteroatoms in the graphitic lattice can be used to modify the original structure and shape by substituting for graphene [55]. 2D and 3D networks give composite-based SCs a high degree of flexibility. When Byun et al. came up with the idea of creating a graphene/h-boron nitride hybrid film, they did so via a simple solution-based technique. The as-fabricated SC, which includes hybrid film electrodes, a Ni current collector, and a PVA-KOH electrolyte, exhibits excellent electrochemical properties as well as flexibility. Upon bending, the flexible SC demonstrated a volumetric capacitance of 95 F cm^{-3} at 50 mV s^{-1} and retained its SC activity. The majority of 3D networks are made up of foams (hydrogels and aerogels), 3D porous frameworks, and sponges. Tran et al. created 3D co-doped hole defect graphene hydrogel (NSHGH) electrodes with a specific capacitance of 536 F g^{-1}, which is significantly higher than the previous record [56]. The NSHGH-based FSSC is a revolutionary design that provides excellent electrochemical performance and mechanical versatility. The SC exhibited excellent cyclability, with 94% of its capacitance remaining after 5,000 cycles at 10 mV s^{-1}, demonstrating its durability. During this time, Wang et al. developed an N-doped porous carbon and graphene aerogel (NPCGA) hybrid electrode with a high specific capacitance of 608 F g^{-1} in 1 M H$_2$SO$_4$ at 0.1 A g^{-1} and a high specific capacitance of 608 F g^{-1} at 0.1 A g^{-1}. Instead, the flexible SC built on NPCGA technology demonstrated greater energy density (12.4 W h kg^{-1}) and power density (2,432 W kg^{-1}), as well as strong cyclability (specific capacitance retention of 92% over 10,000 cycles).

14.7   FLEXIBLE RECHARGEABLE BATTERIES

SCs and batteries are vastly different in terms of power and energy density. It is because of the low ion flow in redox reactions that battery capacity is greater than that of SCs [56]. As a result, improvements in battery rate performance and cycle life, as well as battery safety, are required for FESDs. To meet the growing demand for an electrical energy storage system that is small, light, and safe while also having a high capacity, researchers are experimenting with new electrode materials and rechargeable batteries. LIBs, lithium-sulfur batteries (LSBs), and SIBs. LIBs, LSBs, and SIBs are examples of such batteries. When it comes to energy storage systems, LIBs, LSBs, and SIBs each have their own set of advantages and disadvantages, as seen in Table 14.1.

TABLE 14.1
The Advantages and Disadvantages of Rechargeable Batteries

Graphene	LIBs	SIBs	LSBs
Advantages	High energy density; good stability; (light density, small atomic weight of 2.3 g mol^{-1} low potential (−3.04 V)	High abundance of 2.75%; low cost of sodium; good stability; good rate performance. (Standard potential −2.70 V)	High abundance; low cost of sulfur; high theoretical capacity; (anode: 3,860 mA h g^{-1}; cathode: 1,673 mA h g^{-1}) high energy density (2,600 W h kg^{-1})
Disadvantages	High cost; low abundance of 0.0065%; low capacity. (Theoretical capacity: 372 mAh g^{-1} of LiC6; ~780 mA h g^{-1} of Li2C6; ~1,116 mA h g^{-1} of LiC$_2$)	Low coulombic efficiency. (Larger Na+ size compared to Li, atomic weight of 6.9 g mol^{-1})	Low electrical conductivity; poor stability; low-rate performance. (Large volume changes of sulfur, soluble polysulfide)

14.7.1 LITHIUM-ION BATTERIES

Because of its high operating potential and extended cycle life, LIB technology, which was created in the 1990s, has subsequently become the dominating technology in practical applications.

Because of these two significant problems, the introduction of LIBs has prompted a flurry of research into other battery designs (LIB). The theoretical gravimetric capacity of the graphite negative electrode is $372 \, mA \, hg^{-1}$, which is significantly lower than expected. The limited availability and high cost of Li make it unable to fulfill the increased demand expected in the next years. Furthermore, the increasing expansion of flexible batteries in flexible electronics needs the development of LIBs with high capacity, long cycle life, high power density, high energy density, and flexibility in deformed states [57]. In light of the fact that rechargeable batteries have a limited life span, self-healing is a desired feature.

14.7.2 PURE GRAPHENE

GP has emerged as a possible candidate for energy storage devices like SCs and SIBs, according to recent studies. Individual sheets of 2D planar graphene can be assembled into a GP via flow-directed assembly because of its great mechanical strength and flexibility. To further reduce weight and volume, employing the GP directly as a LIB anode with no binder chemicals or conducting additives is another option. An improvement in overall energy density can be achieved for a device using these electrodes, as opposed to more traditional ones. Graphene nanoplatelets (GNPs) and GO were used to provide a flexible, high-performance, and additive-free anode in LIBs [58]. The composite structure's stability and Li storage capacity can both be improved by using GOs.

Furthermore, the spacing between graphene layers and the presence of small creases in the GP that was created resulted in better lithium storage and flexibility. The GNP/GO article served as the basis for the development of a LIB anode that has a high specific capacity of $694 \, mA \, h \, g^{-1}$, high-rate performance, and stable cycling characteristics. It was discovered that flexible pouch-type batteries, which can be bent and folded, were extremely stable even when the battery was bent and folded. Its excellent flexibility and conductivity were maintained after 750 bends, showing that the material has potential in flexible LIBs.

14.7.3 GRAPHENE/METAL OXIDE

Graphene/metal oxide as a result of its low theoretical capacity, graphene anode materials cannot be profitably used for high-power applications, which have been in significant demand. When it comes to flexible LIBs, the morphologies of 2D hybrid films and 3D networks, such as 3D foams, play an important role in determining the flexibility of the LIBs made of graphene and various TMOs, such as TiO_2, MoO_3, SnO_2, and V_2O_5 [59]. For example, Feng et al. created bendable $mTiO_2$-GS/G composite films [60]. The $mTiO_2$-GS/G electrode retained good charging/discharging capabilities when packed in flexible cells, even when bent. $150 \, mA \, h \, g^{-1}$ capacities at

10°C in flat and bending situations. Moreover, at 400 cycles, flat and bent composite electrodes still have 130 and 150 mA h g^{-1} capacities, respectively. At 20°C, two flexible cells in series could power a commercial 3 V LED, and the LED's brightness did not vary when the cells were bent. Noerochim et al. created a two-step microwave hydrothermal MoO_3 nanobelt/graphene film electrode [61]. Less than half of the discharge capacity of the pure MoO_3 film was produced after 100 cycles, however, the hybrid composites obtained exhibit superior cyclability and rate capability. To make V_2O_5/rGO composite films, Liu et al. used a unique lyotropic liquid crystal assembly process [62]. The flexible LIB with V_2O_5/rGO electrodes and Cu sheets as current collectors has a capacity of 215 mA h g^{-1} at 0.1 A g^{-1} and good flexibility. Instead of hybrid films, Kim et al. created a freestanding ultrathin graphite film with $Li_4Ti_5O_{12}$ nanowire (LTO NW) arrays that can act as both a current collector and anode [63]. After chemical etching of the Ni foam, freestanding ultrathin graphite (FSG) films were developed on the FSG films, and then LTO NW arrays were grown on the FSG films. The 5 m thick FSG films were made up of many stacked graphene layers to keep them freestanding. This resulted in discharge capabilities of 154 mA g^{-1} and consistent cycling performance for 500 cycles at 20°C. A flexible LIB made of composite electrodes also showed good flexibility under bending deformations. 3D metal oxide/graphene hybrid networks are also flexible. Solvent evaporation was used by Ding et al. to create 3D graphene/LiFePO$_4$ structures [63]. The graphene/LiFePO$_4$ nanostructures were electrochemically stable and very flexible. At the 0.1°C and 5°C temperatures, the low graphene composites had a capacitance of 163.7 mA h g^{-1}. The cells with graphene/LiFePO$_4$ cathode and lithium metal anode were also bent at 45°, 90°, and 120°, indicating outstanding flexibility. Anodes made of highly dispersed SnO_2 NPs anchored on 3D non-woven cotton covered with graphene (CGN/SnO$_2$) were made by Zhang et al. [64]. The SnO_2 NPs on the GSs act as spacers, reducing the GSs stack. Even after several cycles, the anode retained good reversible capacity, cycling stability, and rate capability. This approach could be used in flexible and wearable electronic devices. Fe_3O_4-CoO was used to fabricate a coral-like 3D hierarchical heterostructure. Fe_3O_4 NPs are interleaved atop cross-linked CoO NFs perched on graphene foam. After 400 cycles, the anode electrode with coral-like hierarchical structure gave a high reversible specific capacity of 1,200 mA h g^{-1}.

The LIBs can also light a green LED when bent and keep their capacity when bent, demonstrating their remarkable flexibility.

14.7.4 OTHERS FLEXIBLE RECHARGEABLE BATTERIES

In practice, LIBs use graphitic carbon anodes, which have unfulfilled theoretical capacity for LiC_6 and poor rate performance [65]. Metal oxides (Co_3O_4 and GeO_2) and metal sulfides (MoS_2) are excellent electrode materials to replace graphitic carbon [66]. 3D interconnected porous Ni-doped graphene foam with encapsulated Ge quantum dot/Ni-doped graphene yolk-shell nanostructure (Ge-QD@NG/NGF) [66]. This paper compares the flexible Ge-QD@NG/NGF/PDMS electrode to the conventional electrode, underlining the importance of the current collector in the flexible electrode configuration. The specific reversible capacity of the composite battery with Ge-QD@NG/NGF/PDMS anode and lithium foil counter electrode is

1,220 mA h g⁻¹, and the reversible capacity retention is above 96%. The rate performance of the composite battery at 40°C is excellent. Furthermore, the electrode's
flexibility was examined using a Ge-QD@NG/NGF/PDMS yolk-shell electrode. A
2D microstructured metal oxide/carbon composite has significant promise in flexible
LIBs because of its microstructure. Guo and colleagues developed a pie-like GP@
Fe_3O_4@carbon film for use in LIBs [67]. Throughout 1,000 cycle tests at 2 A g⁻¹,
this pie-like electrode with internal array design displayed outstanding cyclability,
with a specific capacity of 852 mA g⁻¹ over the course of the study. As a binder-free
integrated substrate, GP is preferred because it has the ability to speed electron transmission, buffer changes in Fe_3O_4 volume, enhance capacity, and give more flexibility.
A 2D/3D hybrid film structure was produced by Li and colleagues utilizing vacuum
filtering and thermal annealing. The components of the 2D/3D thin film anode work
together to improve overall performance. After 100 cycles at 100 mg⁻¹, its initial discharge specific capacity was 1,656.8 mA h g⁻¹ with a MnO_2 concentration of 56%,
and after another 100 cycles at 56%, it was 1,172.25 mA h g⁻¹. Furthermore, the
damaged LIB showed excellent adaptability. Elementary S is an excellent cathode
material for LSBs in FESDs. S-based materials, on the other hand, offer excellent
electrochemical characteristics and are flexible enough to be used in flexible LIBs.
As well as exhibiting excellent capacity and cyclability, the graphene/S electrode
demonstrated great efficiency (98%) and capacity retention (70%) after 250 cycles
of operation. Two flexible batteries were constructed with the help of the composite
electrode. With its graphene-S cathode and Li metal anode, a flexible LIB can illuminate a green LED when it is bent. When folded, a flexible battery with a $LiMn_2O_4$
cathode and a graphene-S anode could provide 1.95 V of electricity to a red LED.

14.8   CONCLUSION

Industrial applications for lightweight FESDs include flying gadgets, electric cars,
transparent flexible devices such as wearable displays and touch-screens, and transparent flexible devices such as holographic displays. In other words, there is still
opportunity for further growth of FESD in the industrial sector. In order to meet the
rising need for FESDs, the design and engineering of innovative materials will be
encouraged. Optimization of material and device attributes is frequently required to
increase performance. A number of problems must be addressed in order to achieve
high-performance and cost-effective FESD fabrication. First and foremost, selecting
the proper electrode material is critical. Constraints on GS restacking may cause
FESD performance to suffer a considerable decrease. 1D fibers, 2D films, and 3D
networks (foams) with large SSA and high mechanical strength are the best architectures for graphene-based FESDs. Second, it is critical to use flexible substrates
and electrolytes. When used as both current collectors and electrode supports, flexible substrates have the potential to save space and weight. Protection from current
leakage and mechanical loads is supplied by electrolytes containing PVA/H_3PO_4
and H_2SO_4-PVA, respectively, as shown in the diagrams. The following are the
directions in which graphene-based FESDs will go: For FESDs, we require unique
nanostructured flexible electrode materials, substrates, and electrolytes, all of which
are currently unavailable. Graphene-based FESDs with strong electrochemical

and flexible capabilities are already making significant development in the market. High-performance FESDs are projected to make extensive use of graphene-based nanocomposites in the near future.

REFERENCES

[1] K. Sudhakar, M. Winderla, and S. S. Priya, "Net-zero building designs in hot and humid climates: A state-of-art," *Case Studies in Thermal Engineering*, vol. 13, p. 100400, 2019, doi: 10.1016/j.csite.2019.100400.

[2] N. Roslan, M. E. Ya'acob, M. A. M. Radzi, Y. Hashimoto, D. Jamaludin, and G. Chen, "Dye Sensitized Solar Cell (DSSC) greenhouse shading: New insights for solar radiation manipulation," *Renewable and Sustainable Energy Reviews*, vol. 92, no. December, pp. 171–186, 2018, doi: 10.1016/j.rser.2018.04.095.

[3] N. Díez, M. Qiao, J. L. Gómez-Urbano, C. Botas, D. Carriazo, and M. M. Titirici, "High density graphene-carbon nanosphere films for capacitive energy storage," *Journal of Materials Chemistry A*, vol. 7, no. 11, pp. 6126–6133, 2019, doi: 10.1039/c8ta12050a.

[4] M. Bahiraei and S. Heshmatian, "Graphene family nanofluids: A critical review and future research directions," *Energy Conversion and Management*, vol. 196, no. March, pp. 1222–1256, 2019, doi: 10.1016/j.enconman.2019.06.076.

[5] L. Fan, L. Tang, H. Gong, Z. Yao, and R. Guo, "Carbon-nanoparticles encapsulated in hollow nickel oxides for supercapacitor application," *Journal of Materials Chemistry*, vol. 22, no. 32, pp. 16376–16381, Aug. 2012, doi: 10.1039/c2jm32241b.

[6] X. Peng, L. Peng, C. Wu, and Y. Xie, "Two dimensional nanomaterials for flexible supercapacitors," *Chemical Society Reviews*, vol. 43, no. 10. pp. 3303–3323, 2014. doi: 10.1039/c3cs60407a.

[7] H. Jiang, P. S. Lee, and C. Li, "3D carbon based nanostructures for advanced supercapacitors," *Energy and Environmental Science*, vol. 6, no. 1, pp. 41–53, 2013, doi: 10.1039/c2ee23284g.

[8] Z. Niu et al., "A 'skeleton/skin' strategy for preparing ultrathin free-standing single-walled carbon nanotube/polyaniline films for high performance supercapacitor electrodes," *Energy and Environmental Science*, vol. 5, no. 9, pp. 8726–8733, 2012, doi: 10.1039/c2ee22042c.

[9] S. Wang, Y. Chen, Y. Ma, Z. Wang, and J. Zhang, "Size effect on interlayer shear between graphene sheets," *Journal of Applied Physics*, vol. 122, no. 7, 2017, doi: 10.1063/1.4997607.

[10] S. Shiraishi, "Electric double layer capacitors," In *Carbon Alloys: Novel Concepts to Develop Carbon Science and Technology*, pp. 447–457, Jan. 2003, doi: 10.1016/B978-0 08044163-4/50027-9.

[11] N. Nitta, F. Wu, J. T. Lee, and G. Yushin, "Li-ion battery materials: Present and future," *Materials Today*, vol. 18, no. 5, pp. 252–264, 2015, doi: 10.1016/j.mattod.2014.10.040.

[12] L. Zhang and X. S. Zhao, "Carbon-based materials as supercapacitor electrodes," *Chemical Society Reviews*, vol. 38, no. 9, pp. 2520–2531, 2009, doi: 10.1039/b813846j.

[13] Q. Wu, M. Chen, K. Chen, S. Wang, C. Wang, and G. Diao, "Fe$_3$O$_4$-based core/shell nanocomposites for high-performance electrochemical supercapacitors," *Journal of Materials Science*, vol. 51, no. 3, pp. 1572–1580, 2016, doi: 10.1007/s10853-015-9480-4.

[14] Z. Weng, Y. Su, D. W. Wang, F. Li, J. Du, and H. M. Cheng, "Graphene-cellulose paper flexible supercapacitors," *Advanced Energy Materials*, vol. 1, no. 5, pp. 917–922, 2011, doi: 10.1002/aenm.201100312.

[15] H. Nguyen Bich and H. Nguyen Van, "Promising applications of graphene and graphene-based nanostructures," *Advances in Natural Sciences: Nanoscience and Nanotechnology*, vol. 7, no. 2, 2016, doi: 10.1088/2043-6262/7/2/023002.

[16] Y. Huang, J. Liang, and Y. Chen, "An overview of the applications of graphene-based materials in supercapacitors," *Small*, vol. 8, no. 12, pp. 1805–1834, 2012, doi: 10.1002/smll.201102635.

[17] M. Zheng et al., "High-performance flexible solid-state asymmetric supercapacitors based on ordered mesoporous cobalt oxide," *Energy Technology*, vol. 5, no. 4, pp. 544–548, 2017, doi: 10.1002/ente.201600391.

[18] F. Meng, Q. Li, and L. Zheng, "Flexible fiber-shaped supercapacitors: Design, fabrication, and multi-functionalities," *Energy Storage Materials*, vol. 8, pp. 85–109, 2017, doi: 10.1016/j.ensm.2017.05.002.

[19] X. Guo et al., "Nanostructured graphene-based materials for flexible energy storage," *Energy Storage Materials*, vol. 9, pp. 150–169, Oct. 2017, doi: 10.1016/J. ENSM.2017.07.006.

[20] Y. Hu et al., "All-in-one graphene fiber supercapacitor," *Nanoscale*, vol. 6, no. 12, pp. 6448–6451, 2014, doi: 10.1039/c4nr01220h.

[21] G. K. Veerasubramani, K. Krishnamoorthy, P. Pazhamalai, and S. J. Kim, "Enhanced electrochemical performances of graphene based solid-state flexible cable type supercapacitor using redox mediated polymer gel electrolyte," *Carbon*, vol. 105, pp. 638–648, 2016, doi: 10.1016/j.carbon.2016.05.008.

[22] S. Zheng et al., "Arbitrary-shaped graphene-based planar sandwich supercapacitors on one substrate with enhanced flexibility and integration," *ACS Nano*, vol. 11, no. 2, pp. 2171–2179, 2017, doi: 10.1021/acsnano.6b08435.

[23] Y. Xue, L. Zhu, H. Chen, J. Qu, and L. Dai, "Multiscale patterning of graphene oxide and reduced graphene oxide for flexible supercapacitors," *Carbon N Y*, vol. 92, pp. 305–310, 2015, doi: 10.1016/j.carbon.2015.04.046.

[24] J. Y. Kim, J. W. Lee, H. S. Jung, H. Shin, and N. G. Park, "High-efficiency perovskite solar cells," *Chemical Reviews*, vol. 120, no. 15, pp. 7867–7918, 2020, doi: 10.1021/acs. chemrev.0c00107.

[25] H. Zhou, H. J. Zhai, and G. Han, "Superior performance of highly flexible solid-state supercapacitor based on the ternary composites of graphene oxide supported poly(-3,4-ethylenedioxythiophene)-carbon nanotubes," *Journal of Power Sources*, vol. 323, pp. 125–133, 2016, doi: 10.1016/j.jpowsour.2016.05.049.

[26] X. Yu and H. S. Park, "Sulfur-incorporated, porous graphene films for high performance flexible electrochemical capacitors," *Carbon N Y*, vol. 77, no. May, pp. 59–65, 2014, doi: 10.1016/j.carbon.2014.05.002.

[27] J. S. Lee, C. Lee, J. Jun, D. H. Shin, and J. Jang, "A metal-oxide nanofiber-decorated three-dimensional graphene hybrid nanostructured flexible electrode for high-capacity electrochemical capacitors," *Journal of Materials Chemistry A*, vol. 2, no. 30, pp. 11922–11929, 2014, doi: 10.1039/c4ta01695e.

[28] J. Yu et al., "Metallic fabrics as the current collector for high-performance graphene-based flexible solid-state supercapacitor," *ACS Applied Materials and Interfaces*, vol. 8, no. 7, pp. 4724–4729, 2016, doi: 10.1021/acsami.5b12180.

[29] J. L. Shi, W. C. Du, Y. X. Yin, Y. G. Guo, and L. J. Wan, "Hydrothermal reduction of three-dimensional graphene oxide for binder-free flexible supercapacitors," *Journal of Materials Chemistry A*, vol. 2, no. 28, pp. 10830–10834, 2014, doi: 10.1039/c4ta01547a.

[30] J. Liang, C. Jiang, and W. Wu, "Printed flexible supercapacitor: Ink formulation, printable electrode materials and applications," *Applied Physics Reviews*, vol. 8, no. 2. 2021. doi: 10.1063/5.0048446.

[31] S. H. Lee, J. H. Kim, and J. R. Yoon, "Laser scribed graphene cathode for next generation of high performance hybrid supercapacitors," *Scientific Reports*, vol. 8, no. 1, 2018, doi: 10.1038/s41598-018-26503-4.

[32] Z. Wan, X. Chen, and M. Gu, "Laser scribed graphene for supercapacitors," *Opto-Electronic Advances*, vol. 4, no. 7. 2021. doi: 10.29026/oea.2021.200079.

[33] W. Ma et al., "Flexible all-solid-state asymmetric supercapacitor based on transition metal oxide nanorods/reduced graphene oxide hybrid fibers with high energy density," *Carbon*, vol. 113, pp. 151–158, Mar. 2017, doi: 10.1016/J.CARBON.2016.11.051.

[34] W. Ma et al., "Hierarchical MnO2 nanowire/graphene hybrid fibers with excellent electrochemical performance for flexible solid-state supercapacitors," *Journal of Power Sources*, vol. 306, pp. 481–488, Feb. 2016, doi: 10.1016/J.JPOWSOUR.2015.12.063.

[35] F. Z. Amir, V. H. Pham, D. W. Mullinax, and J. H. Dickerson, "Enhanced performance of HRGO-RuO2 solid state flexible supercapacitors fabricated by electrophoretic deposition," *Carbon*, vol. 107, pp. 338–343, Oct. 2016, doi: 10.1016/J.CARBON.2016.06.013.

[36] C. Y. Foo, A. Sumboja, D. J. H. Tan, J. Wang, and P. S. Lee, "Flexible and highly scalable V2O5-rGO electrodes in an organic electrolyte for supercapacitor devices," *Advanced Energy Materials*, vol. 4, no. 12, 2014, doi: 10.1002/aenm.201400236.

[37] J. Bao et al., "All-solid-state flexible thin-film supercapacitors with high electrochemical performance based on a two-dimensional V_2O $_5 \cdot H_2O$/graphene composite," *Journal of Materials Chemistry A*, vol. 2, no. 28, pp. 10876–10881, Jul. 2014, doi: 10.1039/c3ta15293f.

[38] I. K. Moon, S. Yoon, and J. Oh, "Three-dimensional hierarchically mesoporous $ZnCo_2O_4$ nanowires grown on graphene/sponge foam for high-performance, flexible, all-solid-state supercapacitors," *Chemistry - A European Journal*, vol. 23, no. 3, 2017, doi: 10.1002/chem.201602447.

[39] Y. He et al., "Freestanding three-dimensional graphene/Mno_2 composite networks as ultralight and flexible supercapacitor electrodes," *ACS Nano*, vol. 7, no. 1, 2013, doi: 10.1021/nn304833s.

[40] M. Liu and J. Sun, "In situ growth of monodisperse Fe_3O_4 nanoparticles on graphene as flexible paper for supercapacitor," *Journal of Materials Chemistry A*, vol. 2, no. 30, 2014, doi: 10.1039/c4ta01442a.

[41] S. Liu, X. Liu, Z. Li, S. Yang, and J. Wang, "Fabrication of free-standing graphene/polyaniline nanofibers composite paper via electrostatic adsorption for electrochemical supercapacitors," *New Journal of Chemistry*, vol. 35, no. 2, 2011, doi: 10.1039/c0nj00718h.

[42] H. P. Cong, X. C. Ren, P. Wang, and S. H. Yu, "Flexible graphene-polyaniline composite paper for high-performance supercapacitor," *Energy and Environmental Science*, vol. 6, no. 4, 2013, doi: 10.1039/c2ee24203f.

[43] Q. Wu, Y. Xu, Z. Yao, A. Liu, and G. Shi, "Supercapacitors based on flexible graphene/polyaniline nanofiber composite films," *ACS Nano*, vol. 4, no. 4, 2010, doi: 10.1021/nn1000035.

[44] H. Zhou, G. Han, Y. Xiao, Y. Chang, and H. J. Zhai, "Facile preparation of polypyrrole/graphene oxide nanocomposites with large areal capacitance using electrochemical codeposition for supercapacitors," *Journal of Power Sources*, vol. 263, pp. 259–267, Oct. 2014, doi: 10.1016/J.JPOWSOUR.2014.04.039.

[45] K. Shu, C. Wang, C. Zhao, Y. Ge, and G. G. Wallace, "A free-standing graphene-polypyrrole hybrid paper via electropolymerization with an enhanced areal capacitance," *Electrochimica Acta*, vol. 212, pp. 561–571, Sep. 2016, doi: 10.1016/J.ELECTACTA.2016.07.052.

[46] S. Lehtimäki, M. Suominen, P. Damlin, S. Tuukkanen, C. Kvarnström, and D. Lupo, "Preparation of supercapacitors on flexible substrates with electrodeposited PEDOT/graphene composites," *ACS Applied Materials and Interfaces*, vol. 7, no. 40, 2015, doi: 10.1021/acsami.5b05937.

[47] D. Yu et al., "Scalable synthesis of hierarchically structured carbon nanotube-graphene fibres for capacitive energy storage," *Nature Nanotechnology*, vol. 9, no. 7, 2014, doi: 10.1038/nnano.2014.93.

[48] J. Tang, P. Yuan, C. Cai, Y. Fu, and X. Ma, "Combining nature-inspired, graphene-wrapped flexible electrodes with nanocomposite polymer electrolyte for asymmetric capacitive energy storage," *Advanced Energy Materials*, vol. 6, no. 19, 2016, doi: 10.1002/aenm.201600813.

[49] X. Zhang, Y. Lai, M. Ge, Y. Zheng, K. Q. Zhang, and Z. Lin, "Erratum: Fibrous and flexible supercapacitors comprising hierarchical nanostructures with carbon spheres and graphene oxide nanosheets (J. Mater. Chem. A (2015) DOI: 10.1039/c5ta03252k)," *Journal of Materials Chemistry A*, vol. 3, no. 25. 2015. doi: 10.1039/c5ta90128f.

[50] Q. Zhou, C. Jia, X. Ye, Z. Tang, and Z. Wan, "A knittable fiber-shaped supercapacitor based on natural cotton thread for wearable electronics," *Journal of Power Sources*, vol. 327, pp. 365–373, Sep. 2016, doi: 10.1016/J.JPOWSOUR.2016.07.048.

[51] D. Fu, H. Li, X. M. Zhang, G. Han, H. Zhou, and Y. Chang, "Flexible solid-state supercapacitor fabricated by metal-organic framework/graphene oxide hybrid interconnected with PEDOT," *Materials Chemistry and Physics*, vol. 179, pp. 166–173, Aug. 2016, doi: 10.1016/J.MATCHEMPHYS.2016.05.024.

[52] X. Yu and H. S. Park, "Sulfur-incorporated, porous graphene films for high performance flexible electrochemical capacitors," *Carbon N Y*, vol. 77, pp. 59–65, Oct. 2014, doi: 10.1016/J.CARBON.2014.05.002.

[53] X. Wu, Q. Wang, W. Zhang, Y. Wang, and W. Chen, "Nano nickel oxide coated graphene/polyaniline composite film with high electrochemical performance for flexible supercapacitor," *Electrochimica Acta*, vol. 211, pp. 1066–1075, Sep. 2016, doi: 10.1016/J.ELECTACTA.2016.06.026.

[54] K. Vijaya Sankar and R. Kalai Selvan, "Fabrication of flexible fiber supercapacitor using covalently grafted CoFe$_2$O$_4$/reduced graphene oxide/polyaniline and its electrochemical performances," *Electrochimica Acta*, vol. 213, pp. 469–481, Sep. 2016, doi: 10.1016/J.ELECTACTA.2016.07.056.

[55] H. An et al., "Free-standing fluorine and nitrogen co-doped graphene paper as a high-performance electrode for flexible sodium-ion batteries," *Carbon*, vol. 116, pp. 338–346, May 2017, doi: 10.1016/J.CARBON.2017.01.101.

[56] N. Q. Tran, B. K. Kang, M. H. Woo, and D. H. Yoon, "Enrichment of pyrrolic nitrogen by hole defects in nitrogen and sulfur co-doped graphene hydrogel for flexible supercapacitors," *ChemSusChem*, vol. 9, no. 16, pp. 2261–2268, Aug. 2016, doi: 10.1002/cssc.201600668.

[57] H. Pan, J. Ma, J. Tao, and S. Zhu, "Hierarchical architecture for flexible energy storage," *Nanoscale*, vol. 9, no. 20, pp. 6686–6694, 2017, doi: 10.1039/c7nr00867h.

[58] M. Kim, D. Y. Kim, Y. Kang, and O. O. Park, "Facile fabrication of highly flexible graphene paper for high-performance flexible lithium ion battery anode," *RSC Advances*, vol. 5, no. 5, pp. 3299–3305, 2015, doi: 10.1039/c4ra13164a.

[59] L. Noerochim, J. Z. Wang, D. Wexler, Z. Chao, and H. K. Liu, "Rapid synthesis of free-standing MoO$_3$/graphene films by the microwave hydrothermal method as cathode for bendable lithium batteries," *Journal of Power Sources*, vol. 228, pp. 198–205, Apr. 2013, doi: 10.1016/J.JPOWSOUR.2012.11.113.

[60] B. Feng et al., "Free-standing hybrid film of less defective graphene coated with mesoporous TiO$_2$ for flexible lithium ion batteries with fast charging/discharging capabilities," *2D Materials*, vol. 4, no. 1, 2017, doi: 10.1088/2053-1583/4/1/015011.

[61] L. Noerochim, J. Z. Wang, D. Wexler, Z. Chao, and H. K. Liu, "Rapid synthesis of free-standing MoO$_3$/Graphene films by the microwave hydrothermal method as cathode for bendable lithium batteries," *Journal of Power Sources*, vol. 228, pp. 198–205, Apr. 2013, doi: 10.1016/J.JPOWSOUR.2012.11.113.

[62] H. Liu et al., "A lyotropic liquid-crystal-based assembly avenue toward highly oriented vanadium pentoxide/graphene films for flexible energy storage," *Advanced Functional Materials*, vol. 27, no. 12, 2017, doi: 10.1002/adfm.201606269.

[63] S. D. Kim, K. Rana, and J. H. Ahn, "Additive-free synthesis of $Li_4Ti_5O_{12}$ nanowire arrays on freestanding ultrathin graphite as a hybrid anode for flexible lithium ion batteries," *Journal of Materials Chemistry A*, vol. 4, no. 48, 2016, doi: 10.1039/c6ta09059a.

[64] X. Zhang et al., "A free-standing, flexible and bendable lithium-ion anode materials with improved performance," *RSC Advances*, vol. 6, no. 112, 2016, doi: 10.1039/c6ra19347a.

[65] X. Tang et al., "The positive influence of graphene on the mechanical and electrochemical properties of SnxSb-graphene-carbon porous mats as binder-free electrodes for Li+ storage," *Electrochimica Acta*, vol. 186, pp. 223–230, Dec. 2015, doi: 10.1016/J.ELECTACTA.2015.10.170.

[66] Y. Sun et al., "Coral-inspired nanoengineering design for long-cycle and flexible lithium-ion battery anode," *ACS Applied Materials and Interfaces*, vol. 8, no. 14, 2016, doi: 10.1021/acsami.6b02011.

[67] J. Guo, H. Zhu, Y. Sun, L. Tang, and X. Zhang, "Pie-like free-standing paper of graphene paper@Fe_3O_4 nanorod array@carbon as integrated anode for robust lithium storage," *Chemical Engineering Journal*, vol. 309, pp. 272–277, Feb. 2017, doi: 10.1016/J.CEJ.2016.10.041.

15 Effect of Carbon Addition as an Active Material in Fabrication of Energy Storage Devices

Malachy N. Asogwa and Ada C. Agbogu
University of Nigeria

Assumpta C. Nwanya
University of Nigeria
University of the Western Cape Sensor Laboratories

Fabian I. Ezema
University of Nigeria
Northwestern Polytechnical University
iThemba LABS-National Research Foundation
University of South Africa (UNISA)

CONTENTS

DOI: 10.1201/9781003215196-15

15.1   INTRODUCTION

In response to the global energy demand, material scientists have been researching on better ways energy could be harnessed and stored. This led to the development of electrochemical energy storage systems which comprises fuel cells, batteries, and supercapacitors [1,2]. Supercapacitor, also called ultracapacitor because of its high capacitance [1,3–5], is an improvement on the conventional capacitors and battery technologies [3]. For storage devices to be efficient and to be used in high-power applications such as an electric vehicle, electrical grids, etc., it should have high power delivery (high power density), high or moderate energy density, and high operating life time [5]. Supercapacitors possess capacitance that is higher than conventional capacitors although with lesser voltage restrictions. Some of the advantages supercapacitors have over batteries and conventional capacitors include: instant charge, stores more energy (has high energy density) than electrolytic capacitors (but less than batteries), longer lifespan (measured in charge/discharge cycles) above rechargeable batteries but below electrolytic capacitors. Supercapacitors balance energy storage with charge and discharge times, and have a wide range of operating temperatures [2–4].

However, supercapacitors have the disadvantage of low energy density relative to that of batteries. The energy density of a supercapacitor gives the extent or the amount of energy in a given system or region per unit volume or per unit mass. It is low compared with that of batteries. Globally, researchers are making efforts to find ways to improve on the energy density of supercapacitors. They do this by making appropriate choice of electrolytes with high-potential window and high capacitance with a matching improved electrode. Supercapacitors also have self-discharge rate and gradual voltage loss thereby limiting its applications. The rate of discharge of supercapacitors is higher than lithium-ion batteries [6] in terms of voltage loss. Batteries provide a near-constant voltage output until they are used up however, the voltage yield of supercapacitors diminishes linearly with their discharge.

Fabricating a supercapacitor involves the sandwich of current collector, electrode materials, electrolyte, and a separator. The separator reserves ions and keeps the ions temporarily before the passage of the ions during charge–discharge processes. Supercapacitors are of three types. These are electrochemical double-layer capacitors (EDLCs), pseudocapacitors, and the hybrid supercapacitors.

In EDLCs, charges are stored at the electrode/electrolyte boundary and electrochemical double-layer stores energy via non-faradaic process. Non-faradaic process involves no transfer of charge between electrodes and electrolytes; on the other hand, pseudocapacitors transfer charge between electrode and electrolyte and are possible through faradaic means; a reduction–oxidation reactions process [1,3]. It is 'pseudo' because it involves a similar process just like batteries. The hybrid supercapacitors combine the mechanism of EDLCs and pseudocapacitors in the storage principles.

The structure and the properties of the component materials are factors that determine the efficiency of energy storage devices [2,3,7]. This basically depends on the electrodes and the electrolytes of a supercapacitor. Various electrodes have been employed based on the different supercapacitor types. The EDLC electrodes that have been studied include activated carbon, carbon nanotubes, and graphene [8], while conductive polymers and metal oxides are studied for pseudocapacitors [6,9] and then composites, asymmetric, and battery type for hybrid supercapacitors [10]. Electrolyte materials in supercapacitors are of different types. These include solid-state electrolytes, ionic liquids, organic electrolytes (e.g., acetonitrile or propylene carbonate (PC)), and aqueous electrolytes.

Organic electrolytes utilizing the solvents as mentioned above are utilized to dissolve salts like tetraethyl ammonium tetrafluoroborate (TEABF$_4$), etc. Aqueous electrolytes contain water which serves as the solvent for H_2SO_4, KOH, Na_2SO_4, etc. Ionic liquids, on the other hand, comprise pyrrolidinium, aliphatic quaternary ammonium salt, or imidazalium together with anions such as PF$_6$ and BF$_4$. Dry, gel, and polyelectrolyte forms the composition of solid-state polymer electrolytes [11].

Supercapacitors are charged by connecting the positive and the negative terminals to an energy source and hence a potential is applied to the electrodes. Discharging a supercapacitor involves removal of electrons through an external circuit with a load connected to it [12]. Supercapacitors find applications in pulse power systems such as commercial vehicles, cell phones, etc.; in medicine as defibrillators; in military and space as detonators, lasers, regenerative braking, load leveling, smoothing and uninterrupted power systems (UPSs); metro buses, power Siberian trains among other numerous applications [13].

The electrochemical performance is tested using techniques such as cyclic voltammetry, charge and discharge test at constant current also known as constant current charge discharge (CCCD), chrono potentiometry as well as electrochemical impedance spectroscopy. To characterize a supercapacitor, three basic parameters must be taken into consideration, namely, the net capacitance of the device, its potential window (working) and the effective series resistance (ESR). These serves as the evaluative considerations for supercapacitor characterizations [14,15].

It is important to optimize the energy and power densities of supercapacitors in other to make it promising and maximize its efficiency for power applications in the future. Developing advanced electrode and electrolyte materials possessing broad-working potential and a minute ESR (high ion conductivity), using a cost-effective device fabrication method would be paramount in other to achieve these goals [8,11]. Hence, when the electrochemical characteristics of supercapacitors such as cycle life, energy density, and discharge rate and power density are improved, then supercapacitors will serve better than the present battery storage technology and will become a perfect alternative.

In this review, we present an overview of supercapacitors. These include the history, types, structures, and principle of operations of supercapacitors as energy storage devices. The various electrode materials, electrolytes, current collectors, separators that have been studied and used for supercapacitive applications will be reviewed. Finally, the various characterization methods used to study supercapacitors and supercapacitive materials will be presented.

15.1.1 HISTORY OF SUPERCAPACITORS

Scientists started experimenting on supercapacitors as far back as the 1950s, and porous carbon material electrodes were fundamental in the construction of capacitors. For example, the design of fuel cells and rechargeable batteries involved covering low-voltage electrolytic capacitors with permeable electrode, and it was assumed that energy is kept as charge in the carbon electrodes just like it is stored in 'etched' foils in an electrolytic capacitor; H. Becker 1957 [16].

Standard Oil of Ohio (SOHIO) in 1966 recorded another developed version of electrical energy storage device while working on experimental fuel cell designs [17]. It was recognized, patented and registered by Donald L. Boos in 1970 as an electrolytic capacitor having activated carbon electrodes. Aluminum foils (two) utilizing activated carbon formed an early electrochemical capacitor. It has its electrodes embedded in an electrolyte and a thin spongy insulator separates the enclosure. A capacitor with a capacitance of one farad that is significantly higher than the electrolytic capacitor of the same dimensions was achieved from this design. This forms the basis of most electrochemical capacitors.

There was a slight discrepancy between supercapacitors and battery performance as regards electrochemical energy storage. This was pointed out in 1975, 1980, and 1991 by Brian Evans Conway while working on ruthenium oxide electrochemical capacitors [16]. 1982 records the first supercapacitor that was observed with low-internal resistance. This was marketed using the brand name "PRI Ultracapacitor" and subsequently was developed for military applications via the Pinnacle Research Institute (PRI) [16].

The first-generation EDLCs were fabricated around 1987 and had a relatively high internal resistance. The high internal resistance was observed to be a limiting factor to the current produced. They were used to power-static random-access memory (SRAM) for low-current application chips or for data backup [16]. This leads to the availability of enhanced electrode base materials with better capacitance values at the end of the 1980s [16]. Concurrent work was done by the development of electrolytes with better conductivity and this was seen to lower the ESR with an increase in charge/discharge currents [1,3].

Maxwell Laboratories in 1992 progressively adopted the term Ultracapacitor from PRI; the Pinnacle Research Institute termsit "Boost Caps" which signifies its use for power applications. In 1994, David A. combined some features of electrolytic and electrochemical capacitors. Evans was used to develop successfully an electrolytic-hybrid capacitor using an anode of 200V tantalum electrolytic capacitor. The material component works on the high dielectric strength of the anode in a high capacitance value of pseudocapacitive metal oxide (e.g., ruthenium (iv) oxide) cathode from an electrochemical capacitor [18,19]. In 1999, supercapacitors were observed to have partial storage of electrical charge in the Helmholtz double layer. Due to the faradaic reactions with "pseudo capacitance", charge transfer of electrons and protons between electrode and electrolyte was possible. Supercapacitor therefore can be defined in terms of its rise in observed capacitance by surface redox reactions with faradaic charge transfer between electrodes and ions.

Conway in his research greatly utilized the redox reaction, intercalation and electrosorption (adsorption onto a surface) of pseudocapacitors to advance the knowledge of electrochemical capacitors. The energy content of supercapacitors increases with the square of the voltage and because of this; there is a need to look for ways to increase the electrolytes breakdown voltage [2,3].

Developments in 2007 featured lithium-ion capacitors. Fredrick, FDK, pioneered this. It combines a pre-doped lithium-ion electrochemical electrode with an electrostatic carbon electrode. This configuration increases the capacitance value and also reduces the anode potential when pre-doped resulting in a high cell output voltage and as a result leads to enhanced specific energy [4,20].

Researchers at the University of Florida created a prototype supercapacitor battery that takes up a fraction of the space of a lithium-ion cells. It charges and recharges more quickly up to 30, 000 times. Other innovations have been set to improve supercapacitive storage with the use of graphene. It has been shown that lightweight supercapacitors with the specific capacitance between 150 and 550 Fg^{-1} can be created by the utilization of graphene [9,17].

15.2 PRINCIPLE OF OPERATION AND STRUCTURE OF A SUPERCAPACITOR

Before giving details about the operation of a supercapacitor, let us look at the conventional capacitor. A capacitor is an electronic component that stores charge in the form of electric potential having its two parallel plates separated by a small distance, d. The amount of charge that a capacitor store is known as its capacitance. It depends on the distance/separation between plates and the surface area of the electrodes.

Mathematically,

$$C = \frac{\varepsilon_o \varepsilon_r A}{d} \qquad (15.1)$$

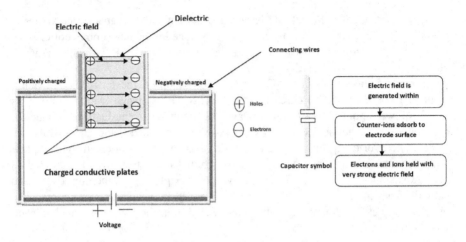

FIGURE 15.1   Schematic of operation of a capacitor.

C, ε_o, ε_r, A, and d are the capacitance, permittivity of free space, relative permittivity, surface area and distance of separation between the electrodes, respectively. The dependence of the capacitance of a capacitor on the distance and the surface area of the electrodes is evident in equation 15.1. While the capacitance decreases with increase in the distance of separation of the plates, it increases with increase in the surface area.

When a capacitor is connected to a power source, an electric field is created across the dielectric causing a positive charge, +q to form on one plate and a negative charge, −q on the other plate. The very intense electric field holds the electrons and the ions in the electrolytes close to one another and counter ions adsorb to the electrode surface. This electric field is where the cell stores its energy as potential energy. The operation of a capacitor follows the chat below (Figure 15.1).

Supercapacitors follow a similar operation just as conventional capacitors. It is a device with two electrodes; positive and negative with a separator in between that prevents physical contact between the two electrodes while facilitating ion transport in the device. Supercapacitors are distinguished by its great capacitance derived from molecular dielectrics. It has nanostructured electrodes with high surface area. In some cases, it has charge transfer reactions similar to those in batteries. Supercapacitors fill the gap between conventional capacitors and batteries. It has great power density (able to discharge so much of its power) but small energy density limiting its applications. The capacitance and energy of a capacitor vary inversely to the distance of separation, d. Since d is very small, large specific capacitance is enabled which translates to a large amount of energy in electrochemical capacitors [8].

15.2.1   Power Density of a Supercapacitor

The power density of a supercapacitor is given as follows:

$$P = \frac{1}{4\,W_{\text{Ts}}R_{\text{cell}}}V^2 \tag{15.2}$$

where P, V and W_{Ts} are the power density, cell voltage and device's weight. The cell weight is the sum of the weights of the electrodes, electrolyte solution, current collectors and others. R_{cell} is the equivalent resistance of the cell.

15.2.2 ENERGY DENSITY OF A SUPERCAPACITOR

The energy density of a supercapacitor is given by

$$E = \frac{1}{2} C_{Ts} V^2 \tag{15.3}$$

C_{Ts} is the total capacitance; from the equations, V, C_{Ts}, W_{Ts} and R_{cell} are the basic variables in determining the performance of an electrochemical capacitor. The energy and power densities of supercapacitors are optimized by increasing the values of V and C_{Ts} and reducing simultaneously the values of W_{Ts} and R_{cell} [11].

15.2.3 CYCLE LIFE

The cycle life of a supercapacitor has been observed to depend on the electrode, electrolyte, cell type, charge/discharge rate and temperature [11]. Therefore, to increase the cycle life of a supercapacitor, these factors should be taken into consideration.

15.2.4 THERMAL STABILITY

It has also been shown that most potential applications of electrochemical supercapacitors are in the range of 30–70°C. Hence, expanding the temperature range can further provide room for many of its applications [11].

15.2.5 SELF-DISCHARGE RATE

This is a limiting factor for electrochemical capacitors applications. When there is self-discharge, the current leaks, causing the reduction in cell voltage. From previous studies, the supercapacitor's self-discharge rate and its mechanism depends on electrolyte type, impurities and the residual gases [11]. When self-discharge rate is avoided in a supercapacitor, the cycle life of a supercapacitor will be optimized [11].

15.3 TYPES OF SUPERCAPACITORS

Electrochemical capacitors are of three types, namely, EDLC, pseudocapacitors and hybrid capacitors. These capacitors are differentiated by their electrode configuration and operational workability [14].

15.3.1 ELECTROCHEMICAL DOUBLE-LAYER CAPACITOR (EDLC)

Electrochemical double-layer capacitors (EDLCs) utilize carbon-based materials as the electrodes. This has been widely employed commercially in supercapacitor

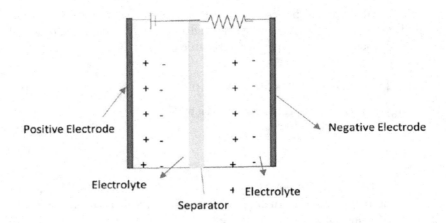

FIGURE 15.2   Electrochemical double-layer capacitor (EDLC).

because of its desirable qualities. It possesses energy density that approximates to 10 Whkg^{-1} when compared to rechargeable batteries and fuel cells. This has attracted a lot of research interest in the past few years and efforts are being made by researchers to further enhance the energy density of EDLCs.

In an EDLC are two carbon-based electrodes, an electrolyte and a separator (Figure 15. 2). Charges are stored electrostatically or non-faradaically which results in overall accumulation of charge on the surface of the electrodes by the contact with other surfaces. Charge exchange occurs in the process whenever any two surfaces contact and separate. These effects of charge exchange are usually only noticed when at least one of the surfaces has a high resistance to electrical flow. This happens because the charges that are transferred get trapped there for a time long enough for their effects to be noticed. These charges then remain on the electrode until they are neutralized by a discharge.

Charge accumulation occurs at the electrode surfaces when a potential difference is applied across the electrodes. This causes the ions in the electrolyte solution to diffuse across the separator into the pores of the electrode of opposite charge. Charge separation, therefore, occurs in EDLC$_S$ when the device is charged. This again enables it to store energy in the double layer at the electrode–electrolyte interface. This is a reversible phenomenon at the electrode–electrolyte interface in preference to the energy storage mechanism that takes place due to chemical reactions within the bulk of the material. This is essential to make an EDLC have high specific power, rapid charging and discharging rates, and long lifespan of thousands up to millions of cycles. These characteristics of an EDLC have attracted growing interests and research toward enhancing the performance and the specific energy of an EDLC device [21–23].

The electrodes are configured to prevent recombination of ions thus forming an electrochemical double layer of charge at each electrode. The double layer of charge at each electrode, the high surface area of the electrodes and decrease in distance between electrodes allows EDLCs to achieve higher energy densities than conventional capacitors. However, if the pores are small, electrolytic ions are restricted leading to high ESR, and low energy and power densities [3,8,9,24].

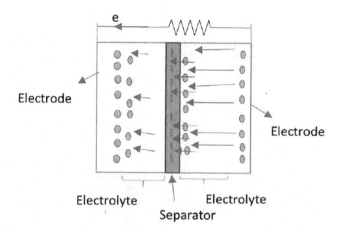

FIGURE 15.3   Schematic of typical pseudocapacitors.

15.3.2   PSEUDOCAPACITORS

These capacitors are called "pseudo" because instead of following a non-faradaic method of storing charge as in EDLCs, they store charge faradaically following redox processes just as batteries. Generally, pseudocapacitors (Figure 15. 3) store charge at the surface of the electrodes through a redox reaction at the electrode/electrolyte interface. This involves physical change during redox reactions. Pseudocapacitors have poor cycling stability compared with EDLCs and exhibitenergy density between 10 and 15 Whkg^{-1} which is larger than that exhibited by EDLCs [3,25].

15.3.3   HYBRID CAPACITORS

Hybrid capacitors (Figure 15.4) are utilizing the features of EDLCs and pseudocapacitors. Usually, they are asymmetric; that is, they combine an electrode-possessing pseudo capacitance and an electrode with double-layer capacitance. This pseudo-capacitor electrode provides its high specific capacitance, while the EDLC materials make the device to be highly reversible. Hybrid capacitors are also preferred to symmetrical EDLCs because of its high rated voltage with the corresponding higher specific energy [10,26].

It has been observed that the mechanism of double-layer capacitance and pseudo capacitance contribute to the total capacitance value of an electrochemical capacitor but with a great variation ratio, considering the design of the electrodes and the composition of the electrolyte. This led to the development of the concepts of supercapattery and supercabattery. This signifies the operational resemblance of hybrid supercapacitor devices with the combinations of supercapacitor and the rechargeable battery [27–30].

15.4   ELECTRODE MATERIALS FOR SUPERCAPACITIVE APPLICATIONS

The choice of good electrode material is paramount when it comes to optimizing the energy density of supercapacitors. The nanomaterials used as electrodes in

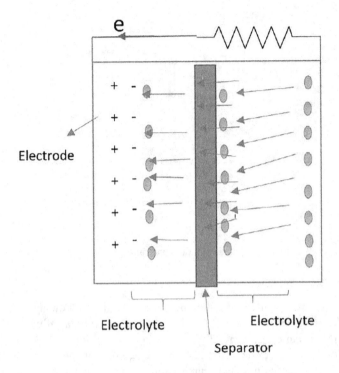

FIGURE 15.4 Schematic of a typical hybrid capacitor.

supercapacitors should have high life cycle, long-term stability, high specific surface area, high electric conductivity, resistance to electrochemical oxidation and reduction as well as high temperature stability among other properties [1,19,20]. Supercapacitor electrodes materials are classified based on the supercapacitor types. Figure 15.5 shows the electrode materials according to the different types of supercapacitors.

15.4.1 EDLCs ELECTRODE MATERIALS

Carbon-based electrode materials are widely used commercially in different forms as carbon aerogels, activated carbons, carbon fibers and carbon nanotubes

15.4.1.1 Carbon Aerogels

Carbon aerogels are a continuous network of nanoparticles of carbon with interfaced mesopores. Amesoporous material is a material containing pores with diameters between 2 and 50 nm, according to IUPAC nomenclature [31,32]. The mesoporous interfaces are utilized in electrodes with no need for binder material. Carbon aerogels electrodes possess a low equivalent series resistance (ESR) and this aids the optimization of supercapacitor power performance [10].

15.4.1.2 Activated Carbon

Activated carbon is also used as electrode in EDLCs. They are known for their large surface area because of their porous nature [33]. When the pores in an activated carbon

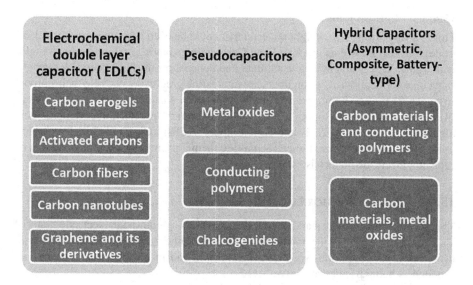

FIGURE 15.5   Supercapacitor types and their different electrode materials.

are high, it could limit its application as an electrode material. Pores that are smaller than electrolyte ions will not contribute to the charge storage. Hence, high porosity may also lead to poor conductivity which consequently can limit maximum power density in an EDLC device [29,30,34]. In addition, a distributed electrolyte resistance might ensue and this restricts charge and discharge rate [35]. Activated carbon is less expensive than other carbon material, thereby can be easily accessed [28].

15.4.1.3 Carbon Nanotubes (CNTs)

CNTs also known as called buckytubes, have capacitance comparable to activated carbon electrodes. This is due to the continuous charge distribution offered by their high surface area. The improved electronic, mechanical, optical and chemical properties, alongside with its high surface area, enable its application in supercapacitors [26]. CNTs are inexpensive, durable and readily available. They possess mesoporous nature, which can permit easy diffusion of ions in electrolytes. This consequently decreases the ESR and hence gives room for maximum power optimization [26].

15.4.1.4 Graphene

Graphene possesses high electrical conductivity and charge transport mobility thereby making it a desirable electrode material for high-power EDLCs. Nanosheets of graphene grown on current collectors exhibit high performance which is comparable to that of conventional dielectric capacitors. Preparation of permeable graphene oxide (as electrode materials) by KOH activation for application in EDLCs has been reported [36]. Heating nano-sized diamond at 1,650°C in a helium atmosphere as well as camphor decomposition over nickel nanoparticles among other methods and thermal exfoliation of graphic oxides [37–40] are being researched for the improved synthesis of graphene for supercapacitor applications.

15.4.1.5 Carbon Fibers

This is a very strong lightweight material that could be used as a supercapacitor electrode. It is also known as graphite fiber. Carbon fibers having a high tensile strength with lightweight is an enviable candidate and hence a promising material for supercapacitive applications. Carbon fibers are interesting precursors because their fibrous structure (i) is very homogeneous and well-defined, ensuring high reproducibility, (ii) is beneficial for developing directly accessible micro porosity, and (iii) provides higher adsorption rates. Activated carbon fibers (ACFs) have good electric conductivities along their fiber axis, thereby making them a promising alternative as supercapacitor electrode materials [41,42].

15.4.2 PSEUDOCAPACITOR ELECTRODE MATERIALS

This includes metal oxides and conductive polymers. Metal oxide electrode materials include ruthenium oxide, manganese oxide, and other transition metal oxides and hydroxides. Chalcogenides such as cobalt sulfide and molybdenum sulfide can be grouped as being pseudocapacitive.

15.4.2.1 Metal Oxides

Metal oxide electrodes are promising also for supercapacitor applications because of its large specific capacitance and long operation time. The capacitance of metal oxides is usually higher than carbon materials. However, they have lower power densities than EDLCs, and their cycle life can be limited by mechanical stress, caused by reduction–oxidation reactions. Metal oxides such as RuO_2, I_rO, MnO_2, V_2O_5, NiO, CO_3O_4, SnO_4, and Fe_2O_3, and some hydroxides such as $Co(OH)_2$, $Ni(OH)$ or their composites have been widely investigated as pseudocapacitive electrode [11]. Several studies have focused on utilizing RuO_2 with metal oxides, such as SnO_2, MnO_2, NiO, VO_x, TiO_2, MoO_3, WO_3, and CaO, which are cheap to form composite oxide electrodes [8,11,43].

The linear dependence on the current versus voltage curve of MnO_2 and RuO_2 and its faradaic behavior makes them good electrodes for pseudocapacitors. The charge storage comes from electron-transfer mechanisms rather than the accumulation of ions in the electrochemical double layer. Faradaic redox reactions that occur inside the active electrode materials are the principle, which makes pseudocapacitance possible. Transition metal oxide such as MnO_2 has been researched greatly because of its cheap price compared to noble metal oxides such as RuO_2.

There are two mechanisms for charge storage of metal oxide. These include the intercalation of protons (H^+) or alkali metal cations (K^+, Na^+, etc.) in the bulk of the material upon reduction and the de-intercalation upon oxidation [44]. This is shown in equation 15. 4) [45] using MnO_2 as an example.

$$MnO_2 + H^+\left(K^+\right) + e^- \rightleftharpoons MnOOH(K) \tag{15.4}$$

Next, adsorption (surface) of electrolyte cations on MnO_2.

$$\left(MnO_2\right)_{surface} + K^+ + e^- \rightleftharpoons \left(MnO_2^-K^+\right)_{surface} \tag{15.5}$$

Not every material that has faradaic behavior is a good electrode for pseudocapacitors. Example, material such as $Ni(OH)_2$;

This is a battery-type electrode, and it has a non-linear dependence on current versus voltage curve [46].

a. **Ruthenium oxide**

Ruthenium oxide is a good candidate for supercapacitive applications, as it offers higher capacitance in its hydrous form than carbon materials and conducting polymers [47–49]. However, ruthenium oxide has disadvantages of high cost and toxicity, hence the need for fabrication methodologies that will improve its quality. In its operation, it uses metal-type conductivity and reversible redox reactions which occurs at the electrode–electrolyte interface [8,23,32,50–52].

b. **Manganese oxide**

Manganese oxide when used as supercapacitor electrode material allows good capacitive performance and is relatively safe and promising and hence has attracted research interests over the years [53]. However, it has the challenge of poor conductivity, and it suffers low material utilization because of its charge–discharge processes which involves surface atoms only. Efforts are being made to address the operational challenges inherent in manganese oxide when used as electrodes by primarily using it in composite form with CNTs and other carbon-based materials [8,54–56].

15.4.2.2 Conductive Polymers

Conductive polymers are among the promising pseudocapacitive electrode. They possess high capacitance, high conductivity and low ESR relatively, and offer high capacitive behavior via redox reactions that happens on the surface and throughout its bulk. The redox processes are highly reversible accompanied by phase transformations. Sometimes, they are doped for better performance and in some cases considered better than carbon materials, such as activated carbon. Conductive polymers include polyacetylene (PA), polypyrrole (PPy), poly-aniline (PANI), and poly-(3, 4-ethylenedioxythiophene) (PEDOT). These have been shown to exhibit specific capacitance comparable to that of metal oxides like ruthenium oxide. The major setbacks for conductive polymer pseudocapacitive electrodes include lack of *n*-doped conductive polymer as well as limited mechanical stability during its charge–discharge cycles owing to mechanical stress [57,58].

PANI and Pay are the most promising pseudocapacitive conductive polymer electrodes because of their low cost, environmental stability and facile synthesis [30–62].

15.4.3 Hybrid Capacitor Electrode Materials

These are electrode materials that make use of the characteristics of EDLCs and pseudocapacitors. They are classified based on their electrode configurations as composite, asymmetric and battery-type hybrid capacitors.

15.4.3.1 Composites

These are formed from carbon-based materials having either metal oxides or conducting polymer in a single electrode. Varying ratios of the carbonaceous material and the metal oxide or conducting polymers could be used to make the composite electrode in order to achieve a particular purpose in a supercapacitor.

15.4.3.2 Asymmetric

This combines non-faradaic and faradaic processes of the EDLCs and pseudocapacitors, respectively. They are engineered in a way that makes carbon the negative electrode, while either metal oxide or conducting polymer serves as the positive electrode.

15.4.3.3 Battery Type

This combines two different electrodes – a supercapacitor electrode and a battery electrode – thereby using the properties of supercapacitors and batteries in a single cell. Better capacitance is achieved in supercapacitors having a combination of these different electrodes [63–65]. Ni_3S_4 microflowers were successfully synthesized, and the electrochemical performance gave 1,797. 5 F g^{-1} at a current density of 0.5 A g^{-1} [66]. Cu_2S–Ag_2S composite synthesized using the SILAR method gave a specific capacity of 772 Cg^{-1} at a scan rate of 10 mVs^{-1} [67].

15.5 ELECTROLYTE MATERIALS FOR SUPERCAPACITIVE APPLICATIONS

Electrolytes are one of the most important constituents of electrochemical supercapacitors. They aid ionic conductivity and charge compensation on both electrodes of the cell. These electrolytes can be classified based on the solvents used:

- **Organic electrolytes**: Organic solvents such as acetonitrile or PCare used to dissolve salts like TEABF$_4$, etc.
- **Aqueous electrolytes**: Water serves as the solvent for various salts such as H_2SO_4, KOH, Na_2SO_4, etc.
- **Ionic liquids**: Pyrrolidinium, imidazalium or quaternary aliphatic ammonium salt together with anions such as BF$_4$, PF$_6$, etc.
- **Solid-state polymer electrolytes**: polymer electrolyte (dry), polymer electrolyte (gel), and polyelectrolyte.

Criteria for choosing electrolytes for applications in supercapacitors include wide potential window, the broad range of working temperature, high ionic conductivity, low viscosity, high electrochemical stability, well-matched with the electrolyte material, environmental friendliness, low cost and low flammability [68].

It is important to note that no electrolyte material can meet all the required conditions. However, H_2SO_4(acid), KOH (alkaline) and KCl (neutral) are characterized with low cost, high ionic conductivity, environmental friendly and high capacitance making them interesting for application in supercapacitors especially in the aqueous system. The high conductivity in aqueous electrolyte could reduce the equivalent series resistance thereby leading to a significant increase in the power density of supercapacitors. Evaluation of the overall performance of aqueous electrolytes requires that the dimensions of hydrated and bare ions, the flow of ions which alters the ionic conductivity, as well as the specific capacitance be taken into consideration.

Appropriate choice of good electrodes with high specific capacitance and electrolytes with a large-operating voltage provides efficient ways of achieving supercapacitors of high energy density. Therefore, electrolytes with a large potential window, high ionic conductivity and large working temperature range should be a top priority of choice for higher performance of supercapacitor in comparison to seeking new electrodes [29,68]. There has been some review on supercapacitor electrolytes [11]; however, Okwundu et al. have presented some trending carbon-based materials with their matching electrolytes for non-faradaic supercapacitors indicating their practical performances such as specific surface area, capacitance, and energy and power densities [69].

15.5.1 ELECTROLYTES FOR ELECTROCHEMICAL DOUBLE-LAYER CAPACITOR (EDLC)

The specific energy of an EDLC can be influenced greatly by the double-layer capacitance and the operating potential of the cell. However, the electrochemical window of the electrolyte solution could limit the overall performance of the cell-operating potential. Therefore, researchers are making efforts to find electrolytes with accommodating electrochemical window within which the solvent is not reduced or oxidized.

Ionic liquids are among the electrolytes used extensively in EDLCs applications due to the relatively large potential windows which exceed 5V. There has not been an assertive finding on the effect of the structure of ionic liquids on the performance of an EDLC despite its advantages as mentioned. However, few research findings has shown that only small amount of ILs or ILs diluted in solvents can affect the electrochemical properties [70,71]. Few studies have shown that the structural variation of an IL has a great influence on the capacitance of an EDLC but doesnot necessarily improve the stability of ILs electrochemically [47,72]. It is paramount, therefore, for an investigation of the effects of double-layer capacitance and the electrochemical stability on the specific energy of a device. Regrettably, very many researchers have focused on either the double-layer capacitance or the electrochemical stability [47,71]. At room temperature, studies have shown that an EDLC with N, N-diethyl-N-Methyl-N-(2-methoxyethyl)ammonium tetrafluoroborate electrolyte has a wide potential window but an EDLC with 1-ethyl-3-methylimidazolium tetrafluoroborate possessed a low discharge when used as the electrolyte [22]. Ionic liquids have been shown to possess a great decomposition temperature which allows wider range of temperature applications above $100°C$ [48,49].

Polymer electrolyte and aqueous solutions are limited to 1 V and organic electrolytes based on acetonitrile or cyclic carbonates, e.g., PC has a potential window limited to 3.5 V owing to the decomposition of the solvent [73,74].

15.5.2 ELECTROLYTES FOR PSEUDOCAPACITORS

Neutral electrolyte salts such as Na_2SO_4 have shown to work well with pseudocapacitor electrodes. Neutral electrolytes have large working potential and less corrosive features. Other types of neutral electrolyte salts include LiCl, Li_2SO_4, NaCl, K_2SO_4, etc. [15].

15.6 SUPERCAPACITIVE CURRENT COLLECTORS

Current collectors are utilized in supercapacitors to connect electrode materials to the supercapacitor's terminals. It assembles and moves electrons from and to the active materials. Achieving these requires materials of high conductivity such as aluminum, copper or stainless steel. The collectors should be made from the same material if the enclosure is made of a metal like aluminum to avoid forming a corrosive galvanic cell. This is because metals corrode easily in non-aqueous electrolytes, are heavy in weight and usually reduce the gravimetric energy density of the whole cell. A resistance that contributes to the internal resistance of the EDLC and limits the power density exists between the metal current collectors and the carbon electrode. Carbon-coated with aluminum foil current collectors has also been explored [75] in order to reduce this resistance. Replacing metal current collectors has also been considered with carbon nanotubes (CNTs), carbon cloth, graphite foil, graphene reinforced with PANI carbon fibers and non-graphite PANI carbon fibers [75]. Carbon fibers derived from electrospun poly (ionic liquid) as the current collector have been shown to be better than carbon-coated with aluminum; when used as current collectors [75].

In aqueous electrolytes such as sulfuric acid current collectors should have a protective coating that is resistant to the electrolytes. The protective material should involve at least a conductive carbon powder (having p-type conductivity) or a polymer binder such as per chlorovinyl-based resin. Some studies have used proposed the use of lead or lead alloy [76] as a protective coating for current collectors when used in acidic electrolytes such as sulfuric acid. This is because lead or lead alloys are very constant with sulfuric acid electrolytes and exhibit high over-voltage of hydrogen and oxygen gassing and are readily obtainable at low cost [76]. Other studies have used corrosion-resistant metallic materials such as to reduce the corrosion [11]. The protective layer should be layered on the metal current collector before the attachment of the electrode and must allow free and efficient charge transfer from the active mass to the current collector. Several layers of the protective material that allows easy adhesion could also be applied.

To reduce the cost of current collectors materials such as indium tin oxide (ITO), carbon-based materials and electrically conductive polymers, ECPs have been advocated for current collectors using strong electrolyte [11]. Researchers have shown that the peak currents in a supercapacitor is enhanced by cooperating a current collector onto the electrode-electrolyte interface [77–79].

15.7 SUPERCAPACITOR ELECTRODE–ELECTROLYTE INTERFACE SEPARATORS

A separator separates the two electrodes in an electrode–electrolyte interface and helps to prevent short circuit. Oftentimes, the pore of separators which is a few hundredths of a millimeter and also permeable to the conducting ions reduces ESR. When separators are chemically inert, the electrolyte's stability and conductivity are maintained. Examples of separators are open capacitor papers, cellulose, glass fibers and polymer membranes. More complicated constructions use nonwoven permeable polymeric film such as kapton, polyacrylonitrileor porous woven ceramic fiber or woven glass fibers [80–82].

The preference of separator material depends on the type of electrode, working temperature and electrochemical supercapacitor cell voltage. For instance, a cellulose separator can function well in organic solvents but may suffer from degradation in an H_2SO_4 electrolyte [11].

15.8 FABRICATION OF SUPERCAPACITORS

Generally, fabrications of supercapacitors require two electrode materials, a current collector and a separator. All of these materials are sandwiched, thereby forming a supercapacitor device (Figure 15.6). The separator serves as a reservoir of ions. Current collectors conduct electrical current from the electrodes. The overall interface enables the flow of ionic current between electrodes while preventing electronic current from discharging the cell [8].

15.9 CHARACTERIZATION METHODS FOR SUPERCAPACITORS

Cyclic voltammetry (CV), galvanostatic charging and discharging (GCD) and electrochemical impedance spectroscopy are employed to characterize, test and diagnose electrochemical supercapacitors in a targeted electrolyte [50,83].

15.9.1 Cyclic Voltammetry (CV)

CV involves the measurement of current response in an electrode or device within a potential range. Voltammetry sweeps include cyclic voltammetry, linear sweep voltammetry, staircase voltammetry, etc. Our interest here is incyclic voltammetry which is greatly used to analyze the electrochemical performances of supercapacitors. In CV experiments, currents are usually generated due to electron transfer between the redox species and the electrodes. This is transported via the solution by the migration of ions. Voltammetry principle requires two electrodes in practice;

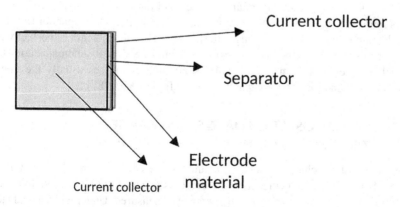

FIGURE 15.6 Schematic of a typical supercapacitor fabrication interface.

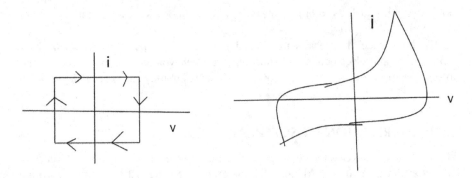

FIGURE 15.7   Voltammogram (cyclic) of (a) ideal capacitor and (b) real capacitor.

however, it is difficult to maintain steady potential while also passing current to counteract the redox events at the working electrode. Therefore, a three-electrode cell (working electrode, counter electrode and reference electrode) is usually employed. This separates the role of referencing the potential applied and balancing the current produced. The working potential of a three-electrode cell needs to be varied, and it depends also on the operating reference potential which is fixed. This maintains the potential difference applied as required for cyclic voltammetry. Furthermore, no current passes between reference and working electrode in cyclic voltammetry experiments. The current is balanced at the counter electrodes by the one at the working electrodes.

Performing cyclic voltammetry requires to linearly sweep the potential between the working and reference electrode until it reaches a preset limit. It is later swept back in the opposite direction, switching potentials. This is done numerous times during a scan, the device in real-time measures the changing current between the working electrode and the counter probes. The plot is known as a cyclic voltammogram [84–86].

The total charge accumulated at the surface of the electrode can be found by integrating the current with respect to time [86].

Lower scan gives high capacitance values and the resultant shape of CV curves gives information on the electrochemical processes in the supercapacitor (for example, the rate of charge and discharge) [87]. An ideal supercapacitor is perfectly rectangular in shape and depends always on the scan rate. A real supercapacitor on the other hand, deviates from a rectangular shape and comprises usually the internal resistance (R) and the overall capacitance (C) (Figure 15.7) [87].

15.10   GALVANOSTATIC CHARGE–DISCHARGE TECHNIQUES (GCD)

The galvanostatic charge–discharge evaluates the electrochemical capacitance of materials in a controlled current condition. This differs from cyclic voltammetry in that while the current is controlled, the voltage is measured. Undeniably, a GCD technique is one of the most used in the field of supercapacitors; it can be applied across a laboratory scale to an industrial one. Chronopotentiometry is another name used to describe GCD and it helps one evaluate capacitance, resistance and cyclability and other electrochemical properties of a capacitor. A pulse current is often applied across the working electrode and the resulting potential is measured; usually against a reference electrode as a function of time. This potential after current input changes suddenly following the internal resistance (IR) loss. This is due to concentration of over-potential felt across the electrodes. This occurs when the concentration of the reactant is exhausted at the electrode surface.

The voltage variation is given by equation 15.6:

$$V(t) = iR + \frac{t}{C}i(V) \qquad (15.6)$$

$V(t)$ is the voltage as a function of time, 'R' is the resistance, 'C' is the capacitance and 'i' is the current. The slope of the galvanostatic charge–discharge curve gives the capacitance of the supercapacitor (from equation 15.6). For pseudocapacitors, equation 15.7 is applied when the $V(t)$ curve profile is not as linear as it should be and C is obtained by integrating the current over the discharge time or charge time

$$C = I\frac{I\Delta t}{\Delta V}F \qquad (15.7)$$

'I' is the set current, 'Δt' is the discharge time (or charge time) and 'ΔV' is the potential window. The series resistance is derived from the voltage drop (V drop) taking place over the current inversion (ΔI). This is given as follows:

$$R = \frac{V_{drop}}{\Delta I} \qquad (15.8)$$

Current inversion makes the voltage drop directly linked to the resistance of the cell. The cyclic stability of supercapacitor cells (EDLCs or redox capacitors) can be obtained by having the parameters of capacitance and resistance measurements repeated over a large number of cycles [87].

15.10.1 IMPEDANCE SPECTROSCOPY

AC impedance spectroscopy involves the measurement of different impedance responses at ac frequencies. The impedance analysis gives details about the interface and kinetics and mechanism of reactions occurring at the interface [87]. This technique can be used to find the double-layer capacitance as well as other electrode processes at complex interfaces.

15.11 SUMMARY AND CONCLUSION

Just as batteries, supercapacitors are also used to store energy for application in electrochemical devices. Supercapacitors are gaining interest in research because of their desirable qualities such as high power density and fast charge and discharge rate. However, they have low energy density limiting their applications. Various electrodes based on materials such as carbonaceous nanomaterials, transitional metal oxides, conducting polymers and chalcogenides in addition to composites of these materials have been tested and applied in supercapacitors. Aqueous, organic and ionic electrolytes have been used in various supercapacitors. While aqueous electrolytes offer good conductivity, non-aqueous electrolytes enhance the operational potential window of the supercapacitor. If the energy densities of supercapacitors are optimized, then supercapacitors can have a wide range of usage. To increase the energy density of supercapacitors, it is better to use electrolytes that have high conductivity, wide operational potential windows and also with high temperature stability. In addition to this, a matching electrode with a large surface area, and matching pore size with the electrolyte ion should be used while maintaining a short distance of separation between the electrodes during its fabrication.

In conclusion, if the electrodes and electrolytes used in a supercapacitor are properly engineered and optimized, supercapacitors could serve as an alternative to battery energy storage technology.

REFERENCES

[1] P. Lu, D. Xue, H. Yang, and Y. Liu, "Supercapacitor and nanoscale research towards electrochemical energy storage," *Int. J. Smart Nano Mater.*, vol. 4, no. 1, pp. 2–26, 2013, doi: 10.1080/19475411.2011.652218.
[2] D. A. Scherson and A. Palencsár, "Batteries and electrochemical capacitors," *Electrochem Soc Interface*, vol. 15, no. 1, p. 17, 1827.
[3] O. Haas and E. J. Cairns, "Electrochemical energy storage," *Annu. Reports Prog. Chem. - Sect. C*, vol. 95, no. 6, pp. 163–197, 1999, doi: 10.1039/pc095163.
[4] K. Naoi and W. Naoi, "New generation supercapacitors in synergy with Li-Ion technology," in *AABC 2014- Advanced Automotive Battery Confernce, LLIBTA Symposium Track A: Cell Materials and Chemistry and Track B: Battery Engineering- Large Lithium Ion Battery Technology and Application and ECCAP Symposium - Large EC Capacitor Technology and Application*, 2014.
[5] J. Lu, T. Wu, and K. Amine, "State-of-the-art characterization techniques for advanced lithium-ion batteries," *Nat. Energy*, 2017, doi: 10.1038/nenergy.2017.11.
[6] V. Augustyn, P. Simon, and B. Dunn, "Pseudocapacitive oxide materials for high-rate electrochemical energy storage," *Energy Environ. Sci.*, 2014, doi: 10.1039/c3ee44164d.
[7] Y.-G. Guo, J.-S. Hu, and L.-J. Wan, "Nanostructured materials for electrochemical energy conversion and storage devices," *Adv. Mater.*, vol. 20, no. 15, pp. 2878–2887, 2008, doi: 10.1002/adma.200800627.
[8] M. Vangari, T. Pryor, and L. Jiang, "Supercapacitors: Review of materials and fabrication methods," *J. Energy Eng.*, 2013, doi: 10.1061/(ASCE)EY.1943-7897.0000102.
[9] J. Zhang, Y. Cui, and G. Shan, "Metal oxide nanomaterials for pseudocapacitors," 2019, [Online]. Available: http://arxiv.org/abs/1905.01766.
[10] A. Muzaffar, M. B. Ahamed, K. Deshmukh, and J. Thirumalai, "A review on recent advances in hybrid supercapacitors: Design, fabrication and applications," *Renew. Sustain. Energy Rev.*, 2019, doi: 10.1016/j.rser.2018.10.026.
[11] C. Zhong, Y. Deng, W. Hu, J. Qiao, L. Zhang, and J. Zhang, "A review of electrolyte materials and compositions for electrochemical supercapacitors," *Chem. Soc. Rev.*, 2015, doi: 10.1039/c5cs00303b.
[12] E. Talaie, P. Bonnick, X. Sun, Q. Pang, X. Liang, and L. F. Nazar, "Methods and protocols for electrochemical energy storage materials research," *Chem. Mater.*, 2017, doi: 10.1021/acs.chemmater.6b02726.

[13] "Electrochemical energy storage: simple definition," 2015. https://www.energie-rs2e. com/en/articleblog/electrochemical-energy-storage-simple-definition (accessed Jun. 09, 2020).

[14] S. Zhang and N. Pan, "Supercapacitors performance evaluation," *Adv. Energy Mater.*, 2015, doi: 10.1002/aenm.201401401.

[15] D. Harrison, F. Qiu, J. Fyson, Y. Xu, P. Evans, and D. Southee, "A coaxial single fibre supercapacitor for energy storage," *Phys. Chem. Chem. Phys.*, 2013, doi: 10.1039/c3cp52036f.

[16] Wikipedia, "Supercapacitor - Wikipedia," *Wikipedia*, vol. 26, pp. 1–47, 2018, [Online]. Available: https://en.wikipedia.org/wiki/Supercapacitor#Evolution_of_components.

[17] Z. S. Wu, X. Feng, and H. M. Cheng, "Recent advances in graphene-based planar micro-supercapacitors for on-chip energy storage," *Natl. Sci. Rev.*, 2014, doi: 10.1093/nsr/nwt003.

[18] A. C. Nwanya, D. Obi, R. U. Osuji, R. Bucher, M. Maaza, and F. I. Ezema, "Simple chemical route for nanorod-like cobalt oxide films for electrochemical energy storage applications," *J. Solid State Electrochem.*, 2017, doi: 10.1007/s10008-017-3520-8.

[19] K. Naoi, "'Nanohybrid capacitor': The next generation electrochemical capacitors," *Fuel Cells*, 2010, doi: 10.1002/fuce.201000041.

[20] J. H. Chae, K. C. Ng, and G. Z. Chen, "Nanostructured materials for the construction of asymmetrical supercapacitors," *Proc. Inst. Mech. Eng. A: J. Power Energy*, 2010, doi: 10.1243/09576509JPE861.

[21] M. Conte, "Supercapacitors technical requirements fornew applications," *Fuel Cells*, 2010, doi: 10.1002/fuce.201000087.

[22] O. Barbieri, M. Hahn, A. Herzog, and R. Kötz, "Capacitance limits of high surface area activated carbons for double layer capacitors," *Carbon*, 2005, doi: 10.1016/j.carbon.2005.01.001.

[23] R. Kötz and M. Carlen, "Principles and applications of electrochemical capacitors," *Electrochim. Acta*, 2000, doi: 10.1016/S0013-4686(00)00354-6.

[24] A. Celzard, F. Collas, J. F. Marêché, G. Furdin, and I. Rey, "Porous electrodes-based double-layer supercapacitors: Pore structure versus series resistance," *J. Power Sources*, 2002, doi: 10.1016/S0378-7753(02)00030-7.

[25] J. W. Park et al., "In situ synthesis of graphene/polyselenophene nanohybrid materials as highly flexible energy storage electrodes," *Chem. Mater.*, 2014, doi: 10.1021/cm500577v.

[26] A. Laforgue, P. Simon, J. F. Fauvarque, J. F. Sarrau, and P. Lailler, "Hybrid supercapacitors based on activated carbons and conducting polymers," *J. Electrochem. Soc.*, 2001, doi: 10.1149/1.1400742.

[27] A. C. Nwanya et al., "Zea mays lea silk extract mediated synthesis of nickel oxide nanoparticles as positive electrode material for asymmetric supercabattery," *J. Alloys Compd.*, vol. 822. 2020, doi: 10.1016/j.jallcom.2019.153581.

[28] L. Yu and G. Z. Chen, "Redox electrode materials for supercapatteries," *J. Power Sources*, 2016, doi: 10.1016/j.jpowsour.2016.04.095.

[29] E. Frackowiak and F. Béguin, "Carbon materials for the electrochemical storage of energy in capacitors," *Carbon*, 2001, doi: 10.1016/S0008-6223(00)00183-4.

[30] P. Simon and Y. Gogotsi, "Materials for electrochemical capacitors," *Nat. Mater.*, 2008, doi: 10.1038/nmat2297.

[31] "Recommendations for the characterization of porous solids (Technical Report)," *Pure Appl. Chem.*, 1994, doi: 10.1351/pac199466081739.

[32] J. Rouquerol et al., "Recommendations for the porous solids," *Pure Appl. Chern*, 1994, doi: 10.1351/pac199466081739.

[33] A. G. Pandolfo and A. F. Hollenkamp, "Carbon properties and their role in supercapacitors," *J. Power Sources*, 2006, doi: 10.1016/j.jpowsour.2006.02.065.

[34] Y. Zhai, Y. Dou, D. Zhao, P. F. Fulvio, R. T. Mayes, and S. Dai, "Carbon materials for chemical capacitive energy storage," *Adv. Mater.*, 2011, doi: 10.1002/adma.201100984.

[35] P. Simon and Y. Gogotsi, "Charge storage mechanism in nanoporous carbons and its consequence for electrical double layer capacitors," *Philos. Trans. R. Soc. A: Math., Phys. Eng. Sci.*, 2010, doi: 10.1098/rsta.2010.0109.

[36] S. Murali et al., "Preparation of activated graphene and effect of activation parameters on electrochemical capacitance," *Carbon*, vol. 50, no. 10, pp. 3482–3485, 2012, doi: 10.1016/j.carbon.2012.03.014.

[37] A. H. Castro Neto, F. Guinea, N. M. R. Peres, K. S. Novoselov, and A. K. Geim, "The electronic properties of graphene," *Rev. Mod. Phys.*, 2009, doi: 10.1103/RevModPhys.81.109.

[38] A. K. Geim and K. S. Novoselov, "The rise of graphene," *Nat. Mater.*, 2007, doi: 10.1038/nmat1849.

[39] Y. Zhu et al., "Graphene and graphene oxide: Synthesis, properties, and applications," *Adv. Mater.*, 2010, doi: 10.1002/adma.201001068.

[40] S. Stankovich et al., "Graphene-based composite materials," *Nature*, 2006, doi: 10.1038/nature04969.

[41] J. F. Snyder, E. L. Wong, and C. W. Hubbard, "Evaluation of commercially available carbon fibers, fabrics, and papers for potential use in multifunctional energy storage applications," *J. Electrochem. Soc.*, 2009, doi: 10.1149/1.3065070.

[42] S. Hu, S. Zhang, N. Pan, and Y. Lo Hsieh, "High energy density supercapacitors from lignin derived submicron activated carbon fibers in aqueous electrolytes," *J. Power Sources*, 2014, doi: 10.1016/j.jpowsour.2014.07.063.

[43] I. Kelpšaite, J. Baltrušaitis, and E. Valatka, "Electrochemical deposition of porous cobalt oxide films on AISI 304 type steel," *Medziagotyra*, 2011, doi: 10.5755/j01.ms.17.3.586.

[44] M. Toupin, T. Brousse, and D. Bélanger, "Charge storage mechanism of MnO_2 electrode used in aqueous electrochemical capacitor," *Chem. Mater.*, 2004, doi: 10.1021/cm049649j.

[45] S.-C. Pang, M. A. Anderson, and T. W. Chapman, "Novel electrode materials for thin-film ultracapacitors: Comparison of electrochemical properties of sol-gel-derived and electrodeposited manganese dioxide," *J. Electrochem. Soc.*, 2000, doi: 10.1149/1.1393216.

[46] T. Brousse, D. Bélanger, and J. W. Long, "To be or not to be pseudocapacitive?" *J. Electrochem. Soc.*, 2015, doi: 10.1149/2.0201505jes.

[47] M. Galiński, A. Lewandowski, and I. Stepniak, "Ionic liquids as electrolytes," *Electrochim. Acta*, 2006, doi: 10.1016/j.electacta.2006.03.016.

[48] D. Wei and A. Ivaska, "Applications of ionic liquids in electrochemical sensors," *Anal. Chim. Acta.* 2008, doi: 10.1016/j.aca.2007.12.011.

[49] N. DeVos, C. Maton, and C. V. Stevens, "Electrochemical stability of ionic liquids: General influences and degradation mechanisms," *ChemElectroChem*, 2014, doi: 10.1002/celc.201402086.

[50] Y. Wang, Y. Song, and Y. Xia, "Electrochemical capacitors: Mechanism, materials, systems, characterization and applications," *Chem. Soc. Rev.*, 2016, doi: 10.1039/c5cs00580a.

[51] J. Yan, Q. Wang, T. Wei, and Z. Fan, "Recent advances in design and fabrication of electrochemical supercapacitors with high energy densities," *Adv. Energy Mater.*, 2014, doi: 10.1002/aenm.201300816.

[52] C. D. Lokhande, D. P. Dubal, and O. S. Joo, "Metal oxide thin film based supercapacitors," *Current Appl. Phys.*, 2011, doi: 10.1016/j.cap.2010.12.001.

[53] M. C. Nwankwo et al., "Electrochemical supercapacitive properties of SILAR-deposited Mn_3O_4 electrodes," *Vacuum*, vol. 158, 2018, doi: 10.1016/j.vacuum.2018.09.057.

[54] M. D. Stoller, S. Park, Z. Yanwu, J. An, and R. S. Ruoff, "Graphene-based ultracapacitors," *Nano Lett.*, 2008, doi: 10.1021/nl802558y.

[55] C. G. Ezema et al., "Photo-electrochemical studies of chemically deposited nanocrystalline meso-porous n-type TiO_2 thin films for dye-sensitized solar cell (DSSC) using

simple synthesized azo dye," *Appl. Phys. A Mater. Sci. Process.*, vol. 122, no. 4, 2016, doi: 10.1007/s00339-016-9965-2.

[56] M. C. Nwankwo et al., "Syntheses and characterizations of GO/Mn$_3$O$_4$ nanocomposite film electrode materials for supercapacitor applications," *Inorg. Chem. Commun.*, vol. 119, p. 107983, 2020, doi: 10.1016/j.inoche.2020.107983.

[57] R. Balint, N. J. Cassidy, and S. H. Cartmell, "Conductive polymers: Towards a smart biomaterial for tissue engineering," *Acta Biomater.* 2014, doi: 10.1016/j.actbio.2014.02.015.

[58] D. Kumar and R. C. Sharma, "Advances in conductive polymers," *Eur. Polym. J.* 1998, doi: 10.1016/S0014-3057(97)00204-8.

[59] J. P. Boudou, M. Chehimi, E. Broniek, T. Siemieniewska, and J. Bimer, "Adsorption of H$_2$S or SO$_2$ on an activated carbon cloth modified by ammonia treatment," *Carbon*, 2003, doi: 10.1016/S0008-6223(03)00210-0.

[60] J. P. Zheng, "A new charge storage mechanism for electrochemical capacitors," *J. Electrochem. Soc.*, 1995, doi: 10.1149/1.2043984.

[61] C. Liu, F. Li, M. Lai-Peng, and H. M. Cheng, "Advanced materials for energy storage," *Adv. Mater.* 2010, doi: 10.1002/adma.200903328.

[62] A. C. Nwanya et al., "Electrochromic and electrochemical capacitive properties of tungsten oxide and its polyaniline nanocomposite films obtained by chemical bath deposition method," *Electrochim. Acta*, vol. 128, 2014, doi: 10.1016/j.electacta.2013.10.002.

[63] K. Jurewicz, S. Delpeux, V. Bertagna, F. Béguin, and E. Frackowiak, "Supercapacitors from nanotubes/polypyrrole composites," *Chem. Phys. Lett.*, 2001, doi: 10.1016/S0009-2614(01)01037-5.

[64] E. Frackowiak, V. Khomenko, K. Jurewicz, K. Lota, and F. Béguin, "Supercapacitors based on conducting polymers/nanotubes composites," *J. Power Sources*, 2006, doi: 10.1016/j.jpowsour.2005.05.030.

[65] H. P. De Oliveira, S. A. Sydlik, and T. M. Swager, "Supercapacitors from free-standing polypyrrole/graphene nanocomposites," *J. Phys. Chem. C*, 2013, doi: 10.1021/jp400344u.

[66] H. Wang, M. Liang, D. Duan, W. Shi, Y. Song, and Z. Sun, "Rose-like Ni$_3$S$_4$ as battery-type electrode for hybrid supercapacitor with excellent charge storage performance," *Chem. Eng. J.*, vol. 350, pp. 523–533, 2018, doi: 10.1016/j.cej.2018.05.004.

[67] S. A. Pawar, D. S. Patil, and J. C. Shin, "Electrochemical battery-type supercapacitor based on chemosynthesized Cu$_2$S[sbnd]Ag$_2$S composite electrode," *Electrochim. Acta*, vol. 259, pp. 664–675, 2018, doi: 10.1016/j.electacta.2017.11.006.

[68] K. Lota, A. Sierczynska, I. Acznik, and G. Lota, "Effect of aqueous electrolytes on electrochemical capacitor capacitance," *Chemik*, vol. 67, no. 11, pp. 1138–1145, 2013.

[69] O. S. Okwundu, C. O. Ugwuoke, and A. C. Okaro, "Recent trends in non-faradaic supercapacitor electrode materials," *Metall. Mater. Eng.*, vol. 25, no. 2, pp. 105–138, 2019, doi: 10.30544/417.

[70] K. Yuyama, G. Masuda, H. Yoshida, and T. Sato, "Ionic liquids containing the tetrafluoroborate anion have the best performance and stability for electric double layer capacitor applications," *J. Power Sources*, 2006, doi: 10.1016/j.jpowsour.2006.09.002.

[71] M. Ue, M. Takeda, A. Toriumi, A. Kominato, R. Hagiwara, and Y. Ito, "Application of low-viscosity ionic liquid to the electrolyte of double-layer capacitors," *J. Electrochem. Soc.*, 2003, doi: 10.1149/1.1559069.

[72] M. Montanino, M. Carewska, F. Alessandrini, S. Passerini, and G. B. Appetecchi, "The role of the cation aliphatic side chain length in piperidinium bis(trifluoromethansulfonyl)-imide ionic liquids," *Electrochim. Acta*, 2011, doi: 10.1016/j.electacta.2011.03.089.

[73] H. Wu et al., "The effects of electrolyte on the supercapacitive performance of activated calcium carbide-derived carbon," *J. Power Sources*, 2013, doi: 10.1016/j.jpowsour.2012.11.014.

[74] Y. Wang, C. Zheng, L. Qi, M. Yoshio, K. Yoshizuka, and H. Wang, "Utilization of (oxalato)borate-based organic electrolytes in activated carbon/graphite capacitors," *J. Power Sources*, 2011, doi: 10.1016/j.jpowsour.2011.08.026.

[75] E. Josef, R. Yan, R. Guterman, and M. Oschatz, "Electrospun carbon fibers replace metals as a current collector in supercapacitors," *ACS Appl. Energy Mater.*, vol. 2, no. 8, pp. 5724–5733, 2019, doi: 10.1021/acsaem.9b00854.

[76] V. A. Kazarov, "(12) United States Patent," vol. 2, no. 12, 2008.

[77] H. M. Mott-Smith and I. Langmuir, "The theory of collectors in gaseous discharges," *Phys. Rev.*, 1926, doi: 10.1103/PhysRev.28.727.

[78] S. T. Myung, Y. Hitoshi, and Y. K. Sun, "Electrochemical behavior and passivation of current collectors in lithium-ion batteries," *J. Mater. Chem.*, 2011, doi: 10.1039/c0jm04353b.

[79] A. H. Whitehead and M. Schreiber, "Current collectors for positive electrodes of lithium-based batteries," *J. Electrochem. Soc.*, 2005, doi: 10.1149/1.2039587.

[80] A. Schneuwly and R. Gallay, "Properties and applications of supercapacitors from the state-of-the-art to future trends," in *Proceeding PCIM*, vol. 2000. Citeseer, 2000.

[81] C. González, J. J. Vilatela, J. M. Molina-Aldareguía, C. S. Lopes, and J. LLorca, "Structural composites for multifunctional applications: Current challenges and future trends," *Prog. Mater. Sci.*, 2017, doi: 10.1016/j.pmatsci.2017.04.005.

[82] X. Lu, M. Yu, G. Wang, Y. Tong, and Y. Li, "Flexible solid-state supercapacitors: Design, fabrication and applications," *Energy Environ. Sci.*, 2014, doi: 10.1039/c4ee00960f.

[83] P. L. Taberna, P. Simon, and J. F. Fauvarque, "Electrochemical characteristics and impedance spectroscopy studies of carbon-carbon supercapacitors," *J. Electrochem. Soc.*, 2003, doi: 10.1149/1.1543948.

[84] N. Elgrishi, K. J. Rountree, B. D. McCarthy, E. S. Rountree, T. T. Eisenhart, and J. L. Dempsey, "A practical beginner's guide to cyclic voltammetry," *J. Chem. Educ.*, 2018, doi: 10.1021/acs.jchemed.7b00361.

[85] N. Elgrishi, K. J. Rountree, B. D. McCarthy, E. S. Rountree, T. T. Eisenhart, and J. L. Dempsey, "Supporting information: A practical beginner's guide to cyclic voltammetry," *J. Chem. Educ.*, 2017, doi: 10.1021/acs.jchemed.7b00361.

[86] N. Elgrishi, K. J. Rountree, B. D. McCarthy, E. S. Rountree, T. T. Eisenhart, and J. L. Dempsey, "A practical beginner's guide to CV," *Osterr. Musik.*, 2017, doi: 10.1021/acs.jchemed.7b00361.

[87] P. Simon and Y. Gogotsi, "Charge storage mechanism in nanoporous carbons and its consequence for electrical double layer capacitors," *Philos. Trans. R. Soc. A Math. Phys. Eng. Sci.*, vol. 368, no. 1923, pp. 3457–3467, 2010, doi: 10.1098/rsta.2010.0109.

Index

Note: **Bold** page numbers refer to tables and *italic* page numbers refer to figures.

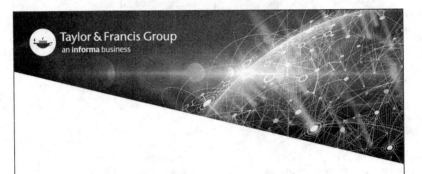

Printed in the United States
by Baker & Taylor Publisher Services